MW00846325

Brief Contents

Contents

7   Fire Alarm Notification   177

8 Fire Alarm System Installation   214

10 Engineering Documents   291

11 Approvals and Acceptance   307

Preface

For more than 100 years the National Fire Protection Association has been work-ing to protect people and property from fire, and part of that effort has been to serve the needs of fire alarm system designers, manufacturers, installers, and users through the publication of its codes, standards, and recommended prac-tices, many of which apply to fire alarms.

The continued rapid evolution of fire alarm systems in the last two decades has completely overshadowed the developments of the previous 100 years. We have progressed from the early fire alarm telegraph systems to signal multiplexing; from the hand ringing of a bell to automatic voice communication systems; from hardwired connections to fiber optics and wireless transmission; from unsupervised smoke detectors to addressable and analog devices that report their status; from relay-based architecture fire alarm control units that were dubbed "solid-state" because of their immense weight and size to today's microprocessor-based fire alarm control units; and from fire detectors that required enormous fires to actuate to smoke detectors that respond to only a few ounces of combustibles.

All of these changes have contributed to the fire alarm systems available today, whether they are single station residential smoke alarms or high-rise computer-controlled combination fire alarm and building management systems.

At the same time, this technical development has contributed to and been influenced by the fire alarm codes and standards used to detail the performance standards, test requirements, and installation of these systems.

Together, the National Fire Protection Association and the Society of Fire Protection Engineers have presented this handbook on *Fire Alarm Signaling Systems* to respond to the need for a guide to acquaint readers with the differ-ent types of fire detection and alarm systems, their application and operation and the requirements for each, as set forth in *NFPA 72®, National Fire Alarm Code®*, as well as Underwriters Laboratories and FM Global Standards.

Fire Alarm Signaling Systems is designed to help the user intelligently apply the fire alarm system technology to comply with owner-based fire protection goals and the applicable codes and standards. The book serves as a how-to-guide that acquaints the reader with the basic requirements for use of each of the fire alarm systems and differentiates between each system. It takes up where the *National Electrical Code® Handbook* leaves off by expanding upon the sec-tions of that handbook that discuss wiring of alarm systems.

Fire Alarm Signaling Systems is *not* intended to substitute for the fire alarm system codes and standards, or to be a summarization of them. Rather, it approaches fire detection and alarm systems from a different perspective: its purpose is to provide information on the basic requirements of fire detection and alarm systems *and* explanations of how fire alarm systems fit and work within those standards. A glossary of terms is provided at the end, and illustrations have been included to expand upon the material presented in the text and to assist the user visualizing actual alarm systems.

This third edition of *Fire Alarm Signaling Systems* contains many significant changes:

- Revisions and updates to reflect the significant changes made to *NFPA 72* over the past nine years
- A new chapter on the history of fire alarm systems
- Updated information to current requirements of the *National Fire Alarm Code*
- Quality control during the installation process
- Prescriptive versus performance design applications
- Intelligibility of voice evacuation fire alarm systems
- Fire alarm systems installation guidelines
- Basic fire alarm system plans review
- Reorganization for better flow and understanding
- Questions to enhance understanding at the end of each chapter
- New sections, which include
  - choosing a fire alarm system
  - establishing fire protection goals
  - matching the fire alarm system to fire protection goals
  - understanding fire alarm system limitations

Other additions to the book include data on fiber optic technology and analog and addressable devices used in fire detection systems.

Until now, there has been no single, consolidated source for the general information provided in this book, though its necessity is evident. The professionals who contributed to this book brought more than 100 years of cumulative experience to the task; their contributions outline the latest advances in fire alarm signaling systems.

The future of fire detection and alarm systems is promising and will continue to grow out of today's basic knowledge. It is hoped that this handbook will assist those in the field to better understand the application concepts and operation of fire alarm systems. The goal of this work is to be the stimulus to take fire alarm systems to their next level of performance to meet our ever-constant need to alert people, in time, for safe evacuation and prompt fire-fighting activities.

The editors would appreciate receiving any comments and suggestions that will improve the contents, as future editions of the book will be revised and expanded to keep the information current.

Acknowledgments

The first edition of *Fire Alarm Signaling Systems* was named *Fire Alarm Signaling Systems Handbook*. It was prepared in 1987 as a joint effort of the National Fire Protection Association and the Society of Fire Protection Engineers. *Fire Alarm Signaling Systems Handbook* was created by a team of talented professionals lead by Charles Zimmerman, who for many years was NFPA's Fire Alarm Signaling Systems Specialist. Charlie was as much a major contributor to the first edition as he was its editor. This edition is dedicated to Charlie for his constant promotion of "doing it right" and his commitment to the fire alarm industry. Additionally, the following technical experts contributed an inestimable amount of their time and expertise to the first edition: Jack Abbott, Joe Drouin, Peter Dubivsky, Al Heim, Vic Humm, Ted Humpel, Stan Kravontka, Bob McPherson, Wayne Moore, Crawley Parris, Mickey Reiss, Jim Roberts, Bill Rogers, Joseph Scheffey, Walter Schuchard, and Max Schulman. Much of what they wrote for *Fire Alarm Signaling Systems Handbook* has been updated and reused in the third edition, and the authors and editors of this edition of *Fire Alarm Signaling Systems* wish to acknowledge their contributions, as well as the contribution of Robert O'Laughlin, who co-authored the first two editions.

The third edition was exclusively prepared by Richard W. Bukowski and Wayne D. Moore.

Special thanks are due to Chuck Durang, Product Manager, Irene Herlihy, Senior Project Editor, Stephanie Lentz of TechBooks and the production services team who pitched in to highball the job through when it really counted.

Wayne D. Moore, P.E.
Richard W. Bukowski, P.E.

CHAPTER

1

The History of Fire Alarm Systems

From the beginning of recorded history, people have understood that early response to a fire is a key to controlling it. In recent centuries, when someone discovered a fire, fire brigades and fire departments were alerted by roving watchmen using hand bell-ringers, church sextons ringing church bells, or people signaling with factory steam whistles. Unfortunately, these systems did not provide very much detail and often directed the fire department to the wrong location. However, with the advent of the telegraph, invented in the early 1840s by Samuel F. B. Morse, fire fighters were given a faster and more accurate fire reporting system.

In 1847, New York became the first U.S. city to begin construction of a municipal fire alarm system required by ordinance "to construct a line of telegraph, by setting posts in the ground . . . for communicating alarms of fire from the City Hall to different fire stations, and [to] instruct the different bell-ringers in the use of said invention."[1] The chief engineer for the city, Cornelius Anderson, "summed up the promise of the proposed fire alarm system in his annual report of 1847: . . . this system would . . . [prevent] nearly all false alarms, and at the same time transmit the alarm in case of fire to all parts of the City with such certainty and promptitude that it would be the means of saving annually to the city a sum nearly equal to the cost of its erection . . . and maintenance."[1]

In March 1851, William Channing, a young doctor and avid fire buff, and Moses G. Farmer convinced the city of Boston to install their version of a municipal alarm system using Samuel Morse's printing register as a major component of the system: "The system consisted of 40 miles of wire to connect the central station to 40 signal boxes and 19 bells in churches, schools and fire engine houses."[1] The system had some technical flaws, but after these were resolved, the system could transmit an electric impulse from a code wheel breaking the circuit and record a Morse code dot or dash on the printing register.

On April 30, 1852, within 24 hours of the systems being placed in service, a fire alarm was transmitted for a fire on Causeway Street.

William Channing presented many papers and lectures on the Boston fire alarm system, and it was during his lecture in March 1855 at the Smithsonian Institute that John N. Gamewell, a South Carolina postmaster and telegraph agent, first heard about the new invention. Gamewell was so impressed with the potential of the Channing/Farmer system that he bought the rights to construct these fire alarm systems, first in the South and the West, and then for the entire country. In 1856, Gamewell joined with his brother-in-law, James M. Gardiner, who had an extensive knowledge of clocks and other intricate mechanisms. By 1871, Gardiner and Gamewell had improved their product line, utilizing spring-driven models of the fire alarm signal boxes and a "noninterference pull" box design developed by Edwin Rogers and Moses G. Crane. In 1880, Gardiner patented a design to eliminate interference between fire alarm boxes transmitting at the same time. Although in the late 19th century, 36 other companies were in the public fire alarm telegraph manufacturing business, the Gamewell Company held a 95% share of the U.S. market.

Extending the public fire alarm telegraph system to stations throughout a building further saved valuable time in fire department notification. The interior manual fire alarm system was first developed by Rogers, who invented what he called the Auxiliary Fire Alarm System. In 1885, George Milliken of Boston developed improvements to the Auxiliary System that allowed for its wider use in buildings throughout U.S. cities.

By 1871, private companies, such as American District Telegraph (ADT), Holmes Electric Protective, and Rhode Island Electric Protective, were founded to serve the public using commercial central stations in districts of a city to provide messenger service and later electric burglar alarm and fire alarm telegraph service to homes and businesses. The early systems used forms of the Auxiliary Fire Alarm System to send one round of code to summon a messenger, two rounds to summon a policeman, and three rounds for a fire emergency. As these telegraph type systems continued to flourish, problems were discovered that reduced their reliability. The major challenge was that a break or ground in the loop serving the signal boxes connected to the central station could disable the entire loop, thus rendering loss of protection to the central station subscribers.

The central stations used runners with jumper cables to repair breaks in the cable caused by fires; they had official-looking badges to gain access to fire scenes to use the jumper cables. The first technological solution to this loop "break and ground" problem came in 1882 when Chauncey F. McCulloh patented a loop design, appropriately labeled the McCulloh Loop.

Fred Gibson of ADT found multiple patents filed by individuals named McCulloh, McCullough, or McCulloch. Lewis H. McCullough and Geoge C. McCullough were issued patents about 5 years apart for a Fire Telegraph System and a Fire or Police Telegraph, respectively. Because both McCullough's had the same address (Richmond, Indiana), it has been assumed, but never proven, they were brothers. Another alarm system patent holder, Alfred H. McCulloch of New England, received his patents in 1895. According to Gibson, Chauncey's patent most closely describes the McCulloh circuit that is still in use today.

In the 1896 *Handbook of the Underwriters' Bureau of New England,*[2] the authors stated, "There is no reason why a thermostat system should not be superior to watchman and clock for a majority of risks." Figure 1-1 shows early heat detectors. The third edition of the *Fire Insurance Inspection and Underwriting Handbook,*[3] published in 1923, stated, "The best service can only be obtained when the watchman is employed for one purpose—that of watching, not when he has to do cleaning nor when he and his family live on the premises, as, in the latter case, more often than not he will be found in his apartment instead of patrolling the building."

In their 1923 standards for Watchman Service and Fire Alarm, the G. A. Insurance Co. stated that the "Watchman must make hourly rounds at approved stations, 6 p.m. to 6 a.m. and also during the day on Sundays, holidays, or when the building is idle. For watchmen, locked lantern burning lard oil, sperm or signal oil is preferred. Electric flashlight [is] also approved." They included a quote attributed to the National Board of Fire Underwriters, that "A fire alarm system when it is an integral part of a central office watchman's supervisory system is considered superior to any other means of transmitting an alarm of fire."

The ninth edition of the National Fire Protection Association (NFPA) *Handbook of Fire Protection*[4] (1941) provided guidance on hiring the right watchman:

> Care should be taken to get the right man. His character, habits, and reliability must be unquestioned. A non-smoker is preferred; in no event should a man be employed who cannot refrain from smoking on his rounds.
>
> His intelligence should be keen; emergencies require quick decisions and good judgment.
>
> The watchman should be a courageous man and his physical condition should be satisfactory. Essentials are unimpaired eyesight, hearing, and sense of smell; two legs and two arms; a sound heart; physical strength. A very young, or an extremely aged watchman should not be employed.

Figure 1-1. Early wooden heat detectors. (Photos courtesy of Wayne Alarm, Lynn, MA.)

The watchman's education should be at least average. He should speak English and be able to write a simple report.

The design of an electric burglar alarm was first patented in 1850 by a Somerville, Massachusetts, man named Augustus Pope. Using the recently developed telephone, Edwin Holmes established a central station and promoted the use of electric burglar alarm systems that transmitted information to his central station.

The first electric fire detector was developed in Brooklyn, New York, in 1863 by Alexander Ross. However, the first electric fire sensor to see commercial use was designed by another New Yorker, William B. Watkins. By the early 1870s, Watkins had developed remotely monitored fire alarm systems using heat detectors and in 1873 formed the first private fire alarm company, which survives to this day as AFA Protective Systems (headquartered in Syosset, New York). Watkins developed the Watkins Thermostat, the precursor to the self-restoring bimetallic heat detector used in fire alarm systems in the 21st century.

Watkins's Automatic Signal Telegraph Company, founded in 1873, was the first business to offer central station fire alarm monitoring. In a brochure published in 1880, Watkins advertised, "The first five minutes at a fire are worth the next five hours." Using heat sensors distributed on each floor of a building connected to a transmitting instrument, Watkin's system could transmit impulses to the receiving station identifying the appropriate building and floor. The system worked well enough to receive the New York Board of Fire Underwriter's approval, with the Board agreeing to grant insurance discounts to their subscribers who had these systems installed in their buildings.

Watkins expanded his operation to Boston by partnering with the Boston Protective Company, a salvage company. After converting one of the Boston Protective dispatching stations to a central station in 1874, Watkins petitioned for and received approval of his systems by the Boston Board of Fire Commissioners.

The 1919 edition of the *Crosby-Fiske-Forster Handbook of Fire Protection* identified 28 electric fire sensors. Watkins developed his alarm-monitoring business to include waterflow alarms in addition to his thermostat alarm systems. Building owners who used these combination systems were able to receive insurance credits that equaled the subscriber's fee. In 1894, Watkins merged with a competitor, the U.S. Electric Fire Signal Company, and renamed the combined company the Boston Automatic Fire Alarm Company. In 1889, the parent company was renamed the Automatic Fire Alarm and Extinguisher Company, shortened to Automatic Fire Alarm (AFA).

Heat detectors were the first devices made available for the protection of residences. Braconier manufactured the first device for this market in about

1921. The device was listed by the Underwriters Laboratories Inc. (UL) and consisted of a 1-in.-diameter steel ball bearing secured in a holder with a low-melting-point solder. When the solder melted due to the heat of a fire, the ball fell from the holder, striking the floor and making a noise.

In the 1930s, several explosive devices were marketed. Two such devices were the Poe firecracker unit (date uncertain) and an Explosive Alarm (1933). Both units consisted of a small metal housing containing a firecracker with one or more exposed fuses.

Also in the 1930s, the first spring-wound bell units appeared. One, manufactured by U.S. Firegard in 1931, consisted of a 5-in.-diameter, mechanically powered bell held in the wound position with a low-melting-point solder. The unit contained a spring-wound clock mechanism, which, when released by the melting of the solder, caused a clapper to spin, which struck the outer gong. Also developed during this period was a unit called Telarm (1933) made by the Lakeside Manufacturing Company in Chicago. This was a spring-wound mechanism actuated by a trip lever connected to a pointer-indicating temperature gauge. The temperature gauge pointer indicated the ambient temperature and the adjustable trip lever could be positioned to trip the alarm anywhere between −40 and +120°F. Thus, this unit served a dual purpose as both a fire detector and ambient temperature gauge.

In 1939, the Gamewell Fire Alarm Company marketed the first electrically powered residential heat detector. This unit consisted of a bimetallic sensor, ac-powered buzzer, and neon pilot light. The unit had a screw shell base that could be screwed into a standard Edison-base light bulb socket. The neon light indicated that power was available; when the bimetal deformed from the heat of the fire, the buzzer was energized.

These very early units apparently did not make any significant impact in the marketplace. By all indications, they were produced for only a short period of time and then discontinued for lack of sales. In fact, prior to 1955, there were very few residential fire detection devices installed in homes in the United States, and those that were used in residential occupancies were predominantly local fire alarm systems, which adapted equipment that had been developed and marketed primarily for commercial buildings.

The start of what might be called today's residential fire detection business began with the UL listing of the first gas-powered fire alarm unit in December 1954, which was presented by William Messik of the Standard Alarm and Signal Company. This unit was quite similar to those still marketed today, consisting of a steel tank containing Freon gas connected to a diaphragm-type air horn by a short length of small-diameter tubing. The tubing was plugged with a low-melting-point solder seal and used a heat collector attached to the tubing at the solder plug location. The unit was listed for a 10-ft space rating. This listing was followed in January 1955 by a listing for the Evergard Fire Alarm Company of their model A-10, also for a 10-ft space rating.

What made these developments significant was not so much the devices themselves (since similar devices were available as early as the 1930s) as the aggressive marketing by these companies. Earlier companies that had long since gone defunct had found that there was no natural market for the device. Although people had an inherent fear of fire, most people felt that they would not have a fire and therefore would not need to spend money on a fire detection device. What these later companies did was to establish distributors who aggressively marketed on a door-to-door basis.

Many things can be said both for and against the sales techniques used in these early years. Cost was high (often averaging $800 per home) and many overzealous salespeople used scare tactics and other high-pressure sales techniques, which were generally effective in both selling the product and instilling fear in the homeowner. On the positive side, however, they were able to achieve some significant market penetration and also did a great deal in the way of educating the public on home fire safety. For example, they were the first to effectively educate the public and promote the concepts of fire drills in the home and escape planning. In fact, as far as cost goes, the $800 per home average was less than the installed prices of residential fire detection systems, which also used heat detection exclusively. Although their market penetration was fairly low (most probably due to the high cost), these devices saved many lives, as witnessed by the number of testimonial letters and fire reports in the files of these companies.

By 1959, the Evergard Fire Alarm Company had released two additional models with 15- and 20-ft space ratings, respectively. This was a significant improvement in the basic sensitivity and response of these units.

Also in 1959, the first spring-wound bell-type heat detector having significant market penetration was listed. This was the model V-25 of Interstate Engineering Company, which had a 25-ft space rating. A spring-wound bell unit had been listed in 1956 by a company called Fyre Fyter, but was offered only for about 1 year and very few were sold. The Interstate Engineering unit was originally marketed in both 6- and 10-in. gong sizes and was marketed in large numbers until it was replaced by a more sensitive model in 1971.

In the early 20th century, ADT, Holmes Protective, AFA, Grinnell, and Automatic Fire Protection (AFP) established contracts with each other to supply detection, sprinkler systems, sprinkler system supervisory equipment, and central station monitoring services. As business in the fire alarm and monitoring fields grew, so did the development of the devices used for detection. George Smith patented the first pneumatic system in 1907, later known as the Aero Automatic Fire Alarm. ADT, AFA, ATMO (from "Atmosphere") manufactured by Kidde, and others developed similar *rate-of-rise*-principle pneumatic heat detectors.

One of the earliest smoke detectors was used on ships. Invented by a merchant seaman named Walter Kidde, the smoke "sniffer" utilized the system of speaking tubes that was being made obsolete by the installation of electric intercoms. These metal tubes ran from the ship's bridge to various compartments. By one account, Kidde was on bridge duty one night when he noticed wisps of smoke emanating from one of the speaking tubes. This gave him the idea to build a box around the tubes in the bridge and install a fan to draw on the tubes and a light to make any smoke visible (see Figure 1-2). Later models incorporated a photocell to automatically detect the smoke and a rotary valve to discharge carbon dioxide down the tube into the compartment from which the smoke had come. Underwriters Laboratories listed the Kidde Model G1 in 1929 (see Figure 1-3) after it had been used in ships for most of the 1920s. One such system was still in use in 1994 on a cruise ship built in 1954 in the Quincy, Massachusetts, Shipyards. Coast Guard regulations still recognize olfactory fire detection systems that blow air from ship compartments at the face of the crewperson steering the ship.

Figure 1-2. The Kidde G1 shipboard smoke detector, showing the numbered speaking tubes to the ship's various compartments.

Figure 1-3. View of the Kidde G1 shipboard smoke detector. It was 6 ft tall and 3 ft wide.

In the late 1930s, ADT developed the first projected-beam photoelectric smoke detector. At about the same time, Walter Kidde Company developed their version of spot-type photoelectric smoke sensors, which were installed by independent alarm companies for the next 12 to 15 years. Also during the 1930s, ionization smoke detector technology was developed in Germany to protect German munitions plants. After the war, scientists working for Cerberus, a Swiss manufacturer, refined the German model and marketed the smoke detectors in America through a Baker Industry company called Pyrene (later called Pyrotronics). The Swiss smoke detectors needed 220 V dc for their power, which of course required that these fire alarm systems be connected to the building's electric power system. This feature made the detector costly and vulnerable to fires that attacked the electric system before the detectors could operate. By the early 1950s, Cerberus was manufacturing and marketing ionization-type smoke detectors for the protection of high-value assets. Faced with high costs and cumbersome licensing requirements in the United States, their task was daunting. For example, early models used radium as the source material, which meant that individual state regulations were involved, which caused chaos in national distribution. Changing the source material to americium brought regulation under the U.S. Atomic Energy Commission (AEC) and a single set of federal requirements. However, the AEC (and its successor, the Nuclear Regulatory Commission, NRC) required that records be kept of every individual detector's installed location and that a wipe test (wiping the exterior with alcohol-soaked gauze and checking that for radiation leakage) be performed on every detector at least annually. Further, every detector bore a label requiring its return to the manufacturer for disposal. Many years later, both of these requirements were dropped.

Other companies began developing improved smoke detection technology. Fire Alert, a company formed in 1964 by Denver Burglar Alarm, developed a detector requiring only 24 V of alternating current to operate. Statitrol, Inc., another Denver-based manufacturer, designed a battery-operated ionization smoke detector. Early photoelectric smoke detector designs by Pyrotector and Electro Signal Lab Inc. (ESL), both Massachusetts companies, refined photoelectric smoke detector technology and brought the cost of these devices down to a mass-marketable price.

Major papers of importance were published in 1962. These were a study by McGuire and Ruscoe[5] (probably the single most-quoted study in fire detection today) and a Federal Fire Council Newsletter, "Needed—Protection for People," published in September 1962.[6] This Fire Council addressed the concern that, until that time, almost all of the fire protection devices in the marketplace were designed for property protection, but were being incorrectly used for protection of people. It discussed the need for systems specifically designed to protect people and the need for research and development. It went on to say that the most logical candidates to conduct this research were municipal, state, and federal governments, as is the case today.

The next major milestone in the residential fire detection business occurred in 1965 with the development of the single-station smoke detector by Pyrotector Company and Aljenik Industries. Both of these units were ac-powered, line-cord-connected, photoelectric units.

When Richard Saltzgaber, the founder of Aljenik, brought his single-station smoke detector for testing to Underwriters Laboratories, they invoked the NFPA requirements for local fire alarm systems by requiring an audible trouble signal upon failure of main operating power and a built-in heat detector. Aljenik chose not to comply with these requirements and began marketing without approval. Pyrotector chose to accept these requirements and included nickel–cadmium batteries and a separate trouble buzzer in their detector. Based on these additions, Pyrotector received the first listing for a single-station residential smoke detector on August 18, 1966. On July 6, 1971, after UL revised their requirements in this area, Aljenik received a listing of its unit.

Even more interesting than the UL experience was Saltzgaber's experience with the NFPA Sectional Committee on Detection and Signaling Systems. In a letter to Saltzgaber dated February 9, 1965, the Chairman of the NFPA Sectional Committee wrote,

> The question at hand is whether or not an electrically operated fire alarm detector utilizing a flexible cord to an electric outlet in a home should receive recognition. The argument in favor is that this type of equipment, utilizing a smoke detection feature, is considered by some to be at least adequate to the protection afforded (by) gas-operated or mechanically operated temperature detecting alarm units which presently bear a UL label.

It is interesting to note that the committee's position was virtually reversed within 5 years.

Throughout the course of U.S. history, the requirements of fire alarm systems have changed as a result of lessons learned from major fires and fire tests. Publication of the results of the Operation School Burning tests conducted by the Los Angeles Fire Department in 1959 provided further impetus to the development of detection technology.[7] This was the first major series of tests employing both heat and smoke detectors. The results had a major impact on the fire protection community, although the full impact was not fully realized until many years later. For example, these tests were the first to conclude that smoke detectors can provide a higher level of life safety in the protection of large, open areas of a building. The results also indicated that detectable quantities of smoke preceded detectable heat levels in most cases. In fact, with Operation School Burning No. 2,[8] detailing a second series of tests and published in 1961, this work represented the only well-accepted definitive study of the comparative response of heat and smoke detection devices for approximately the next 15 years.

Endorsements and regulations fueled the boom in smoke detector sales, which rose from 500,000 units in 1973 to 2.9 million units in 1976. By the late

1970s, there were 130 manufacturers of smoke detectors. However, not all smoke detectors were reliable, and after some embarrassing failures during public tests, many suppliers left the business, with about 50 manufacturers remaining in the early 1980s.

In the mid-1970s, the National Bureau of Standards (now the National Institute of Standards and Technology, NIST) contracted with Illinois Institute of Technology Research Institute and Underwriters Laboratories to obtain data regarding the performance of smoke detectors and their effectiveness in residential environments. The tests were performed in houses located in the Indiana Dunes area and were the first to evaluate detection performance based on the amount of escape time offered when the detector was actuated. Based on the test results,[9,10] smoke detection on every level of a home was proven to provide adequate escape time in roughly 90 percent of the fire scenarios. An independent study in 1976 by the Minneapolis, Minnesota, Fire Department resulted in conclusions nearly identical to the results of the Dunes tests. Other tests in Los Angeles, California, and Ontario, Canada,[11] led to significant changes in the installation, maintenance, and use of smoke detectors.

Watkins and Holmes would be pleased with the systems developed for the 21st century. We now have heat and smoke detectors that continuously transmit what they are sensing in the environment to a computerized fire alarm control panel, which helps to maintain the credibility of the fire alarm system by reducing the responses to nonfire conditions.

BIBLIOGRAPHY

Ditzel, P., *Fire Alarm! The Fascinating Story Behind the Red Box on the Corner,* Fire Buff House Publishers, Conway Enterprises, New Albany, IN, 1990.

Gamewell, *"Emergency Signaling,"* Gamewell Fire Alarm Telegraph Company, New York, 1916.

Meili, E., *My Life with Cerberus, Cerberus* AG, Mannedorf, Switzerland, 1990.

Watkins, *"Protection Against Fire and Burglary, Municipal Fire Alarms,"* Promotional Booklet, Watkins Automatic Signal Telegraph Company, Boston, MA, 1880.

REFERENCES CITED

1. Greer, W., *A History of Alarm Security,* 2nd ed., National Burglar and Fire Alarm Association, Silver Spring, MD, 1991.

2. Crosby, E. U. (Manager), *Handbook of the Underwriters Bureau of New England,* 1896 [Anniversary facsimile edition, NFPA, Quincy, MA, 1986].
3. Dominge, C. C., and Lincoln, W. O., *Fire Insurance Inspection and Underwriting,* 3rd ed., Spectator Company, New York, 1923.
4. Moulton, R. (Editor), *Crosby-Fiske-Forster Handbook of Fire Protection,* 9th ed., NFPA, Boston, MA, 1941.
5. McGuire, J., and Ruscoe, B., The value of a fire detector in the home, Nat. Res. Council. Canada, 1962.
6. Federal Fire Council, "Needed—Protection for People," Newsletter, Federal Fire Council, Washington, DC, September 1962.
7. Los Angeles Fire Department, *Operation School Burning,* NFPA, Boston, MA, 1959.
8. Los Angeles Fire Department, *Operation School Burning No. 2,* NFPA, Boston, MA, 1961.
9. Bukowski, R. W., Waterman, T. E., and Christian, W. J., *"Detector Sensitivity and Siting Requirements for Dwellings:* A Report on the NBS 'Indiana Dunes' Tests," NFPA No. SPP-43, NFPA, Quincy, MA, 1975.
10. Harpe, S. W., Waterman, T. E., and Christian, W. J., *"Detector Sensitivity and Siting Requirements for Dwellings,* Phase 2," ITT Research Institute, Chicago, 1975.
11. Ontario Housing Corporation, "Smoke Detectors in Ontario Housing Corporation Dwellings," Ontario Housing Corporation, Ontario, Canada, January 1978.

2

Establishing Fire Protection Goals and Understanding Fire Signatures

INTRODUCTION

Proper design of a fire alarm system requires an understanding of fire itself—its by-products or *fire signatures,* its effects, how it can be detected, and the importance of installing a fire alarm system. This chapter describes specific fire signatures and their measurement, effects of these signatures on humans, and how fire signatures reach fire detectors. Also treated in this chapter are the importance of establishing fire protection goals and understanding the limitations of fire alarm systems.

CHOOSING A FIRE ALARM SYSTEM

Once the owner of a building decides that he or she needs a fire alarm system, it is the system designer's responsibility to choose a fire alarm system that meets the owner's goals. In many cases, the owner will need guidance in this area, so the designer must be able to articulate what each fire protection goal is and why the owner may or may not wish to establish goals that exceed the requirements of the building code.

ESTABLISHING FIRE PROTECTION GOALS

In order to better understand how best to apply fire detection and fire alarm systems, one must develop the goals that the system must meet. The goals, called fire protection goals, establish the baseline for the detection and alarm system operation. An owner's fire protection goal may be to simply comply with the local building code; *NFPA 5000*™, *Building Construction and Safety Code*™; NFPA *101*®, *Life Safety Code*®; or NFPA 1, *Uniform Fire Code*™. Essentially, when the owner chooses to comply with the prescriptive requirements of a code, he or she defaults to the minimum requirements of that code as the fire protection goal. Most building codes are developed to prevent citywide conflagrations and offer a minimum of occupant protection. In other words, they specify the least one can do to build and occupy a building.

Other fire protection goals include

Life safety
Property protection
Mission protection
Heritage preservation
Environmental protection

Life Safety

When incorporating life safety as a fire protection goal, the owner is acknowledging that the protection of the occupants, both employees and visitors, is the primary reason for installing a fire alarm system. Meeting this goal requires the use of detectors that provide early warning of the fire.

Numerous tests have demonstrated that smoke detectors are the most practical fire detectors for use as an early warning device. One conclusion of the second Indiana Dunes test program[1] stated,

> Supporting the first-year results, the fixed-temperature heat detectors rated for 50-ft (15-m) spacing [135°F (57°C)] used in this test series, in the room of fire origin, provided little life-saving potential. These detectors failed to respond to a majority of the fires and when they did respond they were considerably slower than smoke detectors located remotely from the fire.

The life safety goal can usually be met with the installation of smoke detectors as described in NFPA *101* as "complete smoke detection systems":

> **Complete Smoke Detection Systems.** Where a complete smoke detection system is required by another section of this Code, automatic detection of smoke in accordance with *NFPA 72*®, *National Fire Alarm Code*®;

shall be provided in all occupiable areas, common areas, and work spaces in those environments suitable for proper smoke detector operation.

Additionally, this goal must be met while ensuring the reliability (system will operate when needed) and credibility (no false alarms) of the installed fire alarm system.

Property Protection

A property protection goal is generally developed in concert among the owner, the fire alarm system designer, and the owner's insurance company. Traditionally, heat detection has been used as the detector of choice for property protection; however, depending on the property type and value, other faster responding detectors such as smoke or flame detectors may be used. Generally, the designer will choose an estimated fire size to be detected. He or she will also determine how fast the detection system should operate based on the fire department response time and water supplies available. Issues like these are important considerations for meeting the property protection goal.

Mission Protection

Mission protection can be defined in terms of how the owner of a building or company can answer the question, "Will I be able to continue my operation after the fire?" Some owners do not realize that although they may have insurance, if they take too long to get back into production or operation, their customers may stay with the competitors used during the reconstruction time. Getting those customers back may prove difficult. So once again, the owner and the designer must consider factors such as fire size at detection, water availability, fire department response time, and fire department effectiveness to determine the best form of detection for protecting the owner's mission.

Heritage Preservation

In addition to having the right detector for the application, historic properties often require that the detection be installed in an unobtrusive fashion. The installation of detection should cause as little impact as possible on the historic fabric of the building. More expensive detection or additional concealed detection may be needed to accomplish the dual goals of owners of historic buildings.

Environmental Protection

Protecting the surroundings and understanding the impact of the fire alarm system on the surroundings are both important. There may be noise issues that need to be addressed to ensure the right notification appliance type is chosen. There may be operations that can be performed by the fire alarm control panel that will preserve the surroundings in the space protected by the system. For instance, interfacing the fire alarm system with automatic doors or vents may help to preserve the surroundings during the operation of a suppression system or during a fire condition.

FIRE SIGNATURE FUNDAMENTALS

From the moment of its initiation, fire produces a variety of changes in the surrounding environment. Any of these changes in the ambient condition is referred to as a fire signature and can be monitored by a detection system. The production of smoke, for example, will result in a decrease in visibility that can be detected. To be useful, however, a fire signature should generate a measurable change in some ambient condition, and the magnitude of that change (the *signal*) must be greater than the normal background variations (the *noise*) for the condition. All other factors, for example, hardware costs and detection time, being equal, the preferred fire signature will be the one that can generate the highest signal-to-noise ratio in the earliest period of fire development. Fuel-specific signatures, such as the release of hydrogen chloride (HCl) from polyvinyl chloride (PVC) combustion, may help in detection for specialized applications, but are of little use for general-purpose functions. Individual fire signatures are discussed later. The operational principles of the signatures and their associated detection mechanisms are discussed in a later section of this chapter also.

Aerosol (Smoke) Signatures

The process of combustion releases very large numbers of solid and liquid particles into the atmosphere. The size of the particles ranges from 5×10^4 to 10 micrometers. These particles suspended in air are called aerosols and, when produced by fire, are usually called smoke.

NFPA 72®, *National Fire Alarm Code*®, defines smoke as "the totality of the airborne visible or invisible particles of combustion." In NFPA 90A, *Standard for the Installation of Air Conditioning and Ventilating Systems,* the definition of smoke also includes gases. According to one researcher,[3]

Smoke usually refers to a gaseous disperse system consisting of particles of low vapor pressure produced as a result of incomplete combustion processes. It takes the form of solid carbon particles and minute droplets of high molecular weight carbonaceous matter destructively distilled due to the heat of combustion.

Smoke particulate consists of liquid and solid particles covering a spectrum of particle sizes, depending upon the combustible materials and age of the smoke. As smoke ages and cools, the particles grow in size. This cooling effect relates to the stratification of smoke during smoldering fires and the declining heat release rate from fire plumes.

The essential feature of smoke is its instability. Under the influence of a lively Brownian motion, the particles in a cloud collide with one another and agglomerate. This process goes on continuously until the number of particles has been considerably diminished and their average size greatly increased. The mass concentration of relatively stable smoke is usually low, in the majority of cases, below 1 m^3. High concentration leads to rapid coagulation and loss by sedimentation.

The aerosols resulting from a fire can be classified into two different fire signatures, depending on their ability to scatter light. Particles smaller than 0.3 micrometers in size are usually called *invisible* since they do not scatter visible light efficiently. Particles larger than 0.3 micrometers do scatter visible light and are usually called *visible*.

"Visible" can be a relative term. Objects become visible when they reflect light energy from their surfaces into our eyes. This reflection is a function of the optical characteristics of the surface (color, reflectance). For small particles, the ratio of the wavelength of the light relative to the particle diameter is also important—this ratio is $\alpha = \lambda/d$. For $\alpha < 0.1$, particles will no longer scatter light sufficiently for it to be seen. This means that, given the range of light wavelengths that can be seen, particles below about 0.3 micrometers are invisible in visible light. Therefore, listed smoke-sensing fire detectors are tested only for response to smoke particles and not to gases present in the smoke cloud.

When they first introduced their fire detectors, some manufacturers described the detectors as sensing products of combustion. Underwriters Laboratories Inc. (UL) listed ionization detectors as "combustion products detectors" in the early 1960s. After a UL research program was conducted in early 1968, however, ionization smoke detectors and photoelectric smoke detectors were categorized simply as smoke detectors because both were tested for response to smoke particles only using the same fire scenarios.

The term "products of combustion" can be ambiguous in that aerosols, heat, light energy, and gases are all products of the combustion process.

Table 2-1. Thermal Particulate Point in Air

	Temperature	
Material	°F	°C
Bakelite™	380	191
PVC Insulation	290	142
Acrylan Carpeting	340	169
Wool	360	180
Paper	500	257
Pine Board	320	158
Polystyrene	710	373
Polyethylene	410	208
Motor Oil (SAE30)	310	153

Heating of materials during the preignition stage of a fire produces submicrometer size particles ranging in size from 5×10^{-4} to 1×10^3 micrometers. These particles are generated at temperatures well below ignition temperatures. The temperature at which submicrometer-sized particles are generated from materials is defined as the thermal particulate point (see Table 2-1). These invisible aerosols are generated in very large quantities. A 0.098-ounce (2.8-g) sample of bond paper burned in air in a 7.5-sq ft (2.28-m^2) room with an 8.9-ft (2.71-m) ceiling produces a maximum particle concentration of 3.6×10^{10} particles/cu ft (1.3×10^6 m^3).

As heating of a material progresses toward the ignition temperature, the concentration of invisible aerosols increases to the point where larger particles are formed by agglomeration. As this process continues, the particle-size distribution becomes log normal, with the most frequent particle sizes in the 0.1- to 1.0-micrometer range. Particles less than 0.1 micrometer disappear either by agglomeration or evaporation, and particles greater than 1.0 micrometer are lost through the process of sedimentation, following Stokes's law. Aerosols in this latter size range are remarkably stable and contain both visible and invisible particles. The production of visible aerosols can occur prior to ignition and is usually initiated at temperatures several hundred degrees higher than the thermal particulate point.

Test results from smoldering and flaming combustion of various materials indicate that, generally, smoldering fires produce more large particles than flaming fires.[3] It is important to note that the maximum relative particle concentration for both fire types appears to be in the range of particle sizes smaller than 0.3 micrometer. Particle size seems to affect the optimum detection method.

Energy Release Signatures

In all stages, fire releases energy into the surrounding environment. This energy release produces several useful fire signatures.

IR and UV Radiant Energy: Infrared (IR) and ultraviolet (UV) signatures are usually the earliest energy signatures that can be detected from a fire. With the exception of acetylene and other highly unsaturated hydrocarbons, the infrared emissions from hydrocarbon particles are particularly strong when particles are in the 4.4 micrometer region (due to carbon dioxide, CO_2) and in the 2.7 micrometer region (due to water vapor, H_2O). These emissions account for nearly all the emitted energy. Since the infrared component of sunlight (which is a potential noise, or false signal, source) is reduced in these regions by absorption due to atmospheric carbon dioxide and water, a high signal-to-noise ratio is obtained. The carbon dioxide–water radiation signature can be used effectively for detection, but there is the possibility of noise from manmade infrared sources. The modulation of the energy output level due to flame flicker is another signature that affects the infrared signal from a flame. This flicker is characteristic of flame and has a frequency range of 5 to 30 Hz.

The total infrared signature has been used in the detection of both smoldering and flaming fires. The disadvantage of the infrared signature is that it has a wide range of noise levels from solar and manmade sources. Detectors that use both the carbon dioxide–water vapor and flicker signatures have an excellent signal-to-noise ratio.

The ultraviolet fire signatures appear in flames as emissions from hydroxal (OH), carbon dioxide (CO_2), and carbon monoxide (CO), ranging in size from about 0.27 to 0.29 micrometer. Although the signal-to-noise ratio of the UV signatures is less than that of the IR signatures, UV signatures have been effectively used for flame detection. They are especially effective where magnesium or its alloys may be involved because of the strong signal that burning magnesium causes in the 0.28-micrometer region.

Thermal Energy: Convected thermal energy from a fire causes an increase in the air temperature of the surrounding environment. The time required for release of sufficient energy to produce a significant convected energy signal varies from less than 1 minute for rapidly developing fires to hours for slowly developing/smoldering deep-seated fires. In contrast to aerosol and radiant energy signatures, convected thermal energy signatures often appear well after life-threatening conditions have been reached from excessive aerosol and/or toxic gas concentrations.

Gas Signatures

During a fire, many changes occur in the gas content of the atmosphere. These changes can be found both in the area of fire origin and in areas far removed from the fire. These changes in atmospheric gas content are mostly due to the addition of gases that are not normally present. These changes are termed *evolved gas signatures.*

Another gas signature that occurs during a fire is the reduction of oxygen content; this is termed the *oxygen depletion signature.*

Many gases evolve during a fire, for example, water vapor (H_2O), carbon monoxide (CO), carbon dioxide (CO_2), hydrogen chloride (HCl), hydrogen cyanide (HCN), hydrogen fluoride (HF), hydrogen sulfide (H_2S), ammonia (NH_3), and nitrogen oxides. Most of these evolved gas signatures are fuel-specific, and since they are not associated with a sufficiently large number of fuels, cannot be used for general-purpose detection.

The rate at which any gas is produced in a fire is a relatively constant function of the mass burning rate of the fuel. Referred to as the *yield fraction,* this is the fraction of the burned mass that results in gas. Such yield fractions are a function of the specific fuel, and for oxygenated species depend on the amount of combustion air available.

Carbon monoxide (CO) is present in nearly all fires. Tests have shown a potential use for a carbon monoxide signature in fire detection. The rate of carbon monoxide production varies considerably with the type of fuel, the amount of ventilation, and whether the fire is of the smoldering or open-flaming type. Yield fractions for CO are typically a few percent in fully ventilated fires, but can increase by a factor of 30 to 50 at flashover. Yield fractions in smoldering fires can be just as high, but, due to low burning rates, the quantity of CO produced is low.

When CO is examined as a fire detection signature, it is generally found that detectable levels of CO (75–100 ppm) occur after detectable levels of particulates. There are smoke detectors that use CO in conjunction with particulates as the basis of multimode sensors, which can discriminate unwanted fires from the more common false alarm sources, for example, cooking and steam from showers, which lack the CO component.

FATALITY POTENTIAL FOR FIRE SIGNATURES

Fire signatures create a hostile and often fatal environment. Exposure to signatures produces psychological and physiological effects, both individually and in combination, in humans.

Effects of Aerosols

Aerosols derived from a fire produce several effects on humans. Much of the aerosol consists of carbon particles upon which may be absorbed a number of irritant compounds, such as organic acids, aldehydes, or hydrogen chloride. Studies with dogs exposed to kerosene aerosol consisting largely of carbon soot produced no noticeable respiratory damage. Most of the soot was expelled or coughed out by the dogs within 24 hours. This study further indicated that the aldehydes in wood aerosol were the significant factor in causing respiratory tract swelling and irritation.

Inhalation of these products makes breathing difficult and may be accompanied by severe coughing, a burning sensation in the chest and throat, and irritation of the eyes. In addition, absorption and scattering of light from visible aerosols can reduce visibility to near zero in some cases. Physiological and environmental changes, combined with fear, anxiety, and loss of orientation in a fire situation, often result in panic behavior. Persons who panic in a fire can act irrationally and may become lost or trapped in dead-end corridors or closets. In extreme cases, persons become paralyzed and take no lifesaving action at all.

Visible aerosol concentration is measured in terms of percentage reduction of light per foot (meter) of travel path. In a Japanese study, it was found that persons familiar with their surroundings can tolerate a minimum of 9.2 percent/ft (0.13 OD/m where OD indicates optical density) reduction of light, while those unfamiliar with their surroundings usually only have a 3-percent/ft (0.04 OD/m) tolerance.

Effects of Heat

Heat from fire primarily exposes humans to the potential of burn injury. Experiments with animals have shown that death occurs after 2 minutes of exposure at 212°F (100°C) and after 30 minutes of exposure at 140°F (60°C). Although other tests[4] have shown that the ability to voluntarily endure heat can vary from 2 minutes at 300°F (149°C) to 45 seconds at 740°F (393°C), it is likely that these are extreme levels of tolerance. In addition to surface burns, inhalation of heated gases and aerosols often produces such thermal damage as sloughing of the trachea lining and hemorrhaging in the respiratory track. Exposure to elevated temperatures can also cause shock since blood pools at the body surface as the body attempts to cool itself.[5]

Effects of Toxic Gases

Carbon monoxide, carbon dioxide, and other gases can be generated in sufficient quantity in a fire to produce toxic reactions in humans.

Carbon Monoxide: Carbon monoxide (CO) is present in nearly all hostile fire situations and is an important factor in loss of life from fire. Clinical examination of autopsy reports of fire victims in New York City showed that carbon monoxide poisoning, rather than respiratory tract damage, was the significant factor in death with victims having postburn survival times of less than 12 hours.[6] The usual threshold limit value for exposure to carbon monoxide for an 8-hour working day is 50 parts per million (ppm). In fires with conditions of restricted ventilation, carbon monoxide concentrations as high as 138,000 ppm have been recorded.

Carbon monoxide reduces the oxygen-carrying ability of the blood by combining with hemoglobin to form carboxyhemoglobin. The various effects of carbon monoxide exposure (in order of occurrence) include headache, dizziness, dimness of vision, nausea, increased pulse and breathing rates, confusion and loss of orientation, unconsciousness, reduced pulse and breathing rates, convulsions, and, ultimately, death. Exposure to even moderate levels of carbon monoxide can severely restrict a person's ability to function in a fire situation, especially when combined with the adverse effects of aerosols and elevated temperatures (see Table 2-2). A 1973 study of 106 fire fatalities found that 50 percent involved carbon monoxide poisoning as the cause of death. An additional 30 percent of the fatalities were traced to carbon monoxide poisoning that complicated preexisting effects of heart disease or alcohol use.[7]

Carbon Dioxide: Carbon dioxide (CO_2) is one of the products of complete combustion and is commonly associated with both flaming and smoldering fires in amounts exceeding the threshold limit value of 5000 ppm. Carbon dioxide stimulates the breathing rate; it may also act as an asphyxiant by displacing oxygen. The early symptoms of carbon dioxide intoxication are dizziness and shortness of breath followed by mental excitement and irrational behavior. Concentrations of 9 percent (90,000 ppm) CO_2 in the atmosphere can be fatal in 4 hours due to carbon dioxide—hemoglobin complexing. Although carbon dioxide concentrations up to 6.75 percent have been recorded in bedding test

Table 2-2. **Effects of Carbon Monoxide Exposure**

Concentration (%)	ppm	Exposure Time	Effect
0.02	200	2-3 hr	Mild Headache
0.08	800	45 min	Mild Headache
		2 hr	Possible Death
0.32	3,200	10-15 min	Dizziness
		30 min	Death
0.69	6,900	1-2 min	Dizziness
		10-15 min	Death
1.28	12,800	2-3 breaths	Unconsciousness
		1-3 min	Death

fires, even low concentrations of carbon dioxide will raise the breathing rate and increase the intake of other toxic substances. A 5 percent concentration of CO_2 is used to increase the breathing rate of smoke inhalation victims, speeding the process of elimination of CO found in the blood (see Table 2-3).

Hydrogen Chloride: Hydrogen chloride (HCl) is not found in cellulosic-based fires, but is often associated with the burning of many common plastics, such as vinyls. The threshold limit value is 5 ppm, and a noticeable irritation of the mucous membranes occurs at 35 ppm. Concentrations of 1000 to 2000 ppm produce severe reactions that may be fatal.

Other Gases: Many other gases evolve from burning materials in small but potentially dangerous quantities. Tests of both flaming and smoldering combustion of materials used in aircraft interior finishing produced measurable quantities of toxic

Table 2-3. **Effects of Carbon Dioxide Exposure**

Concentration (%)	ppm	Effect
0.5	5,000	Increased depth of breathing
3.0	30,000	Breathing rate doubles
5.0	50,000	300% increase in breathing rate
10.0	100,000	Possible death, even with sufficient atmospheric oxygen

gases.[8] In the tests, burning polyamides and modacrylics produced hydrogen cyanide. Hydrogen cyanide can also be produced from burning hair, leather, wool, acrylics, polyurethane foams, some dyes, and synthetic fabric finishes. Hydrogen sulfide (H_2S) was produced from chloroprene. Polysulfone plastic produced sulfur dioxide (SO), which can also be produced from burning rubber and wood. Hydrogen fluoride, nitrogen oxides, and ammonia were also detected in the tests.

This is only a small sample of the chemical species that can be present in fire gases. Although usually present only in small, sublethal quantities, these gases can produce additive and synergistic effects within the total system of fire products. These effects are discussed below.

Effects of Oxygen Depletion

Although oxygen depletion is usually confined to the immediate area of a fire, its effects can be felt in remote areas. Oxygen depletion occurs both in a large, open burning fire and, if the building is tightly sealed, smoldering fires of long duration. The physiological effects of oxygen depletion are shown in Table 2-4.

Additive and Synergistic Effects of Toxic Materials

Additive and possible synergistic effects must be considered when dealing with mixtures of toxic materials. In a decade of research, combustion toxicologists have determined that most observed toxic effects in fires are the result of exposure to the short list of toxic species discussed above.

Considering the combined effects of increased breathing rates associated with elevated carbon dioxide and carbon monoxide and decreased oxygen with the irritant effects of fire-generated aerosols and the possible synergistic

TABLE 2-4. Effects of Oxygen Depletion*

Concentration (%)	Exposure Time	Effect
21–17	Indefinite	Respiration volume decrease, loss of coordination, and difficulty in thinking
17–14	2 hr	Rapid pulse and dizziness
14–11	30 min	Nausea, vomiting, and paralysis
9	5 min	Unconsciousness
6	1–2 min	Death within a few minutes

*Approximate figures.

effects of small amounts of other toxic gases, it is clear that persons exposed to these conditions can quickly be rendered helpless. Thus, quick-response detection devices are critical to life safety in fire situations.

TRANSPORT OF FIRE SIGNATURES

A fire detector is a device with a preset sensitivity level. When the area (or conditions) immediately surrounding the detector exceed those preset parameters, a signal is produced. First, fire detectors are designed to respond to matter or energy that is usually but not exclusively associated with combustion. False alarms occur when identical or similar matter or energy is produced within the protected space by controlled combustion or noncombustion sources. Second, all fire detectors have a threshold value, or sensitivity, for sensed parameters that must be exceeded before the detector produces a signal. Third, the detector can only respond to the level of the sensed parameter present at its location, that is, regardless of the amount of matter or energy being released by the fire, only that which actually reaches the detector can be sensed. The mechanisms associated with the transport of matter and energy from the fire to the detector are critical to detector response.

Fire detectors typically sense one of three parameters: (1) convected heat, (2) smoke particles, and (3) radiated energy of mass or energy released by combustion. The generation and transport of each parameter differs, and the differences should be understood for optimum detector application.

Fire is a generic term that refers to flaming combustion, smoldering combustion, or pyrolysis. Flaming combustion usually releases considerable quantities of convected heat, smoke particulate, and radiant energy. Smoldering, however, produces particulate that has considerably different physical properties from particulate produced by flaming combustion. True smoldering only occurs in char-forming materials since the thermal insulating nature of the char is necessary to trap the heat and maintain the glowing reaction zone. Pyrolysis is the thermal degradation process in non-char-forming materials, which produces particulate similar to smoldering but without a glowing reaction zone.

Flaming combustion is easiest to detect because it produces all the signatures to which detectors are sensitive, and the rate of heat release from the combustion is generally sufficient to rapidly transport the particulate to the detector area.

Smoldering and pyrolysis, in contrast, are only adequately detected by smoke detectors. Delays in detection can occur because the limited quantity of heat liberated provides little of the buoyancy needed to drive the particulate to the detector location. The smoke that is produced moves more slowly, is less concentrated in a stratified layer, and is much more subject to the influences of forced convection (heating, ventilating, and air conditioning) systems and other influences.

For detectors that respond to infrared or ultraviolet flame radiation, the energy transport process occurs at the speed of light. The only potential interference with this transport process is blockage of the signal by a physical object. Infrared radiation will reflect off wall surfaces, so even if the detector is not in the direct line of the fire, response can still occur. Response time in this case will be delayed because energy dissipates in the surface reflection process. Ultraviolet-type flame detectors, however, do not respond to reflected energy and must be within the direct line of sight of the flame for response.

Losses are another important factor in the transport process. With thermal energy, the temperature of the fire gases is reduced by entrainment (mixing) of cool air into the fire plume and by convective heat transfer to surfaces. It is often not understood that, when the fire plume first reaches the ceiling and begins to spread outward, as much as 90 percent of the energy contained in the layer is lost to the ceiling. This is one reason why heat detectors are the slowest of fire detectors.

The principal mechanism by which the smoke concentration is reduced is dilution as clean air is entrained into the plume. Although surface losses do occur, the mass loss to surfaces is not significant in reducing the smoke concentration in the upper air layer.

Flame radiation, as with any radiated energy, decreases in intensity with the square of the distance between the radiating source and the target. Some flame detectors can be fitted with concentrators that reduce the effective field of vision and allow the detector to respond at longer distances to the same radiated energy level.

The characteristics of the various fire signatures, with their potential to cause serious fire injury or death, are important considerations in fire alarm system design. Design parameters for signal initiation, transmission, and processing; electrical supervision and installation; and power supply sources are based on fire signatures.

UNDERSTANDING THE LIMITATIONS OF A FIRE ALARM SYSTEM

Unfortunately, many owners expect a fire alarm system to operate before smoke appears or in time for them to use an extinguisher to put out the fire. They are convinced that installing only what the building code requires—most often, smoke detectors installed only in the corridors of their building—will detect a fire in any remote portion of the building. Nothing could be farther from the truth.

If, for instance, the owner's fire protection goal is to detect a smoldering fire and the average ceiling height in the space is 30 ft, the goal will not be met. Smoldering fires lack the force to transport the smoke particles to the detectors located on a high ceiling. Smoke detectors located in a corridor will not respond until the fire in the remote location becomes large enough to force the smoke particles to the corridor ceiling. In this case, by the time detection occurs, the fire will potentially be too large for the fire service to control and extinguish. Unless the owner has made the conscious decision to install the minimum fire alarm system required by the building code, it will be impossible to meet his or her fire protection goals.

The fire alarm system user and designer must consider the chosen detectors' detection capabilities and potential limitations of the design imposed by the building configuration. Although the following is not an all-inclusive list, these items must be considered for their effects on the detection and operation of the fire alarm system:

- Ceiling height
- Air flow
- Ceiling obstructions
- Ambient environment (temperature, humidity, etc.) of the protected space
- Detector type
- Water supply
- Fire department response time
- Fire department capabilities
- Occupancy
- Egress design
- Construction type
- Capabilities of occupants (mobility, hearing, etc.)
- Ambient noise levels

Of these, ceiling height is generally the most important single factor affecting detection. Misapplied detector types or too few detectors or too much faith in a detector's ability to detect all small fires will inevitably lead to disaster.

QUESTIONS

1. As smoke ages and cools, the particles _____ in size:

    a. grow   b. diminish   c. equalize   d. double

2. The cooling of smoke results in _____.

    a. early detection   b. Brownian movement
    c. aging   d. stratification

3. When designing a fire alarm system for life safety of the building occupants, a system of _____ is normally used:

   a. flame detectors   b. heat detectors

   c. smoke detectors   d. manual fire alarm boxes

4. Flame radiation decreases in intensity with _____ between the radiating source and the target:

   a. one-third the distance   b. half the distance

   c. twice the distance   d. the square of the distance

5. The easiest fire signature to detect is:

   a. heat   b. smoke   c. flaming fires   d. smoldering fires

BIBLIOGRAPHY

NFPA Codes, Standards, Recommended Practices, and Manuals (see the latest *NFPA Catalog* for availability of current editions of the following documents):

NFPA 72®, National Fire Alarm Code®.

NFPA 90A, *Standard for the Installation of Air-Conditioning and Ventilating Systems.*

NFPA 101®, Life Safety Code®.

NFPA 1, Uniform Fire Code™.

NFPA 5000™, Building Construction and Safety Code™.

REFERENCES CITED

1. Harpe, S.W., Waterman, T. E., and Christian, W. J., "Detector Sensitivity and Siting Requirements for Dwellings, Phase 2," ITT Research Institute, Chicago, 1975.
2. Haessler, W., "Smoke Detection by Forward Light Scattering," *Fire Technology*, Vol. 1, No. 1, 1965, p. 43.
3. Scheidweiler, A., "New Research in Fire Detection Technology," *International Symposium*, 1968.
4. Dupont, H. B., "How Much Heat Can Firemen Endure?" *Fire Engineering*, Vol. 113, No. 2, 1960, pp. 122–214.
5. Zikria. B.A., *et al.*, "Respiratory Tract Damage in Burns: Pathophysiology and Therapy," *Annals of New York Academy Sciences*, Vol. 150, 1968, pp. 618–626.
6. Zikria, B.A., *et al., Smoke and Carbon Monoxide Poisoning in Fire Victims*, American Association for the Surgery of Trauma, New York, 1971.

7. Johns Hopkins, "Annual Summary Report: I July 72-30," Johns Hopkins University Applied Physics Lab, Baltimore, MD, June 1973.

8. Gross, D., *et al., "Smoke and Gases Produced by Burning Aircraft Interior Materials,"* NBS Building Science Series 18, U.S. National Bureau of Standards, Washington, DC, 1969.

ADDITIONAL READING

Bukowski R. W., "Studies Assess Performance of Residential Detectors," *NFPA Journal,* Vol. 87, No. 1, 1993, pp. 48-54.

Los Angeles Fire Department, "Fire Detection Systems in Dwellings," *NFPA Quarterly,* Vol. 56, No. 3, 1963, pp. 201-246.

Pryor, A.-J., et al., *"Hazards of Smoke and Toxic Gases Produced in Urban Fires,"* Southwest Research Institute, San Antonio, TX, 1968.

3

Fire Alarm Systems Overview

INTRODUCTION

A fire alarm system is a key element of any building's overall fire protection features. Properly designed, installed, operated, and maintained, a fire alarm system can help limit fire losses in buildings, regardless of occupancy. Also, since many of the fire deaths in the United States result from building fires, the use of detection and alarm systems in buildings can reduce the loss of life from fire. This chapter describes the operational characteristics of the various types of fire alarm systems in use.

DESCRIPTION OF A FIRE ALARM SYSTEM

A fire alarm system is a system that is primarily intended to indicate and warn of abnormal conditions, summon appropriate aid, and control occupant facilities to enhance protection of life.

A fire alarm system either automatically or manually links the sensing of a fire condition and the notification of people within and outside of the building that action should be taken to respond to the fire condition. The system might also be designed to initiate that response by initiating operation of a fixed extinguishing system, such as a preaction or deluge sprinkler system; a carbon dioxide, halon, or foam system; and so on. The basic components of a fire alarm system are one or more alarm-initiating-device circuits (IDC) or signaling line circuits (SLC) to which automatic fire detectors, manual fire alarm boxes, water-flow alarm devices, or other alarm-initiating devices are connected. These circuits carry alarm indication to a control panel, which is powered by a primary (main) power supply and a secondary (standby) power supply. Components of a fire alarm system also include: one or more alarm notification appliance circuits to which alarm signals such as bells, horns, speakers, and so on, are connected, or to which an off-premises alarm is connected, or both; and a trouble signal (see Chapter 4).

TYPES OF SYSTEMS

Fire alarm systems are classified according to the type of function they are expected to perform. Their design, installation, maintenance, and use are specified in *NFPA 72®, National Fire Alarm Code®* (see Table 3-1).

Protected-Premises (Local) Fire Alarm Systems

Protected-premises (local) fire alarm systems, as specified in *NFPA 72,* are supervised systems that provide fire alarm and supervisory signals within a facility and produce a signal at the facility only.

The main purpose of a protected-premises fire alarm system is to sound local alarm signals for evacuation of the protected building. A system could be limited to the basic features of a fire alarm system or may have other features such as an annunciator, fire/smoke door control, elevator recall, or voice communication within the protected property (see Figure 3-1).

A protected-premises fire alarm system must be powered by a primary (main) power supply and a secondary (standby) power supply. The secondary supply must be capable of operating the system for a 24-hour period under normal load and subsequently operate the system in alarm for an additional 5 minutes. This power supply requirement covers a normal day of occupancy and then a 5-minute alarm period, if necessary, to evacuate the building (see Figure 3-1).

In a protected-premises fire alarm system, the alarm is not automatically relayed to a fire department; instead, someone must notify the fire department when the alarm sounds. If the building is unoccupied at the time of the alarm, fire department response depends on a neighbor or passerby hearing the alarm and notifying the department (see Figure 3-2).

A protected-premises fire alarm system may carry out several functions in a building, for example, it may serve for automatic fire detection; as a manual fire alarm; in a guard's tour supervisory service; as a sprinkler system waterflow alarm and supervisory signal service; in combination systems (e.g., multipurpose electrical system, burglar alarm, music distribution, and paging systems, whether of voice or coded type); for smoke control; and for control of fire/smoke doors, elevator recall, or air-handling systems actuated by the fire alarm system. A

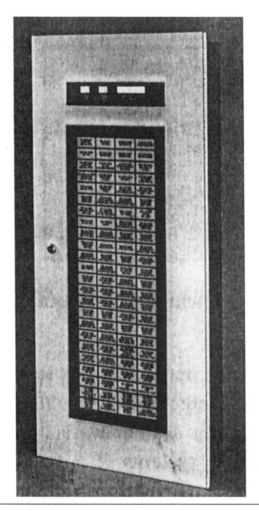

Figure 3-1. Remote annunciator.

protected-premises fire alarm system connected to wireless smoke detectors is shown in Figure 3-3.

A key element of protected-premises fire alarm systems is that all devices, appliances, and equipment in the system must be listed for its intended use. Both Underwriters Laboratories Inc. and FM Global (Norwood, MA) have listed equipment for this service. Other testing laboratories may be acceptable to the authority having jurisdiction (AHJ). The listed fire alarm control units are

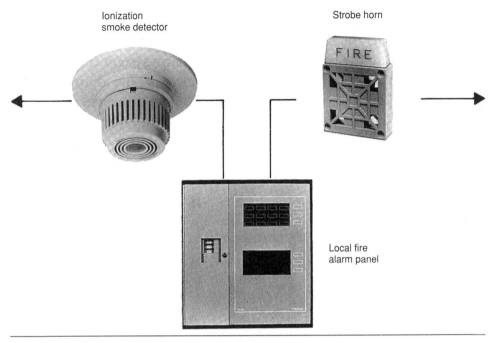

Figure 3-2. Typical equipment used in a protected-premises fire alarm system.

normally tested for ordinary indoor use only. The installation of fire alarm control units in an outdoor environment is not allowed unless it is listed for that environment.

The protected-premises fire alarm system may perform automatic control functions, such as release of fire or smoke doors and smoke dampers or deenergizing computer equipment and heating, ventilating, and air conditioning (HVAC) systems. Automatic controls should not impede building lighting, as this may hamper building evacuation and endanger lives.

Alarm signals can be coded or noncoded. Coded alarm signals consist of at least three complete rounds of the number transmitted. Coded supervisory signals usually have a single round of the number transmitted.

Use of a standard fire alarm evacuation signal is required by the *NFPA 72* for building evacuation. The signal is a uniform Code 3 temporal pattern, using a sound keyed ½ to 1 second ON, ½ second OFF, ½ to 1 second ON, ½ second OFF, ½ to 1 second ON, and 2½ seconds OFF. The signal should be repeated for at least 3 minutes. This signal is for evacuation purposes, and its use should be restricted to occupant evacuation. This fire alarm evacuation signal pattern is

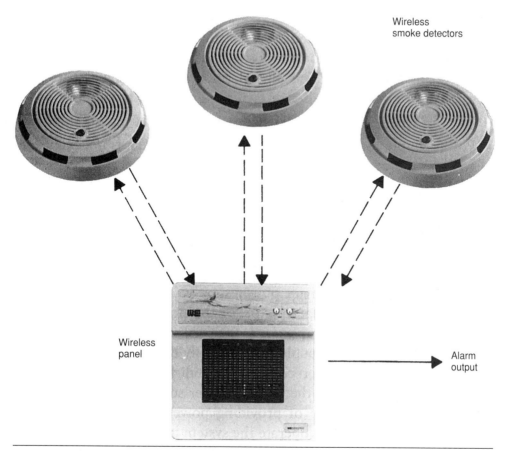

Figure 3-3. Typical equipment used in a wireless local system.

normally *not used* in health-care facilities and penal institutions unless complete evacuation is necessary.

The power for a protected-premises fire alarm system is normally supplied from the commercial power and light service because of its high reliability. As for all alarm systems, a dedicated branch circuit that is mechanically protected should be connected to the power and light service of the facility and be marked "fire alarm circuit control." Engine-driven generators can also provide this power to the system.

Standby, or secondary, power supplies are required for protected-premises systems. Failure of the primary power supply or failure of an electrically monitored-for-integrity circuit is required to cause a trouble condition at the fire alarm control unit. Dry-cell batteries are not permitted for use as the secondary power supply due to their unreliability over time and their lack of recharging capability. Secondary power supplies should be capable of providing power under normal system load for 24 hours and then of operating the system in alarm condition for at least 5 minutes.

Auxiliary Fire Alarm Systems

Auxiliary fire alarm systems are protected-premises fire alarm systems in a protected facility that transmit alarm signals via a municipal fire alarm system. The system is intended to automatically call or notify a fire department following activation of an automatic or manual fire alarm. The municipal system then transmits an alarm to the municipal communications center.

The alarms from an auxiliary fire alarm system and the municipal fire alarm boxes located in the community both transmit their signals with the same equipment and communication method. With an auxiliary system, the owner of the protected facility is responsible for system operation, maintenance, and testing, whereas the municipality is responsible for its system.

An auxiliary fire alarm system circuit connects the alarm-initiating devices to a municipal fire alarm system. The connection is made through a nearby transmitter, a master fire alarm box, or a dedicated telephone line run directly to the municipal connection center switchboard. The signal received by the fire department is identical to that received by the department upon manual activation of a municipal fire alarm box, often called a "street box." Because fire department personnel know which municipal boxes are auxiliarized, responding fire fighters can check for an alarm originating within the protected facility (see Figure 3-4).

There are three types of auxiliary systems recognized in *NFPA 72:* (1) local-energy type, (2) shunt type, and (3) parallel-telephone type.

The local-energy type of system (Figures 3-5 and 3-6) is defined in *NFPA 72* as being electrically isolated from the municipal alarm system and having its own power supply. Tripping of the transmitting device does not depend on the current in the municipal system. In a wired circuit, receipt of the alarm by the municipal communication center (if the municipal circuit is accidentally opened) depends on the design of the transmitting device and the associated municipal communication center equipment, that is, whether or not the municipal system is designed to receive alarms through manual or automatic ground operational facilities. In a radio-box type of system, receipt of the alarm by the municipal communication center depends upon the proper operation of the radio transmitting and receiving equipment.

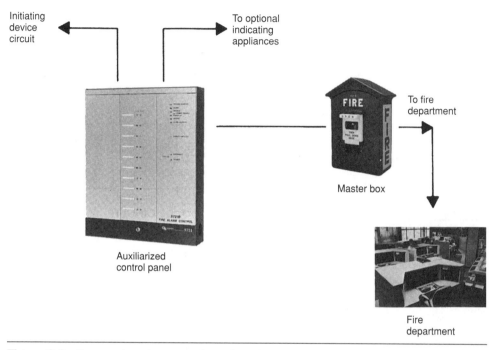

Initiating
device
circuit

To optional
indicating
appliances

To fire
department

Master box

Auxiliarized
control panel

Fire
department

Figure 3-4. Connections in an auxiliary fire alarm system.

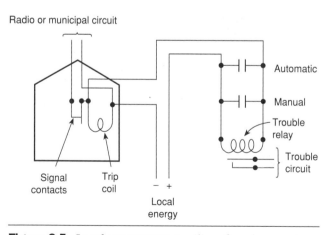

Radio or municipal circuit

Automatic

Manual

Trouble
relay

Trouble
circuit

Signal
contacts

Trip
coil

Local
energy

Figure 3-5. Local energy-type auxiliary fire alarm system.

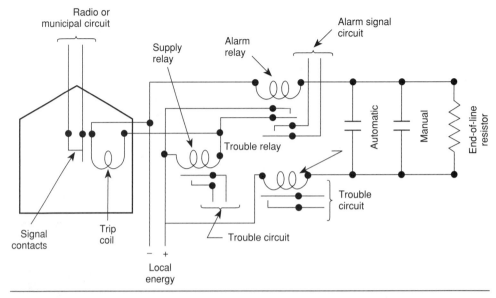

Figure 3-6. Local energy-type auxiliary fire alarm system with local alarm.

The shunt-type system, as described in *NFPA 72 is* electrically connected to and is an integral part of the municipal fire alarm system (see Figures 3-7 and 3-8). A ground fault on the auxiliary circuit is a fault on the municipal circuit, and an accidental opening of the auxiliary circuit will send a needless (or false) alarm to the municipal communication center. An open circuit in the transmitting device trip coil will not be indicated at either the protected property or the municipal communication center; also, if an alarm-initiating device is operated,

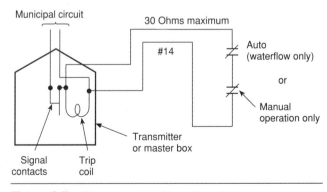

Figure 3-7. Shunt-type auxiliary fire alarm system.

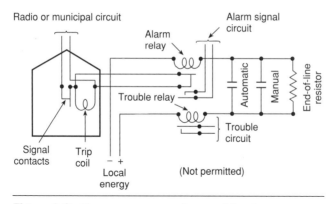

Figure 3-8. Shunt-type master box and local energy-type auxiliary fire alarm system.

the alarm will not be transmitted, but an open-circuit indication will be given at the municipal communication center. If a municipal circuit is open when a connected shunt-type system is operated, the transmitting device will not trip until the municipal circuit returns to normal, at which time the alarm will be transmitted.

NFPA 72 specifies that shunt-type systems are to be used for manual fire alarm and waterflow alarm service only. Automatic fire detection is not permitted on shunt-type systems. This system is noncoded with respect to any remote electric tripping or actuating devices.

Additional design parameters for shunt-type systems are as follows:

1. All conductors of the auxiliary circuit must be in conduit or electrical metallic tubing in accordance with NFPA 70, *National Electrical Code® (NEC)*.
2. Both sides (conductors) of the auxiliary circuit must be in the same conduit.
3. Maximum permitted shunt loop length is 750 ft (228 m) in conduit.
4. The smallest conductor for the auxiliary circuit is 14 AWG (*NEC*).
5. A protected-premises fire alarm system auxiliarized by the addition of a relay coil energized by a local power supply and whose normally closed contacts trip a shunt-type master box is not permitted (see Figure 3-8).
6. Where a municipal transmitter or master box is located within a private premise, it must be installed in accordance with NFPA 1221, *Standard for the Installation, Maintenance, and Use of Emergency Services Communications Systems.*

A parallel-telephone type of auxiliary alarm system is described in *NFPA 72* as a system in which alarms are transmitted over a circuit directly connected to the annunciating switchboard at the municipal communication center and terminated at the protected property by an end-of-line resistor or equivalent. Such auxiliary systems are for connection to municipal fire alarm systems of the type in which each municipal alarm box annunciates at the municipal communication center switchboard by individual circuit (see Figure 3-9).

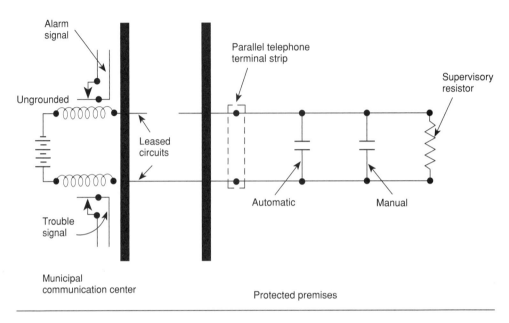

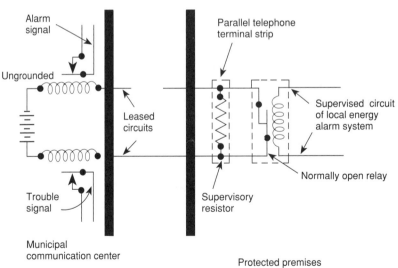

Figure 3-9. Parallel-telephone type of auxiliary fire alarm system.

There are major differences among these three types of auxiliary systems:

1. The shunt-type system is a series closed loop for the alarm-initiating circuit. The local-energy type and parallel-telephone types are open-contact parallel circuits for the alarm-initiating circuit.
2. An accidental open circuit on a shunt-type system activates a fire alarm. An accidental open circuit on a local-energy type or parallel-telephone type of system activates a trouble alarm.
3. A single ground fault on the alarm-initiating circuit of either the shunt or the parallel-telephone type of system extends into the municipal box wired circuit or switchboard.
4. A single ground fault on a local-energy type of system that prevents normal operation on the alarm-initiating circuit causes a trouble signal.

Some of the basic requirements for auxiliary fire alarm systems include:

1. One municipal transmitter or master box that can serve up to a maximum 100,000 sq ft (9290 m^2) fire area
2. Means to manually initiate an alarm signal when automatic fire detectors and alarm service and waterflow alarm services are provided
3. An annunciator located near the transmitting device

Specific power sources are required by *NFPA 72* for these auxiliary systems: the municipal fire alarm system for the shunt-type system, a primary (main) *and* secondary (standby) power supply for the local-energy type of system, and the municipal communications center as the power supply for the parallel-telephone type of auxiliary system. The standby power supply for all types of auxiliary fire alarm systems is required to operate the system for 24 hours of normal operation, followed by 5 minutes of alarm.

The auxiliary alarm systems do not require the use of audible alarm signals within the facility. If fire alarm evacuation signals are needed for the protected property, the system should also comply with the additional provisions of *NFPA 72*.

A design engineer should work closely with the municipal fire department communication center to ensure that the auxiliary fire alarm system is compatible with the municipality's system. This compatibility is critical to efficient auxiliary system operation.

Remote-Station Fire Alarm Systems

Remote-station fire alarm systems are supervised systems in one or more protected facilities that transmit alarm, trouble, or supervisory signals to the remote station through a direct connection (usually a leased telephone line) from each protected property as described in *NFPA 72*. The remote stations are attended by trained personnel 24 hours a day, often at a fire department communications center. Properties under the same or different ownership can be protected by a remote station (see Figures 3-10 and 3-11).

When permitted by the authority having jurisdiction, the remote station may be in a location other than a fire department communications center, where personnel are on duty 24 hours a day and are trained to receive the alarm signal and immediately retransmit it to the fire department. System trouble signals usually are not automatically transmitted to the remote station. (A remote-station system may or may not have an evacuation alarm.)

A remote station differs from a central station primarily in that central station personnel are principally responsible for furnishing and maintaining supervised signaling service and the central station is a privately owned, for-profit operation.

Acceptable methods of transmitting alarm signals from a remote station to the fire department (assuming the fire department communications center is not the remote station) include the following:

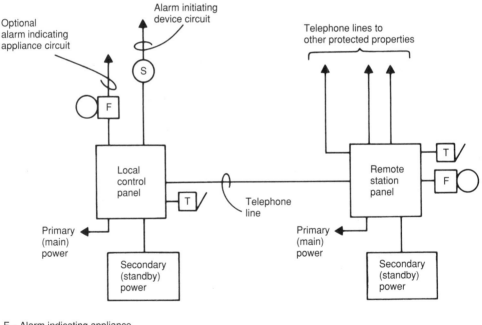

F = Alarm indicating appliance
T = Trouble signal

Figure 3-10. Typical remote-station fire alarm system.

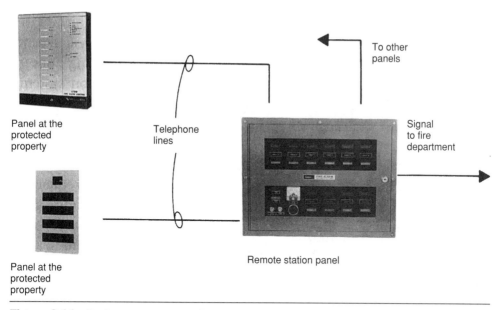

Figure 3-11. Basic components of a remote-station fire alarm system.

1. A dedicated circuit independent of any switching network using voice communications or coded signals
2. A one-way (outgoing only) telephone using a commercial dial network
3. A radio system with fire department frequency (where permitted)

The use of a dedicated independent circuit is the preferred transmission method. Upon receipt of a supervisory or trouble signal, the remote-station operator on duty is responsible for notifying the owner or owner's representative immediately.

NFPA 72 requires an independent secondary (standby) power supply for the remote-station control unit and, if needed, at the protected premises. The secondary supply power source usually can be any permitted for a fire alarm system, but it must have the capacity to operate the system for 24 hours, followed by 5 minutes of alarm. Since a remote station fire alarm system, like an auxiliary fire alarm system, is intended primarily for property protection, an evacuation system is not required. If an evacuation system is installed, it should meet the additional requirements of *NFPA 72*.

Remote-station fire alarm systems may have coded or noncoded operation. Coded systems can protect a maximum of five facilities and may share common connections to the remote station. A noncoded system is used for alarm annunciation if only one facility is protected.

Proprietary Fire Alarm Systems

A proprietary fire alarm system is a protected-premises fire alarm system located within a protected property and operated under single ownership, which provides emergency building monitoring and control services. The proprietary fire alarm system has a supervising station staffed 24 hours a day. An alarm received in the supervising station is transmitted to the fire department, and a permanent record of all alarms is maintained at the supervising station (see Figure 3-12).

The proprietary fire alarm system is a widely used type of system in large commercial and industrial occupancies. Many proprietary fire alarm systems

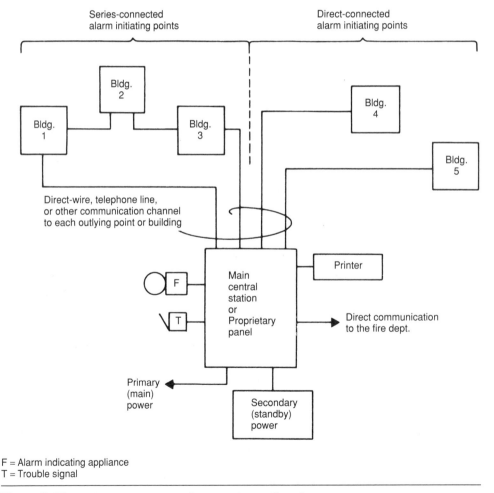

F = Alarm indicating appliance
T = Trouble signal

Figure 3-12. Basic components of a proprietary fire alarm system.

have separate initiating-device circuits for each building zone or subsection, similar to the protected-premises, auxiliary, and remote-station fire alarm systems. However, due to the increasing use of electronics, newer proprietary systems for large buildings often have addressable or addressable analog fire alarm systems (see Figure 3-13). These systems receive all signals throughout the building over one or more pairs of wires and determine the exact location of the fire by use of digitally coded information transmitted over the wires.

Although a proprietary system's fire alarm control unit is similar to the other fire alarm control units, the receiving console can be quite different. A proprietary receiving console consists of individual lights, a digital display, or a cathode ray tube (CRT) visual display terminal and screen indicating the alarm point. An audible alarm to alert the console operator and a hard-copy printer can also be included.

Proprietary control units are required by *NFPA 72* to transmit trouble and alarm signals to the central supervising station. The control unit at the central supervising station, as well as remotely located control equipment, must have a secondary (standby) power supply that will operate the system for 24 hours under normal traffic conditions. Since operators are constantly on duty with a proprietary system, 24 hours of standby power is considered sufficient.

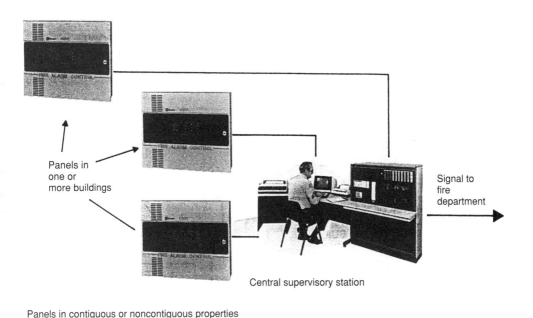

Panels in one or more buildings

Signal to fire department

Central supervisory station

Panels in contiguous or noncontiguous properties under one ownership

Figure 3-13. Single-ownership properties can be protected by a proprietary system.

Large proprietary multiplex- and computer-controlled systems usually have functions in addition to alarm indication and initiation. These systems often provide for smoke control within the building by automatically closing and opening dampers in heating, ventilating, and air conditioning systems and turning on exhaust fans. They can also adjust elevator controls so elevators bypass fire floors and automatically return to the lobby floor for fire department use, as required by ANSI A17.1, *The Safety Code for Elevators and Escalators*. In addition to increasing flexibility, use of addressable fire alarm systems greatly reduces the amount of building wiring. Computer-based proprietary systems often include energy management capabilities that result in energy savings, which has become a major factor in the increased use of these systems in large buildings.

The supervising station for proprietary systems must be staffed with competent personnel who are adequately trained to perform their duties. The supervising station must be constantly attended by a minimum of two staff members responsible for monitoring signals and operating the system. Having "dedicated personnel" is a main feature of a proprietary system, and staff should not be assigned additional duties that may interfere with these responsibilities.

Proprietary systems are used in facilities where a system more complex and elaborate than a local fire alarm system is needed. The supervising station must be located in a separate fire-resistive building or in a fire-resistive enclosed room that is entirely separate from hazardous operations in the facility.

Fire alarms from the supervising station to a fire department or emergency response team must be transmitted promptly and, preferably, automatically. Two-way telephone communication should also be established between the supervising station and the fire department communication center.

Output signals from the supervising station can control certain life safety and property protection functions in the building during emergencies. Control of building operations can include elevator operation, unlocking stairwell and exit doors, releasing fire and smoke dampers, initiating and monitoring fire extinguishing systems and components, hazardous gas shutoffs, air pressurization of stairwells and means of egress, and control of lighting, HVAC systems, and manufacturing processes.

Central Station Fire Alarm System

NFPA 72 defines a central station fire alarm system (see Figure 3-14):

A system or group of systems in which the operations of circuits and devices are transmitted automatically to, recorded in, maintained by, and supervised from a listed central station that has competent and

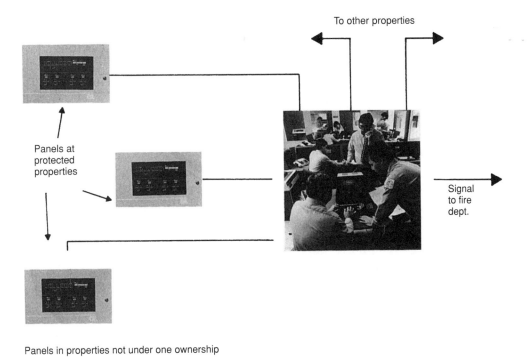

To other properties

Panels at
protected
properties

Signal
to fire
dept.

Panels in properties not under one ownership

Figure 3-14. Multiowner properties connected in a central station fire alarm system.

experienced servers and operators who, upon receipt of a signal, take such action as required by this code. Such service is to be controlled and operated by a person, firm, or corporation whose business is the furnishing, maintaining, or monitoring of supervised fire alarm systems. (*NFPA 72*, 2002 edition, paragraph 3.3.67.2)

The central supervisory station that receives the fire alarm signal in a proprietary fire alarm system is operated by an individual with a proprietary interest in the protected buildings. The central supervisory station of a proprietary system is generally located in a guard office within or near the building or group of buildings protected by the system. On the other hand, the supervising station in a central station fire alarm system is staffed by operators who perform the service for a fee and have no proprietary interest in the protected buildings.

According to *NFPA 72*, communication methods that may be used in a central station fire alarm system include the following:

1. Active multiplex circuits that are part of a supervising station, including systems derived channels

2. Digital alarm communicator systems, including digital alarm radio systems
3. McCulloh systems (not allowed after June 30, 2003)
4. Two-way radio frequency (RF) multiplex systems
5. One-way radio alarm systems
6. Directly-connected noncoded systems

However, the most commonly used method of transmission of the alarm, trouble and supervisory signal is a Digital Alarm Communicator Transmitter (DACT) installed at the protected property and transmitting to a Digital Alarm Communicator Receiver (DACR) located at the Central Station. Commercial properties that have contracted for central station service must employ one of the following combinations of transmission channels when using a DACT, according to *NFPA 72*:

(1) A second telephone line (number)
(2) A cellular telephone connection
(3) A one-way radio system
(4) A one-way private radio alarm system
(5) A private microwave radio system
(6) A two-way RF multiplex system
(7) A transmission means complying with 8.5.4

Exception: One telephone line (number) equipped with a derived local channel or a single integrated services digital network (ISDN) telephone line using a terminal adapter specifically listed for supervising station fire alarm service, where the path between the transmitter and the switched telephone network serving central office is monitored for integrity so that the occurrence of an adverse condition in the path shall be annunciated at the supervising station within 200 seconds.

The following requirements shall apply to all combinations listed in 8.5.3.2.1.4(A):

(1) Both channels shall be supervised in a manner approved for the means of transmission employed.
(2) Both channels shall be tested at intervals not exceeding 24 hours.

Exception No. 1: For public cellular telephone service, a verification (test) signal shall be transmitted at least monthly.

Exception No. 2: Where two telephone lines (numbers) are used, it shall be permitted to test each telephone line (number) at alternating 24-hour intervals.

(3) The failure of either channel shall send a trouble signal on the other channel within 4 minutes.

(4) When one transmission channel has failed, all status change signals shall be sent over the other channel.

Exception: Where used in combination with a DACT, a derived local channel shall not be required to send status change signals other than those indicating that adverse conditions exist on the telephone line (number).

(5) The primary channel shall be capable of delivering an indication to the DACT that the message has been received by the supervising station.

(6) The first attempt to send a status change signal shall use the primary channel.

Exception: Where the primary channel is known to have failed.

(7) Simultaneous transmission over both channels shall be permitted.

(8) Failure of telephone lines (numbers) or cellular service shall be annunciated locally.

8.5.3.2.1.5 DACT Transmission Means. The following requirements shall apply to all digital alarm communications transmitters:

(1) A DACT shall be connected to two separate means of transmission at the protected premises.

(2) The DACT shall be capable of selecting the operable means of transmission in the event of failure of the other means.

(3) The primary means of transmission shall be a telephone line (number) connected to the public switched network. (*NFPA 72,* 2002 ed.)

When a signal is received at a central station, an alarm is immediately retransmitted to the public fire service communications center, and the appropriate authorities are informed. The central station dispatches someone to the protected premises when equipment must be reset manually. Central station systems require secondary (standby) power supplies that will operate the system for a period of 24 hours.

Emergency Voice/Alarm Communications System

As defined by *NFPA 72,* the purpose of an emergency voice/alarm communications system is to provide for the transmission of information and instructions pertaining to a fire emergency to the building occupants and the fire department.

This system also allows communications with those persons remaining in the building. An emergency voice/alarm communications system can be used to partially evacuate a building in emergency conditions while permitting some building occupants to remain in safe areas. Such a system should always be provided whenever the emergency plan includes anything other than immediate and total evacuation of the building.

This system supplements a protected-premises, auxiliary, remote-station, central station, or proprietary fire alarm system. Its standby power supply must operate the voice/alarm signaling service for a fire or other emergency condition for 24 hours without the primary operating power and then be capable of intermittently operating the system in alarm condition for 15 minutes. Because of the emergency nature of a voice/alarm communications system, special requirements in *NFPA 72* also cover the survivability of the system so that fire damage to one notification zone will not result in loss of communication to another.

There are two basic types of emergency voice/alarm communications systems:

1. A stand-alone manually controlled system. Voice or tone signals for this type of system can be selected and distributed by the operator.
2. A system that is integrated into a fire alarm system. Voice or tone signals can be selected and distributed manually or automatically for the integrated system.

The voice/alarm system consists of a series of high-reliability speakers located throughout the building (see Figure 3-15). They are connected to and controlled from the fire alarm communications console located in an area of the building designated as a fire command center. From the fire command center, voice messages can be provided to individual speaker zones or the entire building with specific instructions to the occupants. Some systems have fire warden stations, or fire zones, on each floor from which a fire warden assumes local command and passes on specific evacuation instructions. Operation of the command center sometimes is assumed by a trained building employee until the fire department arrives, at which time the fire officer in charge takes over. The system also can be used during fire-fighting operations for communication with the fire fighters.

A one-way system permits emergency personnel to give voice instructions on either a selective or an "all-call" basis via a microphone and a loudspeaker system. Its use began with the advent of high-rise buildings and other structures where immediate evacuation of the entire building might be impractical and the

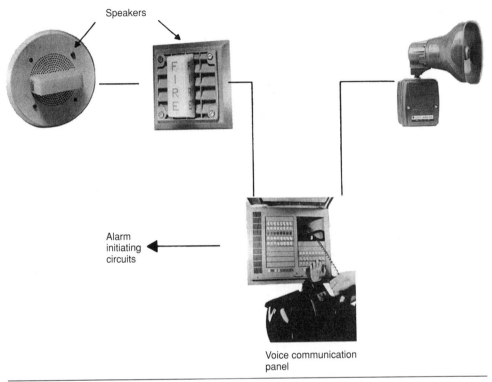

Speakers

Alarm
initiating
circuits

Voice communication
panel

Figure 3-15. Typical speakers connected to a voice panel in an emergency voice/alarm system.

fire plan called for selective partial evacuation or the relocation of the building occupants from the fire area.

A two-way system using telephone or intercom techniques permits communication between fire service personnel while they are investigating or fighting a fire in the building. Most two-way systems use telephones located in fire stairwells so fire service personnel can be assured of reliable communication from any floor to the fire command station. Use of these telephones is also permitted for other building occupants and they are frequently installed on all floors for use in emergency circumstances. A distinctive notification signal, different from any other alarm or trouble signal, is required at the fire command center to indicate an off-hook condition of any telephone on the system.

One important aspect of a voice/alarm communication system is that, since complete building evacuation is not always feasible, occupants can be instructed to relocate to other; nonfire areas of the building and wait for further instructions from the fire department (see Chapter 7, Fire Alarm Notification).

SIGNAL PROCESSING

Processing is the dynamic link between signal input and system response. Signal processing is an interpretive function governing the response of a unit or an entire fire alarm system to the various inputs received by that unit or system. A fire alarm control panel will typically receive numerous, varied signal inputs; outputs can vary as widely depending on the application of the alarm system.

In a simplified application, such as for a protected-premises fire alarm system or a dwelling fire-warning system, the fire alarm control panel will energize the general evacuation output signals. With a more complex system, for example, auxiliary, remote-station, central station, or proprietary fire alarm system, several response (output) modes may need to be activated for certain types of inputs. The interpretive processing routines for matching appropriate outputs to inputs can be established through fixed electric or electronic circuitry or controlled with computer programs.

TYPES OF SIGNALS PROCESSED

Manual Fire Alarm Signals

Requirements for manual fire alarm signals are addressed in *NFPA 72* and in NFPA *101*®, *Life Safety Code*®. Manual fire alarm signals in a fire alarm system are typically initiated from manual fire alarm boxes located throughout the protected premises. Signals can also be initiated outside the fire alarm system, for example, from a municipal fire alarm box. Processed outputs for manual signals vary according to the type of fire alarm system.

Evacuation signals, whether audible (tonal or voice-generated) or visible, are signal outputs in the protected-premises and emergency voice/alarm systems. Various prerecorded tone and voice messages can also be considered as emergency system outputs.

Manual initiation of an alarm signal in an auxiliary or remote-station fire alarm system results in the transmission of signals to a remote point for subsequent fire department interpretation and action. Signals can be transmitted indirectly, or directly to the fire department from a protected-premises fire alarm system.

Manual signals are processed in a similar manner for central station and proprietary fire alarm systems, except that the signal is processed through the private receiving system equipment with possible retransmission to a public system. Unless system design dictates, no distinction is made between manual and automatic fire alarm signals.

Some of the differences among manual system responses are as follows:

1. An auxiliary fire alarm system does not normally use a coded alarm signal. The rationale for this is that auxiliary systems have historically been simple noncoded latching circuits that did not transmit signals in a serial method to a public fire department.
2. Manual signal initiation is not normally part of a dwelling fire alarm system. Since most fires occur while the occupants of the household are sleeping, initiation from automatic detection is the only necessary function of a dwelling fire alarm system.
3. *NFPA 72* and NFPA 1221 assume that other system units or interfaces have processed manual signals prior to retransmission. Noncoded manual alarm inputs are processed in a manner similar to that for automatic alarm inputs.

Automatic Fire Alarm Signals

Automatic fire alarm signals for each system are addressed in *NFPA 72*. Automatic alarms are typically initiated by smoke, heat detectors, waterflow switches, or other suppression-system alarm-monitoring devices, which can be installed on separate circuits or integrated into the same circuit of a fire alarm system. Installation of devices and appliances as well as testing requirements for all devices, appliances, fire safety functions, and control panel features are addressed in *NFPA 72*.

The nature of automatic alarm signal generation and processing depends on the type of fire alarm system used. An evacuation alarm signal is required for building occupants when a protected-premises fire alarm system is specified. Incoming automatic signals for central station and proprietary fire alarm systems are required to be identified and processed by the protected zone or other system subdivision within the protected premises.

Primary power, when restored, must also restore systems to their normal supervisory condition. Central station and proprietary systems have a large amount of information contained within a common system. A rapid, more discriminating means of identifying changes in system status is required in central station systems since identifying and interpreting signals is important to public fire department response.

TYPES OF PROCESSING

Guard's Tour Supervisory Service

In guard's tour service, a supervisory service signal is typically initiated through mechanical key-operated tour stations. These signals can also be generated electronically or optically for encoding and decoding. *Encoding* is defined as the processing of a signal such as that initiated by a switch contact and subsequent transfer of the signal into information that can be interpreted by a machine. *Decoding* is the translation of this information back into signals understandable by human beings.

Guard's tour supervisory signals are not transmitted to public fire departments. Guard's tour supervisory signaling equipment generally tracks all stations within the tour schedule; all stations must report in this type of service. The beginning and end of a tour schedule must register at some remote point, and intermediate nonrecording signals must operate in fixed succession. A guard's tour system can have an exception-reporting package in which all stations must report; such a system also has a delinquency signal to identify any station that has not been operated.

Sprinkler System Waterflow Alarm and Supervisory Signal Service

In general, a sprinkler system waterflow alarm and supervisory signal service in any fire alarm system must be able to interpret the type of system signal and the particular element that has operated. Additionally, the supervisory service must identify restoration of the element to its normal position.

Waterflow supervisory signals consist of one or more of the following: gate valve supervision, pressure supervision, water level supervision, water temperature supervision, and pump supervision. The waterflow alarm signals are the only signals that can be transmitted and processed by auxiliary fire alarm systems. If additional supervisory services are desired, another type of fire alarm system should be specified.

Trouble Signals

To varying degrees, all systems interpret as *trouble signals* those conditions that inhibit alarm transmission or annunciation. Many functions and/or circuits are monitored for integrity against loss of capability, depending on the nature of the system and the complexity of the information processed by that system. In some cases, automatic or manual control of various system functions permits emergency operation under faults that could debilitate the entire fire alarm system.

Fire alarm systems share certain characteristics concerning trouble signals. All systems interpret loss of primary (main) power as a trouble condition, which is usually indicated by an audible signal. Loss of initiating circuits such as an open conductor is also a trouble condition, and fire alarm systems process this signal information accordingly.

Loss of notification-appliance circuits is a trouble condition in protected-premises and emergency voice/alarm communications systems. When remote-station, central station, or proprietary systems have a notification alarm signal, these integrated circuit systems must comply with *NFPA 72* concerning trouble signals for loss of notification-appliance circuits.

Trouble signals for signaling line circuits (SLC) are covered in *NFPA 72*. This standard requires that the remote-station, central station, or proprietary systems recognize a single open or single ground fault as an abnormal (trouble) condition, but not as an alarm or fire.

MATCHING SYSTEM TYPE WITH FIRE PROTECTION GOALS

Owners are encouraged to develop fire protection goals for their facilities. If an owner chooses not to develop his or her own fire protection goals, then by default he or she accepts those goals established in the building and life safety codes. Typically, fire protection goals are based on the following:

- Life safety
- Property protection
- Mission protection
- Heritage preservation
- Environmental protection

Once these goals are established, the fire alarm system type, application, and operation should be integrated with the fire protection goals. The type of system installed will vary depending on building occupancy, location, operations conducted in the facility, and local or state requirements. Typically, the fire protection goals will also establish the extent and type of detection devices used as well as the type of off-premises connection.

CONCLUSION

The type of fire alarm system to be used depends on the protection needs and goals with regard to life safety, property protection, mission protection, and/or heritage protection of each facility or individual property. Although there are requirements basic to all fire alarm systems, specific features of each system—type of primary and secondary power supply, trouble signals, methods of signal transmission, system supervision, and so on—must be carefully considered by the designer for each fire alarm system. *NFPA 72* should be consulted when the designer is matching the fire alarm system to the protection goals desired.

QUESTIONS

1. When an alarm signal is received by a protected-premises fire alarm system:
   a. the notification appliances operate throughout the facility
   b. the fire department is automatically notified
   c. the central station responds
   d. the service company is notified

2. When a commercial building fire alarm system is connected to an off-premises monitoring station (e.g., central station, remote station) using a DACT:

   a. the system connection to the DACT must be installed in the raceway
   b. the DACT must be installed inside the fire alarm control panel
   c. the DACT must be connected to two phone lines
   d. the DACT must be red

3. A remote station most typically:
   a. is operated by a for-profit company
   b. is used only for residential fire alarm systems monitoring
   c. is located at a fire department communications center
   d. uses direct circuit wiring to the fire department

4. Fire protection goals are typically developed by:
   a. the general contractor   b. the fire chief
   c. the owner   d. the building official

5. A signaling line circuit (SLC) is used to interconnect the fire alarm control panel to:
   a. fire department signals   b. addressable devices
   c. multiple DACTs   d. nonaddressable smoke detectors

BIBLIOGRAPHY

NFPA Codes, Standards, Recommended Practices, and Manuals (see the latest *NFPA Catalog* for availability of current editions of the following documents):

NFPA 70, *National Electrical Code®*.

NFPA 72®, National Fire Alarm Code®.

NFPA *101®, Life Safety Code®*.

NFPA 1221, *Standard for the Installation, Maintenance, and Use of Emergency Services Communication Systems*.

ADDITIONAL READING

ANSI A17.1, *The Safety Code for Elevators and Escalators,* American National Standards Institute, New York, 2000.

4

Fire Alarm System Components and Circuits

INTRODUCTION

Electric supervision of a fire alarm system is critical to smooth, uninterrupted system operation. Power supplies, system circuits, and equipment can be supervised so that an open or ground fault condition will activate a trouble signal. This chapter describes the components of a system that should be supervised, the individual circuit and system approaches to supervision, and the relation of device circuits to the fire alarm system. Additional information on supervision can be found in Chapter 3, Fire Alarm Systems Overview, and Chapter 7, Fire Alarm Notification.

The basic components of a fire alarm system include notification appliances, automatic fire detectors, manual fire alarm boxes, waterflow alarm devices, or other alarm-initiating devices connected to a fire alarm control unit (FACU), using circuits described as notification appliance circuits (NAC), initiating-device circuits (IDC), and signaling line circuits (SLC). The FACU is powered by a primary (main) power supply and a secondary (standby) power supply. The fire alarm system can also be connected to an off-premises monitoring station or directly to the fire department responsible for responding to an alarm (see Figure 4-1).

FIRE ALARM CONTROL UNITS

The FACU is designed to be the central operating point of any fire alarm system. Early FACUs were simple relay configurations. The current fire alarm control units range from solid-state conventional types to addressable or addressable/analog types, each providing different features that may or may not be needed in the design application.

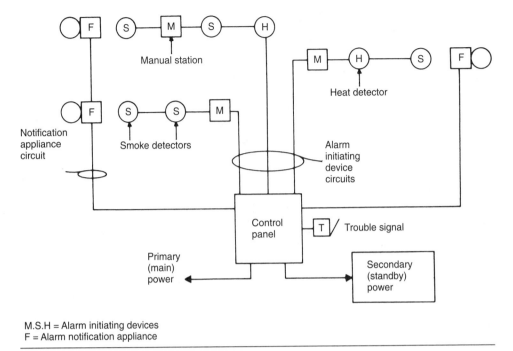

M.S.H = Alarm initiating devices
F = Alarm notification appliance

Figure 4-1. Basic components of a fire alarm system.

Conventional Fire Alarm Control Units

These FACUs are typically designed to accept detection devices that will provide either alarm or normal status information. That is, they are either on or off. Additional information such as exact location (point ID) or alarm levels are not provided by conventional devices or systems. Typically, the devices are arranged in a zone, such as all of the devices on a floor, and when an alarm condition occurs, the zone in alarm will be reported at the FACU. Conventional systems are used when detailed information regarding detector status or location is not critical to report. Many applications fall into this category: small stores, office buildings, and storage facilities.

Addressable Fire Alarm Control Units

Often, especially in larger systems, more information is needed or desired from the fire alarm system by the owner, designer, or authority having jurisdiction. This information could be as basic as wanting to know the times and exact locations from which detectors are reporting. This will enable the fireground commander to track the development of the fire or enable a technician to efficiently locate the source of a system problem or false alarm. The unit is still basically an on/off device, but its on/off status and exact location are transmitted to the addressable FACU.

Addressable/Analog Fire Alarm Control Units

In an effort both to utilize the addressable features of new microprocessor-based FACUs and improve the performance and reliability of system-connected smoke detectors, the addressable/analog system was developed. Addressable/analog devices report their exact location and alarm threshold status. This latter feature allows the FACU to monitor a smoke detector's cleanliness to determine whether or not the detector needs to be serviced *before* it alarms needlessly.

In addition, with an addressable/analog FACU, the actual response level of a smoke detector can be individually set to accommodate the environment in the detector's location.

POWER SUPPLY REQUIREMENTS FOR FIRE ALARM SYSTEMS

Regardless of the type of alarm system, *NFPA 72*®, *National Fire Alarm Code*®, specifies similar requirements for fire alarm system power supplies.

Primary (Main) Power Supply

The primary (main) power supply may be provided by commercial light and power supply or an engine-driven generator. The commercial supply can con sist of one phase of a three-wire supply or a two-wire supply. The connection to light and power supply must be on a dedicated branch circuit, with the disconnecting means marked "fire alarm circuit control," and accessible only to authorized personnel.

An alternate primary power supply is an engine-driven generator consisting of (1) a single engine-driven generator with either a trained operator on duty 24 hours a day or an engine-driven generator and a storage battery with 1-hour capacity, or (2) multiple engine-driven generators one of which must be arranged for automatic starting.

The primary power supply must be supervised and its failure indicated by a distinctive trouble signal.

Secondary (Standby) Power Supply

The secondary (standby) power supply must provide power to the complete system within 30 seconds in event of primary supply failure or whenever the primary power supply drops below 80 percent of rated value. The secondary supply must have the capacity to supply the fire alarm system for 24 hours during normal nonalarm operation and for 5 minutes of alarm operation at the end of that 24 hours ("24/5" capacity).

The amount of battery standby needed can be calculated based on the power requirements of the devices and equipment used. Most manufacturers provide simple methods or programs to calculate the size of the batteries needed to meet the 24/5 requirements. It is important to remember that as batteries age, they lose their ability to supply the original power calculated. To compensate for aging and other environmental effects such as temperature, it is common practice to include a safety factor, arbitrarily ranging from 10 to 20 percent, in the battery load calculations.

Emergency voice/alarm communications systems are required to have 24 hours of standby during normal operation and 5 minutes of emergency operation (alarm notification appliances operating) left at the end of that period.

Acceptable sources of standby power are (1) a storage battery and charger, (2) an engine-driven generator and 4-hour capacity storage battery, and (3) multiple engine-driven generators, one of which must be arranged for automatic starting.

Trouble-Signal Power Supply

A trouble-signal power supply for fire alarm systems operates a trouble signal when the primary (main) power supply fails. The trouble-signal power supply must be independent of the primary supply (but can be the secondary power supply) and cannot use dry cells.

Notification Appliance Power Supplies

Often the main FACU cannot supply the power necessary for all of the visible and audible notification appliances. To allow for system design flexibility and to avoid large gauge wire connections returning to the FACU, remotely located notification appliance circuit power supplies are used to extend the FACU to the floors or areas needing the additional power.

SUPERVISION OF POWER SUPPLIES, SYSTEM CIRCUITS, AND EQUIPMENT

Electric supervision in fire detection and alarm systems monitors the circuit integrity of interconnecting conductors so that when a single-open or a single-ground condition occurs that would prevent normal operation, the condition is automatically transmitted and indicated in the appropriate location. Electric supervision does not normally include monitoring electric conductors for a short-circuit fault or multiple-ground conditions. Short-circuit faults and multiple-ground conditions typically generate an alarm signal that occurs on an initiating-device circuit or signaling line circuit. Use of double-loop circuits or multiple-path conductors instead of electric supervision is not permitted by *NFPA 72.*

Fire alarm systems provide several distinct types of audible signals. To eliminate confusion between *supervised* and *supervisory, NFPA 72,* uses the term *monitoring integrity of installation conductors* to reference the supervision of the system wiring. A *trouble signal* is required to sound upon the occurrence of a single open or single ground condition that would prevent normal operation of the system. A *supervisory signal* indicates the off-normal position, or condition, of some part of a sprinkler or other fire protection system connected to the fire alarm system.

All fire alarm and process-monitoring alarm supervisory systems should be electrically supervised so that the occurrence of a single-open or a single-ground fault condition in the installation wiring that prevents the required normal operation of the system or causes failure of the primary (main) power supply source is indicated by a distinctive trouble signal.

Power supplies, system circuits, and certain equipment in a fire alarm system are supervised. Fault conditions on supplementary circuits need not be indicated by a distinctive trouble signal. However, a fault on these circuits should not affect system operation nor cause an alarm condition at the FACU.

Power Supply Supervision

All sources of energy, that is, power supply, in a fire alarm system should be supervised, except for a power supply that:

1. Operates trouble signal circuits and appliances.
2. Is used as an auxiliary means of maintaining normal system operation following a trouble signal indicating when the primary (main) supply source is interrupted.
3. Supplies supplementary devices and appliances. Failure of this supplementary power supply will not prevent system operation and alarm signal initiation.

System Circuit Supervision

Initiating-Device Circuits, Signaling Line Circuits, and Notification Appliance Circuits: Installation wiring for all appliances or devices that initiate or transmit signals, either manually or automatically, must be supervised. Typical devices are fire alarm boxes, heat and smoke detectors, and automatically operated transmitters. A noninterfering shunt circuit does not need to be supervised, provided that a fault condition of the shunt wiring does not affect essential operating features of a system. Supervision is also not essential for pneumatic continuous-line-type rate-of-rise systems, where the device wiring terminals are connected in parallel across electrically supervised circuits.

Audible and Visible Alarm Notification Appliance Circuits: All installation wiring for operating audible and visible alarm notification appliance circuits are electrically supervised. A short or fault in the notification appliance circuit will be indicated by a trouble signal. Supervision is not essential for the installation wiring of the following:

1. Audible alarm notification appliance that is installed in the same room as a fire alarm control unit, provided the circuit conductors are installed in conduit or have equivalent protection against mechanical injury or tampering.
2. Supplementary alarm notification appliances, provided that a fault condition does not prevent required system operation.
3. Supplementary signal annunciator or other supplemental visible alarm notification appliance, provided that a fault condition does not prevent required system operation.

Equipment Supervision

Speaker amplifier and tone-generating equipment is supervised where speakers are used to produce audible fire alarm signals. An audible trouble signal will result upon failure of the audio amplifier and/or tone-generating equipment. Tone-generating equipment and amplifying equipment need not be supervised if they are enclosed as integral parts and serve only one listed audible notification appliance.

Installation-wiring connections to alarm notification appliances, initiating devices, and supervisory initiating devices are supervised. Supplementary circuits that operate fan motor stops or similar equipment do not need to be electrically supervised, provided that a fault condition does not affect required system operation. The trouble signal circuit does not need to be supervised.

Circuit Performance: Initiating-device circuits, signaling line circuits, and notification appliance circuits can be designated either as Class A or Class B depending on their performance during nonsimultaneous single-circuit-fault conditions. Circuits capable of transmitting an alarm signal during a single-open or a nonsimultaneous single-ground fault on a circuit conductor are designated as Class A.

Circuits incapable of transmitting an alarm beyond a single-open or a nonsimultaneous single-ground fault on a circuit conductor are designated as Class B.

Initiating-device, notification appliance, and signaling line circuits are permitted to be designated by style so as to describe requirements in addition to the requirements shown for Class A or Class B according to the following:

1. An initiating-device circuit is permitted to be designated as either Style A, B, C, D, or E depending on its ability to meet the alarm and trouble performance requirements shown in Table 4-1 during a single-open, single-ground, wire-to-wire short, and loss-of-carrier fault condition.
2. A notification appliance circuit is permitted to be designated as either Style W, X, Y, or Z depending on its ability to meet the alarm and trouble performance requirements shown in Table 4-2 during a single-open, single-ground, and wire-to-wire short fault condition.
3. A signaling line circuit is permitted to be designated as either Style 0.5, 1, 2, 3, 3.5, 4, 4.5, 5, 6, or 7 depending on its ability to meet the alarm and trouble performance requirements shown in Table 4-3 during a single-open; single-ground; wire-to-wire short; simultaneous wire-to-wire short and open; simultaneous wire-to-wire short and ground; simultaneous open and ground; and loss-of-carrier fault conditions. Classifying circuits by class allows a designer to know circuit performance before system specifications are written.

It is not intended that the styles be construed as *grades.* One style system is not necessarily better than another; in fact, a simple style may provide more adequate and reliable signaling for an installation than a complex style. In previous editions of the *National Fire Alarm Code,* the tables contained quantities tabulated under each style to imply that one style was better than the other. The increased quantities for the higher style numbers were based on the ability to signal an alarm during an abnormal condition in addition to signaling the same abnormal condition. When the capacities were at the maximum allowed, the overall system reliability was considered to be equal from style to style.

The number of automatic fire detectors connected to an initiating-device circuit should be limited by good engineering practice. If a large number of detectors is connected to one-initiating device circuit covering a widespread area, pinpointing the source of alarm becomes difficult and time-consuming. With certain types of detectors, a trouble signal results from faults in the detector. When this occurs with a large number of detectors on an initiating-device circuit, locating the faulty detector also becomes difficult and time-consuming.

Table 4-1. Performance of Initiating-Device Circuits*

Class: Style:	B A			B B			B C			A D			A Eα		
Abnormal Condition	Alm	Trbl	ARC	Alm	Trbl	ARC	Alm	Trbl	ARC	Alm	Trbl	ARC	Alm	Trbl	ARC
	1	2	3	4	5	6	7	8	9	10	11	12	13	14	15
Single open	—	X	—	—	X	—	—	X	—	—	X	R	—	X	R
Single ground	—	X	—	—	X	R	—	X	R	—	X	R	—	X	R
Write-to-wire short	X	—	—	X	—	—	—	X	—	X	—	—	X	X	—
Loss of carrier (if used)/ channel interface	—	—	—	—	—	—	—	X	—	—	—	—	—	X	—

*Alm = Alarm; Trbl = trouble; ARC = alarm receipt capability during abnormal condition; R = required capacity; X = indication required at protected premises and as required by *NFPA 72*; α = style exceeds minimum requirements of Class A.

Source: National Fire Alarm Code®, NFPA, 2002, Table 6.5.

TABLE 4-2. Performance of Signaling Line Circuits*

Class:	B			B			A			B			B		
Style:	0.5			1			2α			3			3.5		
	Alm	Trbl	ARC	Alm	Trbl	ARC	Alm	Trbl	ARC	Alm	Trbl	ARC	Alm	Trbl	ARC
Abnormal Condition	1	2	3	4	5	6	7	8	9	10	11	12	13	14	15
Single open	—	X	—	—	X	—	—	X	R	—	X	—	—	X	—
Single ground	—	X	—	—	X	R	—	X	R	—	X	R	—	X	—
Wire-to-wire short	—	—	—	—	—	—	—	—	M	—	X	—	—	X	—
Wire-to-wire short and open	—	—	—	—	—	—	—	—	M	—	X	—	—	X	—
Wire-to-wire short and ground	—	—	—	—	—	—	—	X	M	—	X	—	—	X	—
Open and ground	—	—	—	—	—	—	—	X	R	—	X	—	—	X	—
Loss of carrier (if used)/channel interface	—	—	—	—	—	—	—	—	—	—	—	—	—	X	—

Class:	B			B			A			A			A		
Style:	4			4.5			5α			6α			7α		
	Alm	Trbl	ARC	Alm	Trbl	ARC	Alm	Trbl	ARC	Alm	Trbl	ARC	Alm	Trbl	ARC
	16	17	18	19	20	21	22	23	24	25	26	27	28	29	30
Single open	—	X	—	—	X	R	—	X	R	—	X	R	—	X	R
Single ground	—	X	R	—	X	—	—	X	R	—	X	R	—	X	R
Wire-to-wire short	—	X	—	—	X	—	—	X	—	—	X	—	—	X	R
Wire-to-wire short and open	—	X	—	—	X	—	—	X	—	—	X	—	—	X	—
Wire-to-wire short and ground	—	X	—	—	X	—	—	X	—	—	X	—	—	X	—
Open and ground	—	X	—	—	X	—	—	X	—	—	X	R	—	X	R
Loss of carrier (if used)/channel interface	—	X	—	—	X	—	—	X	—	—	X	—	—	X	—

*Alm = Alarm; Trbl = trouble; ARC = alarm receipt capability during abnormal condition; M = may be capable of alarm with wire-to-wire short; R = required capability; X = indication required at protected premises and as required by *NFPA 72*; α = style exceeds minimum requirements of Class A.

Source: *National Fire Alarm Code®*, NFPA, 2002, Table 6.6.1.

TABLE 4-3. Notification-Appliance Circuits*

Class:	B		B		B		A	
Style:	W		X		Y		Z	
	Trouble indication at protected premises	Alarm capability during abnormal conditions	Trouble indication at protected premises	Alarm capability during abnormal conditions	Trouble indication at protected premises	Alarm capability during abnormal conditions	Trouble indication at protected premises	Alarm capability during abnormal conditions
Abnormal Condition	1	2	3	4	5	6	7	8
Single open	X	—	X	R	X	—	X	R
Single ground	X	—	X	—	X	R	X	R
Wire-to-wire short	X	—	X	—	X	—	X	—

* X = Indication required at protected premises; R = required capability.

Source: National Fire Alarm Code®, NFPA, 2002, Table 6.7.

Monitored Circuits: Each fire alarm system has a number of circuits connected to the fire alarm control unit. Each circuit fulfills a general system function and can be arranged in different ways. Major circuits found in fire alarm systems are basic initiating-device circuits, notification appliance circuits, and signaling line circuits.

Basic Initiating Circuits: The basic alarm-initiating-device circuit (typically Style A, B, or C) consists of a two-wire circuit with an end-of-line resistor. A separate circuit in the control panel monitors the field wiring between the panel and the initiating devices. Initiating devices have normally open contacts and are connected in parallel (see Figure 4-2). As shown, a small supervisory current normally flows through both relays (or an equivalent solid-state circuit), the field installation wires, and the end-of-line resistor. The current is sufficient to energize T_1, but not A_1. Operation of an initiating device (detector) shunts the resistor, increasing the current in the circuit, which energizes A_1 to initiate the alarm. A short between the wires or a ground fault on both wires also causes an alarm.

A single open on the circuit deenergizes T_1 and initiates a trouble signal; it also renders inoperable all initiating devices that are electrically beyond the open as shown in Figure 4-3. When an open fault occurs, audible and visible trouble indications operate at the control unit.

When an automatic fire detector has integral trouble contacts, contacts must be wired between the last detector and the end-of-line resistor. This prevents

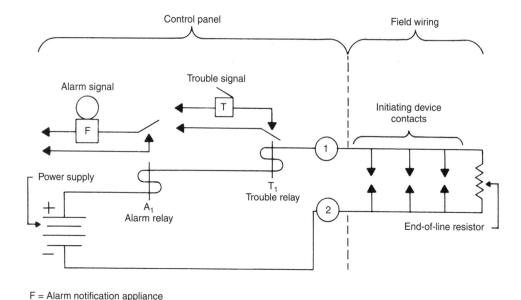

F = Alarm notification appliance

Figure 4-2. Typical Style A, B, or C initiating-device circuit.

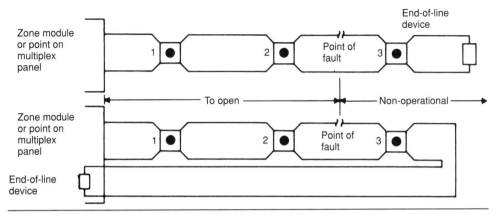

Figure 4-3. Typical Style A, B, or C initiating-device circuit with open fault.

the opening of a trouble contact from impairing the alarm operation of other fire detectors (see Figure 4-4).

To operate the system with an open-circuit fault (Style D or E) on the initiating circuit, wiring is typically arranged as shown in Figure 4-5. The circuit is connected to a control panel designed to receive the signal despite an open on the field wiring circuit. Since the end-of-line device is at the fire alarm control panel in the form of relay T_1, it deenergizes when an open on the wire occurs, sounding the trouble signal T. The panel is then conditioned manually or automatically by

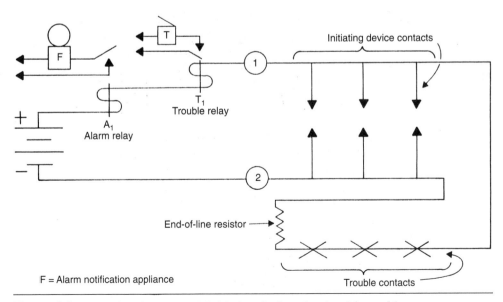

Figure 4-4. Typical Style A, B, or C initiating-device circuit with trouble contacts.

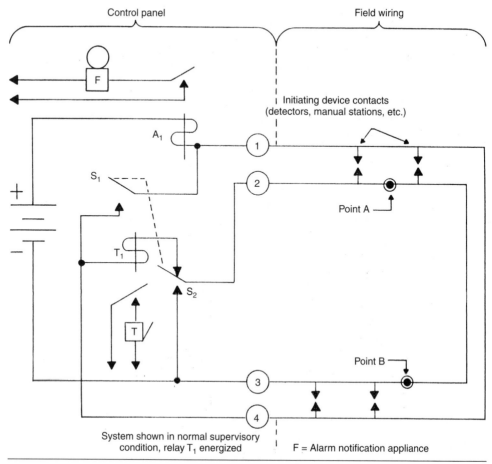

Figure 4-5. Typical Style D or E initiating-device circuit.

the operation of S_1 and S_2 to connect terminals 1 and 4, and 2 and 3 together. The system can then operate for an alarm despite an open circuit on either or both of the initiating circuit wires. Should two opens occur on a single wire, however, such as at point A and B, the device to the right of point A would not be able to operate the system.

Alarm-initiating circuits should be wired as shown in Figure 4-6 to monitor all interconnecting wires. If it is incorrectly wired (Figure 4-7), the fire alarm system would set up as if it were in the normal supervisory (monitored) condition, but an open occurring in riser 1 or 2 would not be indicated at the control unit because the supervisory current would still flow through riser 3 and the end-of-line device.

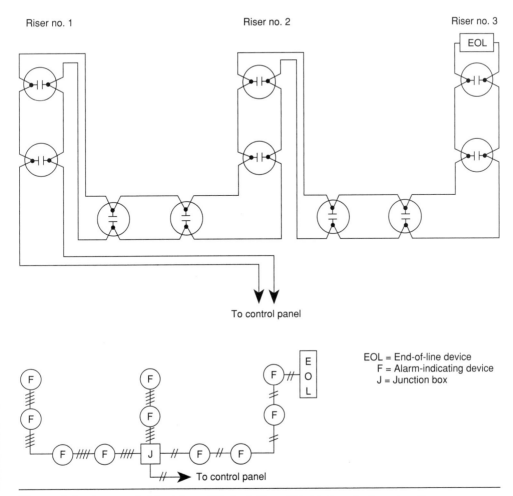

Figure 4-6. Correct field wiring of Style A, B, or C initiating-device circuit.

Notification Appliance Circuits: Notification appliance circuits can also be basic supervised circuits or capable of being conditioned to operate despite an open circuit on the installation wiring.

A typical basic notification appliance circuit is shown in Figure 4-8. In the normal supervisory condition, current flows from the negative side of the power supply through the normally closed contact (A_{1b}) of the alarm relay A_1, then exits terminal 2 and flows through the end-of-line diode D_1, back through terminal 1, through the normally closed contact (A_{1a}), and through the supervisory relay coil T_1 to the positive side of the power supply.

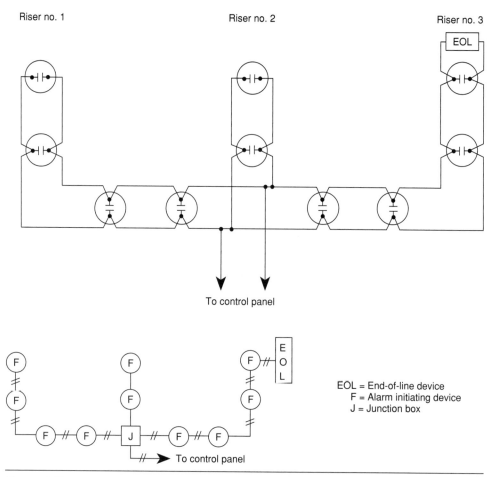

Figure 4-7. Incorrect field wiring of Style A, B, or C initiating-device circuit.

In this supervisory condition, the D_2 diodes block the flow of current through the alarm appliances F, preventing their operation. Sufficient current flows through relay T_1 to energize it as shown. Should an open occur in the notification appliance circuit, or D_1, relay T_1 will deenergize, activating the trouble signal T.

When an alarm occurs and alarm relay A_1 is energized, contacts A_{1a} and A_{1b} transfer from the upper, normally closed position to the lower, normally open position. This changes the potential at terminal 1 from plus ($+$) to minus ($-$), and the opposite occurs at terminal 2. Diode D_2 then blocks the flow of current through the end-of-line device, and the D_2 diodes conduct current to F, causing the indicating appliances to operate. A circuit can also be provided to condition the notification appliance circuit to operate despite an open circuit.

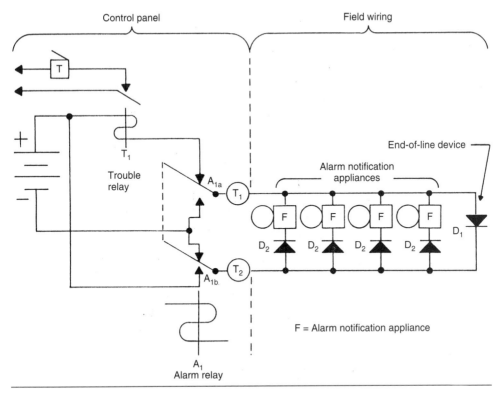

Figure 4-8. Notification appliance supervised circuit.

The parallel supervision circuit is similar to the end-of-line device circuit, except that the panel contains additional circuitry in the form of a bridge circuit (see Figure 4-9). After all the appliances are connected, the bridge circuit is balanced at the panel. Subsequently, an open circuit in the wiring, or the coil or contacts of a single appliance, unbalances the circuit and causes a trouble signal. In addition to supervising the installation wiring, use of the parallel circuit allows all of the notification appliance coils and contacts to be supervised.

Combination Initiating-Device and Notification Appliance Circuits: A hybrid circuit that combines both alarm-initiating devices and alarm notification appliances on a single pair of wires along with supervision of the field wiring is shown in Figure 4-10.

In the normal supervisory condition, the circuit polarity is such that there is no current flowing through the alarm notification appliances [terminal 1 on the control panel is minus ($-$) and terminal 2 is plus ($+$)]. A diode in each alarm notification appliance blocks the flow of current. In the alarm condition, the

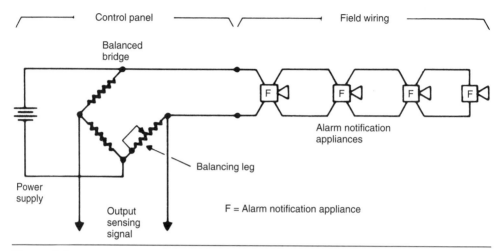

Figure 4-9. Indicating-appliance parallel supervision circuit.

polarity is reversed as shown in Figure 4-8, and current flows through the alarm notification appliance.

The alarm-initiating device must be polarized or contain a current-limiting means or both so the device does not short the circuit when in the alarm condition. Alarm-initiating devices must also withstand the full circuit alarm voltage.

Signaling Line Circuits: Signals are transmitted over signaling line circuits from the protected premises to a supervisory station in all fire alarm systems, except local systems. Although these circuits can take different forms depending upon the type of system, all should be supervised (monitored) in some manner.

The McCulloh circuit, one of the oldest forms of signaling line circuit, is a coded circuit that is still used in many central station systems. These circuits

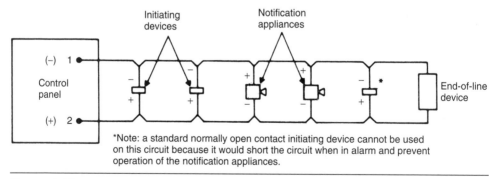

Figure 4-10. Combination alarm-initiating device and notification appliance circuit.

allow the system to receive an alarm even though the fire alarm circuit may be open because of a fault in the box circuit wiring. An alarm can be received via the remaining good leg of the circuit and ground return. Response to the fault at the control panel can be either automatic or manual. When response is automatic, relays perform the conditioning function; when it is manual, a switch must be operated. The automatic grounding-type system is free of circuit grounds until a break occurs. A relay is then deenergized and both sides of the line are tied together (via the relay contacts) at the positive terminal of the power supply. Concurrently, the negative side of the power supply is grounded. In the event of an alarm under these conditions, the fire alarm box mechanism will transmit its signal from the positive line via the signaling contacts, to the signal-responsive devices at the control panel, through ground return to the negative power source. Both boxes and the central control panel must be designed for and be compatible with this type of operation under an open-circuit fault.

In the basic circuit operation, the circuit operates from a grounded positive battery supply (see Figure 4-11). The ungrounded negative side feeds through

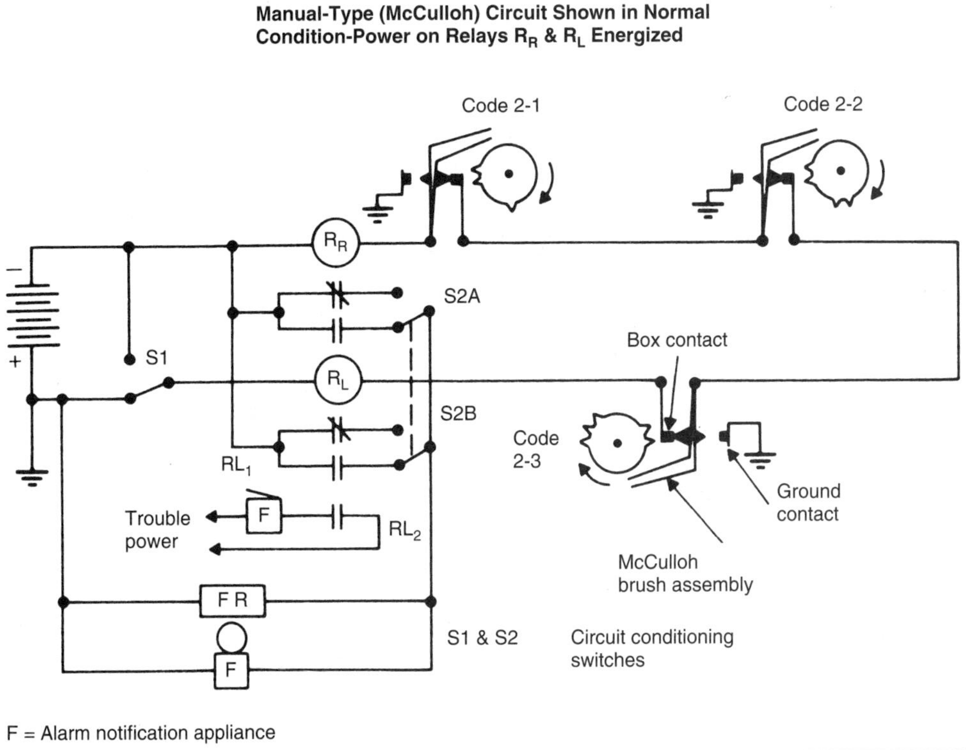

Manual-Type (McCulloh) Circuit Shown in Normal Condition-Power on Relays R_R & R_L Energized

F = Alarm notification appliance

Figure 4-11. McCulloh circuit normal supervisory condition.

the right-hand circuit relay R_R (one leg of a closed series loop), out through the signal line to the alarm box signaling contacts, back through the left-hand ground relay R_L, and returns to ground. This feed forms the normally closed box circuit, which exists during normal operation.

Under an open-fault condition, both the right- and left-hand circuit relays R_R and R_L are released, initiating a trouble signal (see Figure 4-12). The circuit at this moment is either automatically or manually switched to feed battery out through both relays to the open fault. This forms two normally open circuits, which will receive signals from the ground contact of the fire alarm boxes or transmitters.

Under a ground-fault condition, the left-hand relay (R_L) releases, initiating a trouble signal (see Figure 4-13). The system is then conditioned by switching S1 to feed to minus ($-$) potential through both relays, out through the circuit to the ground fault, and back to the battery through the grounded plus ($+$) side. Upon operation of a fire alarm box, the coded pulse is transmitted each time the brush assembly opens the line. Each box or transmitter is equipped

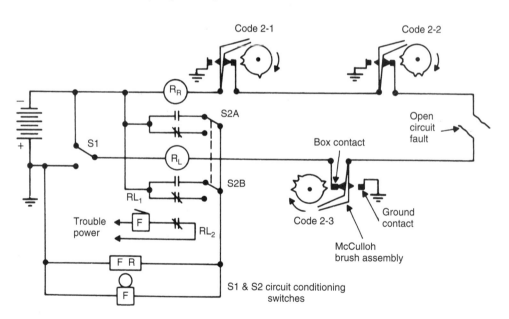

**McCulloh Circuit Shown with Open Fault
Condition-Power on Relays R_R & R_L Deenergized
S_1 & S_2 Manually Transferred**

F = Alarm notification appliance

Figure 4-12. McCulloh circuit with open-fault condition.

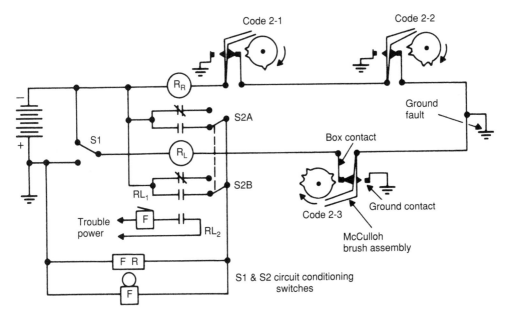

Figure 4-13. McCulloh circuit with ground-fault condition.

with a set of contacts known as a McCulloh brush, which consists of several contact fingers held together to form a closed and an open circuit. Two of the contact fingers are electrically common but are arranged to flex separately.

For proper performance, the opening and grounding functions of the signaling line circuit must be independent of each other. The open and ground contact blades are staggered so that each code wheel tooth will open, close, and then ground the line. If the line does not close before the ground contact is made, the ground signal will not return both ways, and the signal will be lost under some conditions. Signaling line circuits require the McCulloh brush-type contact action at the fire alarm transmitter to be compatible with its associated control unit.

REMOTE-STATION FIRE ALARM SYSTEM

Another method of connecting a fire alarm system directly to the municipal communication center is found in a remote-station fire alarm system, which does not use municipal fire alarm circuits. In this system, a separate pair of telephone wires is leased from the telephone company between each property and the municipal communication center. Generally, these systems are noncoded, and actuation of an individual pilot light identifies the property where the alarm originated (see Figures 4-14 and 4-15).

The system in Figure 4-14 is shown with power on, both A relays deenergized, T relays energized, and in supervisory (monitored) condition. Operation of an initiating device energizes A_1, switching A_{1A} and A_{1B}, which causes a reverse flow of current through the remote station and energizes the system. An open or short on the signaling line circuit deenergizes T_2, initiating a trouble signal at the remote supervisory station.

There are two types of noncoded remote-station fire alarm systems. The first is the reverse polarity type, in which power for the leased circuit originates at the fire alarm control at the protected property. In the event of an alarm, the polarity of the leased line is reversed, for example, through a relay, so that a relay is energized in a receiver at the municipal communication center. A light goes on, identifying the location from which the alarm originated and actuating an alarm notification appliance. Lines are electrically supervised.

The second type of noncoded system utilizes differential current relays or equivalent circuits. A small supervisory current flows through the leased wire

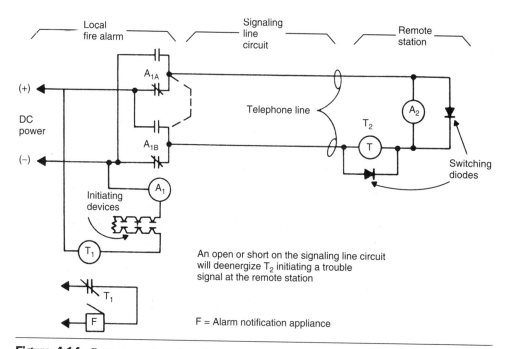

Figure 4-14. Remote-station fire alarm system with reverse polarity circuit.

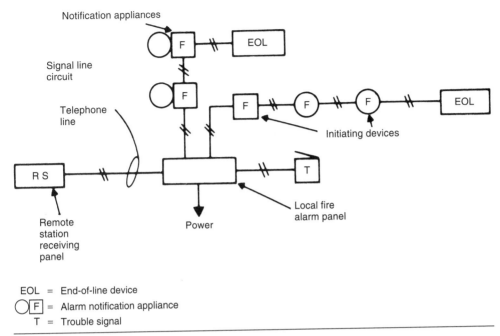

Figure 4-15. Remote-station fire alarm system riser diagram.

and an end-of-line resistor at the protected property. When an alarm is actuated, a relay contact shorts out the end-of-line resistor, causing the current to increase sufficiently to pick up or energize an alarm relay at the municipal communication center. This, in turn, lights a pilot light and sounds the alarm. This system can be powered from either end of the line. In essence, the remote supervising station is the control panel and the signaling line circuit is the field wiring between terminals 1 and 2 and the first initiating device. The initiating devices are located in the protected property.

ADDRESSABLE ALARM-INITIATING DEVICES

Another method of monitoring the interconnection of alarm-initiating devices is with a fire alarm system that is designed to use addressable detectors. In this system, the fire alarm control unit periodically sends an individually coded signal to each addressable alarm-initiating device and receives a signal response in return. A trouble signal is initiated if the control unit does not receive a response or (in some systems) receives a return signal indicating that a smoke detector, for example, exceeded its calibrated limit. Since the devices are addressed individually, there is no need for an end-of-line device or a supervisory current.

When certain addressable systems are used, the alarm-initiating circuit can be wired with branch circuits to other addressable devices. The end-of-line device would be omitted here, and fewer wires are required in risers 1 and 2.

ANALOG ALARM-INITIATING DEVICES

Analog smoke detectors provide three main features: (1) identification of the fire area, (2) sensitivity setting of the devices, and (3) verification times for each device in the system. The analog value from each smoke detector is the percentage of smoke obscuration detected, and is continuously stored and evaluated by the controller. When the value continuously but slowly increases over an extended period of time, an "alert" message is generated and displayed. This indicates the need for detector cleaning and maintenance, thereby reducing false alarms due to dirty detectors. When the analog value from the detector increases rapidly, the device, if identified for alarm initiating, will wait until the predetermined verification time has expired. The device will then determine the new analog value from the detector and place the system into alarm if a sufficient value for the alarm condition is present.

NONSUPERVISED PARTS OF FIRE ALARM SYSTEMS

The internal components of alarm-initiating devices and alarm notification appliances are not required to be supervised, although the power supply and the installation wiring that connect the alarm-initiating devices and the alarm notification appliances to the control panel should be supervised.

Testing laboratories review design and performance of the internal components of a fire alarm system under test conditions to determine whether they can be listed for fire alarm service. This ensures some minimum level of reliability. Even with this review, the known failure rate of electronic components necessitates regular testing of a fire alarm system. Testing a fire alarm system is the only way to determine if the system has failed. More information on system testing can be found in Chapter 12, Fire Alarm System Testing.

QUESTIONS

1. The basic components of a fire alarm system include:

   a. notification appliances

   b. automatic heat detectors

   c. manual fire alarm boxes

   d. a and c

   e. all of the above

2. A fire alarm system can be connected to:

   a. a central station   b. the fire department   c. a remote station
   d. a and b   e. a, b, and c

3. Circuits capable of transmitting an alarm signal during a single-open or a nonsimultaneous single-ground fault on a circuit conductor are designated as:

   a. Class B   b. Class A   c. Style A   d. Style B

4. Electric supervision in fire detection and alarm systems monitors the circuit integrity of interconnecting conductors so that when a single-open or a single-ground condition occurs that would prevent normal operation:

   a. the fire department is automatically notified
   b. the maintenance person is automatically called
   c. the condition is automatically transmitted and indicated in the appropriate location
   d. the alarm sounds throughout the building

5. To compensate for aging and other environmental effects such as temperature, it is common practice to include a safety factor in the battery load calculations:

   a. as high as 25 percent
   b. ranging from 10 to 20 percent
   c. 5 percent on conventional systems
   d. ranging from 15 to 30 percent on voice/alarm communication systems

BIBLIOGRAPHY

NFPA Codes, Standards, Recommended Practices, and Manuals (see the latest *NFPA Catalog* for availability of current editions of the following documents):

NFPA 72®, National Fire Alarm Code®.

NFPA 1221, *Standard for the Installation, Maintenance, and Use of Emergency Services Communications Systems.*

ADDITIONAL READING

ANSI A1, 7.1. *The Safety Code for Elevators and Escalators,* American National Standards Institute, New York, 2000.

5

Signal Initiation

INTRODUCTION

Fire alarm signals can be initiated by several types of devices (either manual or automatic) or through a systems approach in which alarm signals are combined with functions of fire extinguishing systems. Systems or individual devices are activated (and a signal initiated) by heat, smoke, energy radiation, or other detectable by-products of a fire.

An automatic initiating device must be able to differentiate between normal environmental fluctuations, that is, nonfire conditions, and prefire conditions or changes. A signal is initiated when certain preset conditions (usually related to temperature, gas, light, or energy) are reached or exceeded.

Initiating devices can be used in a facility in combination with a fire extinguishing system. Extinguishing system capacity is determined by the area to be protected and the speed of detection desired. Extinguishing systems that are connected to alarm systems require supervisory devices to detect abnormal conditions that might compromise the extinguishing system operation, such as closed valves, low water pressure, low water tank temperature, and water level changes. Signals can also be initiated via a guard's tour supervisory service.

For additional information on audible signals see Chapter 7, Fire Alarm Notification, and on design parameters for heat and smoke detectors, see Chapter 9, Fundamentals of Fire Detection System Design; also refer to Chapter 8, Fire Alarm System Installation.

ALARM-INITIATING DEVICES

Alarm-initiating devices for fire alarm systems are either manual fire alarm boxes (stations) or automatic fire detectors. Both are used to initiate an alarm on a fire alarm system.

General Requirements for Manual Stations: Mounting and Distribution

Manual fire alarm boxes should be approved for the particular application and can be used for fire protective signaling purposes only. However, combination fire alarm and guard's tour boxes are acceptable if so listed. Each box should be securely mounted with the operable part of the box not less than 3.5 ft (1.1 m) or more than 4.5 ft (1.37 m) above the floor level.

Manual fire alarm boxes should be distributed throughout the protected area so they are unobstructed, readily accessible, and located in the normal path of exit from the area, as follows:

1. At least one box should be provided at each exit and on each floor.
2. Additional boxes should be provided so that travel distance to the nearest box will not exceed 200 ft (61 m).

Usually, manual fire alarm boxes are located in the egress corridor adjacent to the door into the stairway. This makes it easy for occupants to activate the box as they enter the stairway.

Manual Fire Alarm Boxes

Manual stations can be one of four types: (1) noncoded or coded, (2) presignal or general alarm, (3) breakglass or nonbreakglass, and (4) single- or double-action.

Noncoded or Coded Stations: A noncoded manual station contains a switch that is housed within a distinctive enclosure. Once actuated, the station contacts transfer. This condition is maintained until the station is reset to normal. Older coded stations contain a mechanically or electrically driven motor that, when activated, turns a code wheel, causing contacts to momentarily open or close to reproduce the code of the station. By counting the code that is sounded out on the system's audible appliances, one is able to determine the location of the alarm. Newer designs are electronic, but function identically. The station is required to repeat its code a minimum of three times. Both coded and noncoded stations can be surface, semiflush, or flush-mounted.

Presignal or General Alarm Stations: Presignal stations initially cause alarm signals to sound only in specific areas such as guard's offices or in the principal's office in a school. Actuation of a key switch on the station or the control panel causes an evacuation signal to sound. A general alarm station is the most common manual alarm system. When actuated, a general alarm station causes evacuation signals to sound immediately.

Breakglass or Nonbreakglass Stations: The term *breakglass* is applied to both noncoded and coded stations. For a breakglass station, actuation of the device to cause an alarm requires the initial action of breaking a glass or other breakable element. This is thought to deter false alarms, and is also intended to address the requirement that, once activated, manual stations cannot be reset without use of a special tool. Some codes specify that the operation of a manual station shall result in the fracture of a breakable element. Early breakglass stations used a glass plate covering the operating switch. The plate was to be broken with a small hammer attached to the station by a chain, but the hammers were frequently missing. Later designs incorporated a glass rod and some placed the rod inside the station. There are occasional reports of the glass rods being mistakenly replaced with other materials, which sometimes results in a station that cannot be activated. Stations without this feature are classified as *nonbreakglass* (see Figure 5-1).

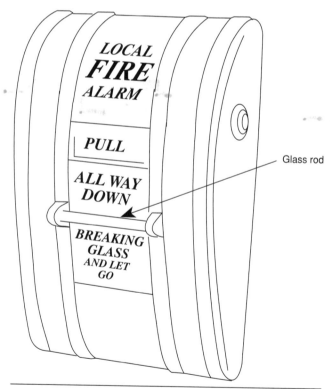

Figure 5-1. Breakglass station.

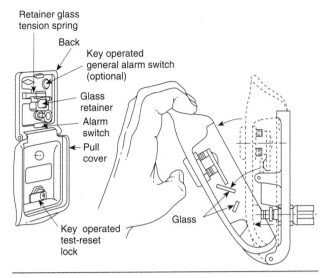

Figure 5-2. Single-action station.

Single- or Double-Action Stations: A single-action station requires only one action by the user to initiate an alarm. The required action is usually breaking a glass element or actuating a lever or other movable part of the station (see Figure 5-2).

A double-action station initiates an alarm as a result of two actions taken by the user. Typically the user must break a glass, open a door, or lift a cover to gain access to a switch or lever, which must then be operated as the second action to initiate an alarm. This is also thought to reduce the incidence of inadvertent actuation. (See Figure 5-3.)

Automatic Fire Detectors

People are excellent but unreliable fire detectors. Various mechanical, electric, and electronic devices in use mimic the human senses of smell, sight, hearing, taste, and touch to detect the environmental changes created by fire.

The most common elements of a fire that can be detected are heat, smoke (aerosol particulate), gas, and light radiation. However, not all fires produce all of these elements, and nonfire conditions can produce similar ambient conditions. The designer of a fire alarm signaling system must be able to differentiate among those elements that might be expected from a fire. The designer must also know the similar ambient conditions that might result from nonfire situations.

Figure 5-3. Double-action station.

Even if all of the elements of fire—heat, smoke, and light—are present in a given fire, the magnitude of each element must exceed some theoretical basic level during fire development. It should be noted that one element will usually appear first; this is especially important if the system is designed primarily for life protection.

Automatic fire detectors are classified in the initiating-devices chapter of *NFPA 72®, National Fire Alarm Code®*, as heat, smoke, and radiant energy detectors and "other" fire detectors. A heat detector is a device that detects abnormally high temperature or rate of temperature rise. Smoke detectors are devices that detect particles of combustion. A radiant energy (flame) detector detects the infrared, ultraviolet, or visible radiation produced by a fire. "Other" fire detectors are devices that detect phenomena other than heat, smoke, or flame produced by a fire.

Detectors are further divided into three types: (1) line type, (2) spot type, and (3) air-sampling type. In a line-type detector, detection is continuous along a path. Typical examples of line-type devices are rate-of-rise pneumatic tubing detectors, projected-beam smoke detectors, and heat-sensitive cable.

In a spot-type detector, the detecting element is concentrated at a particular location. Typical examples of spot-type detectors are bimetallic or fusible alloy heat detectors, certain pneumatic rate-of-rise detectors, most smoke detectors, and thermoelectric detectors.

An air-sampling-type detector consists of a system of piping through which air samples are continuously drawn to a central location where a high-sensitivity analysis system evaluates the air sample for a fire signature. Sampling ports can be located in any pattern, at specific hazards, or both. Current examples employ either photoelectric (scattering) or cloud chamber principles.

Automatic detectors have two operating modes: nonrestorable and restorable. A nonrestorable detector is a device whose sensing element is designed to be destroyed in the process of detecting a fire. In a restorable detector, the sensing element is not ordinarily destroyed in the process of detecting a fire. Restoration of the element may be manual or automatic.

In addition, detectors that respond in more than one way to a fire condition are called combination detectors. Combination detectors either respond to more than one of the fire phenomena (heat, smoke, flame, etc.) or employ more than one operating principle to sense one of these phenomena. Two typical examples of combination detectors are a combination of a heat detector with a smoke detector and a combination rate-of-rise and fixed-temperature heat detector. Combination detectors are also called multimode detectors.

Multimode Detectors: Any type of fire detector or multimode detector can employ a multicriterion mode of operating. Historically, fire detectors were threshold devices: When the signal produced exceeded some preset threshold value, the device went into an alarm state. Today, fire detectors can apply logic to make an alarm decision. The logic used is called a decision algorithm and may range in complexity from a simple truth table (IF the alarm is smoke, sound local audibles; IF the alarm is smoke AND heat OR manually activated, sound local AND notify fire department) to complex calculations such as NIST's sensor-driven fire model. Decision algorithms can be located within individual detectors or within the fire panel to which the detectors are connected. In the latter case, analog data continuously transmitted from one or more fire sensors provide the required input.

Analog Detectors: Any type of fire detector can be arranged as a sensor that transmits data continuously or at intervals to the fire panel for processing. Thus, an analog detector does not initiate an alarm; instead, this decision is made at the panel based on the data from one or more analog detectors. Some analog detector systems can convert the data into engineering units for display, for example, actual temperatures from a heat detector or concentration from a gas detector.

Heat Detectors: Heat detectors are the oldest type of automatic fire detection device, beginning with the development of automatic sprinklers in the 1860s and continuing to the present with various types of devices. A sprinkler can be considered a combined heat-activated fire detector and extinguishing device when the sprinkler system is provided with waterflow indicators connected to the fire alarm control panel. Waterflow indicators detect either the flow of water through the pipes or the subsequent water pressure change upon actuation of the system. Heat detectors were originally used with sprinklers to pinpoint the location of the fire in order to limit water damage.

Heat detectors that only initiate an alarm and have no extinguishing function are still in limited use. Although they are the least expensive of fire detectors and have the lowest false alarm rate of all automatic detecting devices, they also are the slowest in detecting fires. A heat detector is best suited for fire detection in a small, confined space where rapidly building high-heat-output fires are expected, in areas where ambient conditions would not allow the use of other fire detection devices, and where speed of detection is not a prime consideration.

Heat detectors are generally located on or near the ceiling and respond to the convected thermal energy of a fire. They respond either to a specified rate of temperature change or when the detecting element reaches a predetermined fixed temperature. In general, heat detectors are designed to operate when heat causes a specific change in a physical or electrical property of a material or gas such as by melting or expanding.

Operating Principles of Fixed-Temperature Heat Detectors: Fixed-temperature heat detectors are designed to initiate an alarm when the temperature of the operating element reaches a specific point. The air temperature at the time of alarm is usually considerably higher than the rated temperature because it takes time for the temperature of the operating element to increase to its set point. This condition is called thermal lag. Fixed-temperature heat detectors are available to cover a wide range of operating temperatures from approximately 100°F (38°C) and higher (see Table 5-1). Higher temperature detectors are necessary so that detection can be provided in areas normally subjected to high ambient (nonfire) temperatures or in areas zoned so that only detectors in the immediate fire area operate.

TABLE 5-1. Operating Temperature Ranges for Fixed-Temperature Heat Detectors[a]

Temperature Classification	Temperature Rating Range (°F)	Maximum Ceiling Temperature (°F)	Color Code
Low[b]	100–134	20 below rated[c]	Uncolored
Ordinary	135–174	100	Uncolored
Intermediate	175–249	150	White
High	250–324	225	Blue
Extra High	325–399	300	Red
Very Extra High	400–499	375	Green
Ultra High	500–575	475	Orange

[a] The difference between the rated temperature and the maximum ambient temperature should be as small as possible to minimize the response time.

[b] Intended only for installation in controlled ambient areas. Units are marked to indicate maximum ambient installation temperature.

[c] Maximum ceiling temperature has to be 20°F or more below detector-rated temperature.

For SI Units: °C = (5/9) (°F −32).

Nonrestorable Fusible Element-Type Detectors: Eutectic metals (alloys of bismuth, lead, tin, and cadmium that melt rapidly at a predetermined temperature) can be used as operating elements for heat detection. The most common such use for these metals is the fusible soldered joint element in an automatic sprinkler (see Figure 5-4). These elements are nonrestorable since fusing (melting) the element allows the valve cap covering the sprinkler orifice to fall away, water to flow in the system, and the alarm to be initiated.

A closed-head sprinkler can be used as a release device on a fixed-temperature detector for pneumatic operation of a hydraulic differential deluge valve. The closed-head sprinklers are attached to 1/2-in. (12.7-mm)-diameter air pilot lines at approximately 10-ft (3-m) spacing. The dry pilot actuation release system is pressurized with compressed air or nitrogen. Upon activation of a sprinkler, release of pressure from the pilot lines activates the deluge valve. To increase the speed of operation of the deluge valve, an accelerator can be installed on the pilot line detection system. Typical applications of the pneumatic pilot line detection system are transformer protection and cooling tower protection. The pneumatic detection system can also be utilized in those areas where hazardous electric operations are performed requiring deluge system protection.

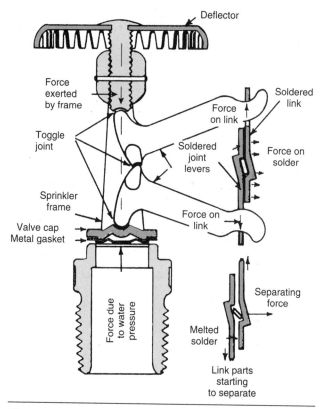

Figure 5-4. Representative arrangement of a soldered link and lever automatic sprinkler.

A eutectic metal may also actuate an electric heat detector and is often used as a solder to secure a spring under tension. When the element melts, the spring action closes contacts and initiates an alarm (see Figure 5-5, items D, F, and G). Devices using eutectic metals cannot be restored; either the device or its operating element must be replaced following operation.

Restorable Bimetallic-Type Detectors: In a restorable bimetallic-type detector, two metals with different coefficients of thermal expansion are bonded together and then heated. Differential expansion causes bending or flexing toward the metal with the lower expansion rate. This action closes a normally open circuit and initiates an alarm. The low-expansion metal commonly used is Invar™, an alloy of 36 percent nickel and 64 percent iron. Several alloys of

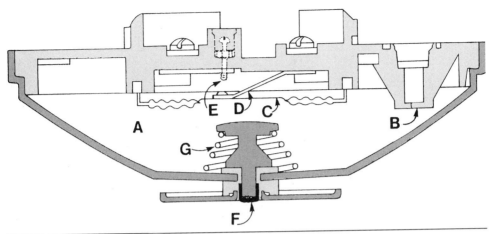

Figure 5-5. A spot-type combination rate-of-rise, fixed-temperature device. The air in chamber A expands more rapidly than it can escape from vent B. This causes pressure to close electric contact D between diaphragm C and contact screw E. Fixed-temperature operation occurs when fusible alloy F melts, releasing spring G, which depresses the diaphragm, closing contact points.

manganese–copper–nickel, nickel-chromium–iron, or stainless steel may also be used for the high-expansion component of a bimetal assembly. Bimetals are used for the operating elements as either strips or snap discs for fixed-temperature detectors.

Figure 5-6 shows a bimetal strip detector, where metal A changes its length when heated. A different metal, metal B, has a higher coefficient of expansion and will expand more than metal A when the temperature rises. When these two metals are bonded together and the temperature increases, metal B (which is on the bottom) will expand more than metal A, causing the entire element to bend upward as shown in the lower part of the figure. This bimetallic action closes an electrical circuit as shown in Figure 5-7 and is similar in operation to that of a bimetallic thermostat controlling the furnace in a home.

A bimetal strip, as it is heated, deforms in the direction of the contact point. With a given bimetal, the width of the gap between the contacts determines the operating temperature; the wider the gap, the higher is the operating point.

The disk-type bimetallic element (Figure 5-7, bottom) is also used extensively in fire alarm systems. The disk is formed of bimetallic material and is prestressed in an arch shape. When the temperature of the bimetallic material rises

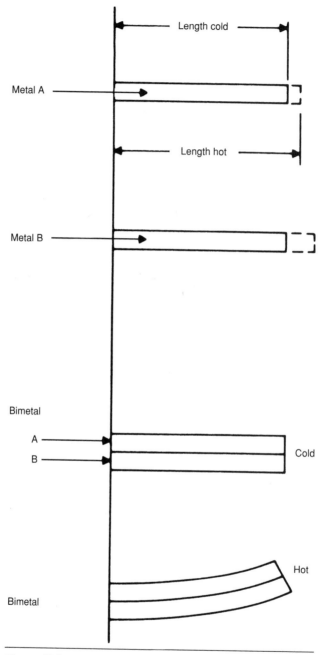

Figure 5-6. Bending action with two dissimilar metals.

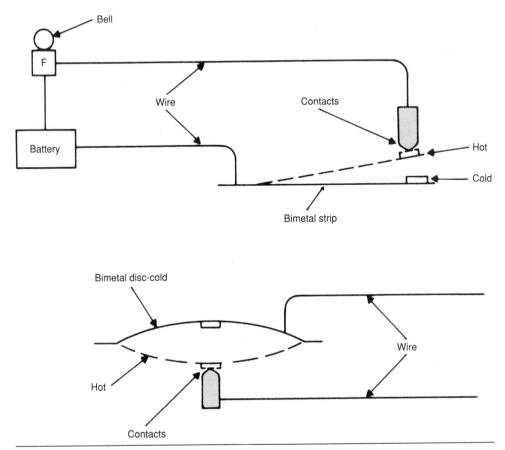

Figure 5-7. (Top) Bending action in bimetallic strip detector; (bottom) bimetallic disk-type detector.

to a predetermined point, the expansion of the metals causes the disk to change from a concave to a convex position in a snap action (as shown by the dashed line). Because of the snap action, the force applied to the contact points is greater on initial contact than the force applied by strip-type bimetallic elements.

Bimetallic detectors are usually designed to restore themselves to their original condition (and the disk to its concave position) after the heat has been removed (see Figures 5-8 and 5-9). A heat collector usually is attached to the frame of the detector to transfer heat more quickly from the room air to the bimetal.

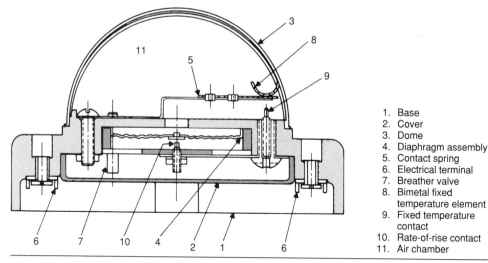

1. Base
2. Cover
3. Dome
4. Diaphragm assembly
5. Contact spring
6. Electrical terminal
7. Breather valve
8. Bimetal fixed
   temperature element
9. Fixed temperature
   contact
10. Rate-of-rise contact
11. Air chamber

Figure 5-8. Combination rate-of-rise and fixed-temperature detector (with resetting fixed-temperature element).

Heat detectors with bimetallic elements are automatically self-restoring after operation when the temperature drops below the operating (set) point. With a bimetallic strip element, the detector will usually be designed such that the detector restores when the element temperature drops just 1 degree or so below the set point. However, if the element is a snap disk, the ambient temperature around the detector may need to be lowered well below the set point for the disk element to snap back to its concave position. For this reason, and the 15°F (8.3°C) operating tolerance permitted for a typical 135°F (57.5°C) ordinary degree fixed-temperature heat detector, a detector of this temperature rating should not be installed where the ceiling temperature might exceed 100°F (37.8°C). (See Table 5-1 for other detector temperature ratings.)

Line-Type Heat Detectors: Various methods of line-type detection have been developed as alternatives to spot-type fixed-temperature detection, such as the heat-sensitive cable, continuous line-type detector. The detector shown in Figure 5-10 uses a pair of steel wires in a normally open circuit. The conductors are held apart by heat-sensitive insulation. The wires, under tension, are enclosed in a braided sheath to form a single cable assembly. When the temperature limit is reached, the insulation melts, the two wires contact, and an alarm is initiated. The fused section of the cable must be replaced following an alarm to restore the system.

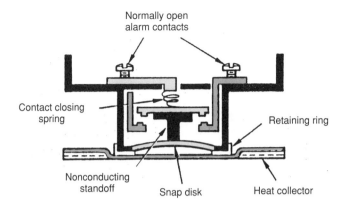

Normally open
alarm contacts

Contact closing
spring

Retaining ring

Nonconducting
standoff

Snap disk

Heat collector

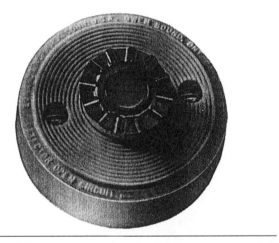

Figure 5-9. Spot-type fixed-temperature snap-disk detector.

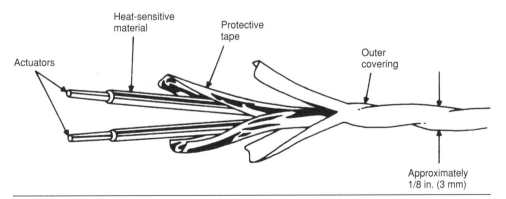

Heat-sensitive
material

Protective
tape

Outer
covering

Actuators

Approximately
1/8 in. (3 mm)

Figure 5-10. Line-type heat detector.

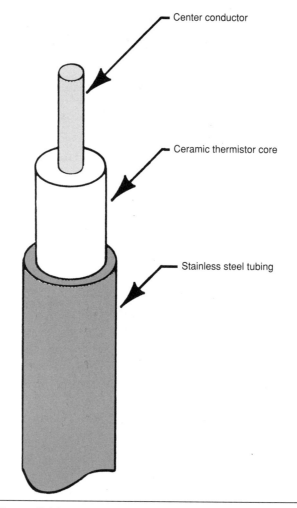

Center conductor

Ceramic thermistor core

Stainless steel tubing

Figure 5-11. View of the construction of the continuous thermal sensor showing outer tubing, ceramic thermistor core, and center wire.

A similar alarm device utilizing a semiconductor material and a stainless steel capillary tube has been used where mechanical stability is a factor in heat detection (see Figure 5-11). The capillary tube contains a coaxial-center conductor separated from the tube wall by a temperature-sensitive semiconductor material. Under normal conditions, a small current (below the alarm threshold) flows in the circuit. As the temperature rises, the resistance of the semiconductor thermistor decreases, allows more current flow, and initiates the alarm.

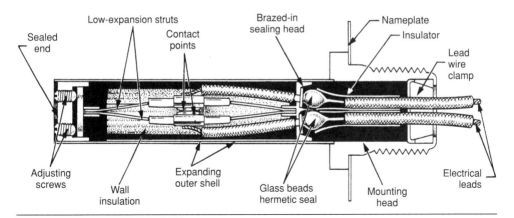

Figure 5-12. Section of spot-type rate compensation detector.

Rate Compensation-Type Detectors: A rate compensation detector is a device that responds when the temperature of the surrounding air reaches a predetermined level, regardless of the rate of temperature rise (see Figure 5-12).

A typical example of a rate compensation detector is a spot-type detector. This type of detector has a tubular casing of a metal that tends to expand lengthwise as it is heated and an associated contact mechanism that will close at a certain point in the elongation. A second metallic element inside the tube exerts an opposing force on the contacts, tending to hold them open. The forces are balanced in such a way that on a slow rate of temperature rise, heat takes longer to penetrate to the inner element. This inhibits contact closure until the total device has been heated to its rated temperature level. However, on a fast rate of temperature rise, there is not as much time for heat to penetrate to the inner element, which therefore exerts less of an inhibiting effect. This contact closure is obtained when the total device has been heated to a lower level, compensating for thermal lag.

Rate-compensating detectors are also automatically self-restoring after operation when the ambient temperature drops to some point below the operating point.

Rate-of-Rise-Type Detectors: One effect that a flaming fire has on the surrounding area is to rapidly increase air temperature in the space above the fire. Fixed-temperature heat detectors will not initiate an alarm until the air temperature near the ceiling exceeds the design operating point. The rate-of-rise detector, however, will function when the rate of temperature increase exceeds a predetermined value, typically around 12°F to 15°F (7°C to 8°C) per minute.

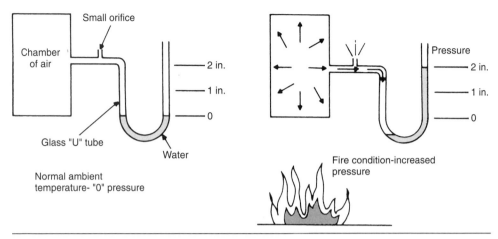

Figure 5-13. Operating principle for a pneumatic fire detector.

Rate-of-rise detectors are designed to compensate for the normal changes in ambient temperature [less than 12°F (7°C) per minute] that are expected under nonfire conditions.

In a pneumatic fire detector, air heated in a tube or chamber expands, increasing the pressure in the tube or chamber (see Figure 5-13). This exerts a mechanical force on a diaphragm, which closes the alarm contacts. If the tube or chamber were hermetically sealed, slow increases in ambient temperature, a drop in the barometric pressure, or both would cause the detector to initiate an alarm regardless of the rate of temperature change. To overcome this, pneumatic detectors have a small orifice to vent the higher pressure that builds up during slow increases in temperature or a drop in barometric pressure. The vents are sized in such a manner that when the temperature changes rapidly, as in a fire situation, the rate of expansion exceeds the venting rate and the pressure rises. When the temperature rise exceeds 12°F to 15°F (7°C to 8°C) per minute, a flexible diaphragm converts the pressure to mechanical action. Pneumatic heat detectors are available for both line- and spot-type detectors. Rate-of-rise operation is shown in Figure 5-14 for a line-type pneumatic heat detector and in Figures 5-5, 5-9, and 5-15 for a spot-type detector.

Most spot-type rate-of-rise detectors are equipped with a fixed-temperature backup feature so that should the temperature rise be slower than 15°F (8°C) per minute, the detector will operate when the detecting element has reached the predetermined fixed point. Spot-type rate-of-rise detectors are installed in the same manner as fixed-temperature detectors.

The line-type pneumatic heat detector consists of metal tubing in a loop configuration attached to the ceiling or wall near the ceiling of the area to be

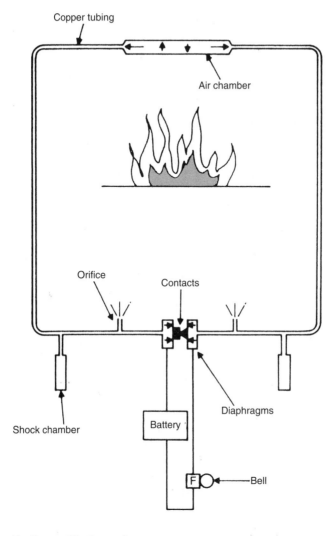

F = Alarm notification appliance

Figure 5-14. Operating principle for a line-type pneumatic heat detector.

protected. Lines of tubing are normally spaced not more than 30 ft (9.1 m) apart, not more than 15 ft (4.5 m) from a wall, and with no more than 1000 ft (305 m) of tubing on each circuit. Also, a minimum of at least 5 percent of each tube circuit or 25 ft (7.6 m) of tube, whichever is greater, must be in each protected area. Without this minimum amount of tubing exposed to a fire condition, insufficient pressure would build up to achieve proper response.

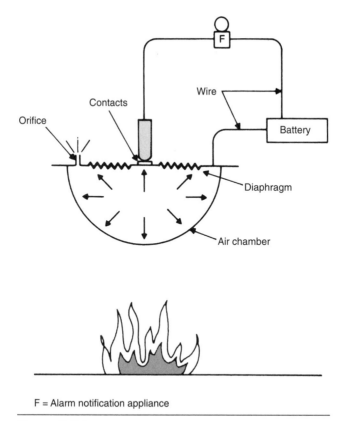

F = Alarm notification appliance

Figure 5-15. Operating principle for a spot-type pneumatic heat detector.

Shock chambers tend to compensate for additional transient increases in pressure that might be caused, for example, by the opening of an oven door. This reduces the possibility of a false alarm.

In small areas, where line-type detectors might have insufficient tubing exposed to generate sufficient pressures to close the alarm contacts, air chambers or rosettes of tubing are used. These units act like a spot-type detector by providing the volume of air required to meet the 5 percent or 25-ft (7.6-m) requirement. Since a line-type rate-of-rise detector is an integrating detector, it will actuate either when a rapid heat rise occurs in one area of exposed tubing or when a slightly less rapid heat rise takes place in several areas where tubing on the same loop is exposed. Pneumatic tubing systems are no longer manufactured, but still may be found in service.

The pneumatic principle is also used to close contacts within spot-type detectors. The difference between line- and spot-type detectors is that the

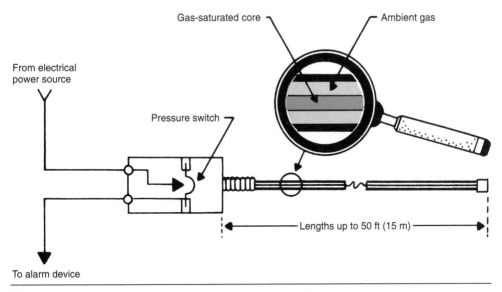

Figure 5-16. Sealed line-type fixed-temperature heat detector.

spot-type detector contains all of the air in a single container rather than in a tube that extends from the detector assembly to the protected area(s).

Sealed Pneumatic Line-Type Detectors: The sealed pneumatic line-type heat detector is another fixed-temperature heat detector for special applications. It is not vented to the air and consists of a capillary tube containing a special salt that is saturated with hydrogen gas. At normal temperatures, most of the hydrogen is held in the porous salt and the pressure in the tube is low.

As the temperature increases at any point along the tubing, hydrogen is released from the salt, which increases the internal pressure and eventually trips a diaphragm pressure switch. The integrity of the capillary tube is supervised by a second pressure switch, which monitors the low pressure present at normal temperatures (see Figure 5-16).

Thermoelectric Effect Detectors: A thermoelectric effect detector is a device with a sensing element consisting of a thermocouple or thermopile unit that produces an increase in electric voltage in response to an increase in temperature. This potential is monitored by associated control equipment, and an alarm is initiated when the voltage increases at an abnormal rate (see Figure 5-17).

Thermopile devices, which operate in the voltage-generating mode, use two sets of thermocouples. One set is exposed to changes in the atmospheric temperature. When temperature rapidly changes as in a fire situation, the

Figure 5-17. Spot-type thermoelectric effect heat detector.

temperature of the exposed set increases faster than the temperature of the unexposed set, generating a net voltage. The voltage increase associated with this potential operates the alarm circuit. Since the thermopile units are connected in series, this voltage need not be within only one detector; the small voltages produced at each unit on a circuit can cumulatively produce an alarm. The sensitivity of a thermopile detector is directly related to the number of thermocouple junctions within the device; the sensitivity increases as the number of junctions increases. Sensitivity can also be increased by designs that focus radiative energy on the exposed junctions. Thermopile effect detectors are no longer manufactured, but still may be found in service.

Thermistor Type Heat Detectors: Most system-connected heat detectors use a heat-sensing element called a thermistor, which increases electrical resistance with increasing temperature. These devices are simple and reliable and have a small mass, which makes them responsive to changes in the surrounding air temperature. With the appropriate electronics, they can respond in either a fixed-temperature or rate-of-rise mode or both. They can also operate as analog sensors sending data to the fire panel, and their signal can be converted to actual temperature readings.

Smoke Detectors: A smoke detector will detect most fires much more rapidly than a heat detector. Smoke detectors are identified by their operating principle; two of the most common operating principles are ionization and photoelectric. As a class, smoke detectors using the ionization principle provide somewhat faster response to high-energy (open flaming) fires since these fires produce large numbers of smaller smoke particles and the ionization sensor response is proportional to the number of particles per unit volume of air. As a class, smoke detectors operating on the photoelectric principle respond faster to the smoke generated by low-energy (smoldering) fires, since these fires generally produce more of the larger smoke particles. A photoelectric sensor's response is proportional to the mass of particles per unit volume of air. However, each type of smoke detector is subjected to, and must pass, the same test fires for listing by testing laboratories.

Smoke detectors can also initiate control of smoke spread in four different ways: (1) by prevention of the recirculation of dangerous quantities of smoke within a building by the mechanical ventilation system, (2) by selective operation of equipment to exhaust smoke from a building, (3) by selective operation of equipment to pressurize smoke compartments, and (4) by operation of doors to close the openings in smoke compartments.

Ionization-Type Smoke Detectors: An ionization-type smoke detector uses a small amount of radioactive material in the sensing chamber to ionize the air, rendering the air conductive and permitting a current flow between two charged electrodes. This gives the sensing chamber an effective electrical conductance. When smoke particles enter the sensing area, they decrease the conductance of the air by attaching themselves to the ions, causing a reduction in ion mobility. The detector responds when the conductance is below a predetermined level.

The theory of ionization detector operation is based on alpha-particle properties. Alpha particles, which are actually helium nuclei, ionize air, dissociating the air molecules into positive ions and negative electrons. If this ionized air is introduced into an electric field, the electrical charge on the ions and the electrons causes them to move, which results in a current. As shown in Figure 5-18, a potential from battery B is applied to plates P_1, and P_2. The air between the plates is ionized by the alpha emitter A, and the charged particles move in the direction indicated by the arrows. A galvanometer measures the current; the value of the current depends upon the strength of the alpha emitter and to a lesser extent on the voltage of battery B. The current in an ionization chamber depends upon the composition of the gas between the electrodes. Ion production is

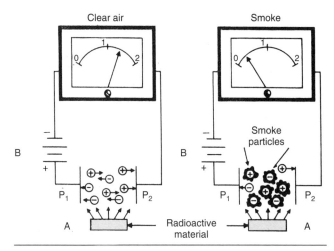

Figure 5-18. Principle of operation of an ionization smoke detector.

dependent on the number and size of the gas molecules. The rate of drift of the ions is closely related to their size and mass, and is not based on gas composition. The alpha particles travel only about 1 in. from the source, after which they capture a free electron and turn into a molecule of helium. Strict government licensing requirements assure that there are no radiation hazards from ionization-type smoke detectors.

Conditions are quite different when the smoke particles enter the ionization chamber. These particles attach themselves to the ions, causing a sharp drop in current, which effectively increases the resistance of the chamber. This, in turn, sounds the alarm through other electrical circuits. A cross-section view of an ionization detector is shown in Figure 5-19.

Photoelectric-Type Smoke Detectors: The suspended smoke particles generated during the combustion process affect the propagation of a light beam passing through the air. This effect can be utilized to detect the presence of a fire in two ways: obscuration of light intensity over the beam path and scattering of the light beam from its path of travel.

Smoke detectors that operate on the light obscuration principle consist of a light source, a light-beam-collimating system, and a photosensitive device, all arranged in a straight line. When dense smoke obscures part of the light beam, or less-dense smoke obscures more of the beam, the amount of light reaching the photosensitive device is reduced by a combination of absorption and scattering, initiating the alarm (see Figure 5-20). The light source is usually a light-emitting

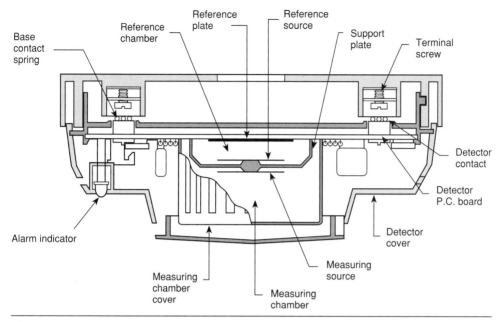

Figure 5-19. Cross-section view of an ionization smoke detector.

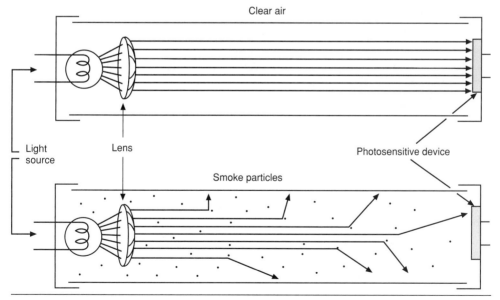

Figure 5-20. Principle of operation of a photoelectric obscuration smoke detector.

Table 5-2. Projected Beam Using Mirrors

Number of Mirrors	Maximum Allowable Beam Length	Beam Setup
0	Listed length L	
1	$(2/3)L = (2/3)(a + b)$	
2	$(4/9)L = (4/9)(c + d + e)$	

Example: Maximum allowable length of a beam for a detector listed at 300 ft, when used with two mirrors is $(4/9) \times 300$ ft, or 133 ft.

For SI Units: 1 ft = 0.30 m.

diode (LED), which is a reliable long-life source of illumination with a low current requirement. Pulsed LEDs can generate sufficient light intensity while operating at low overall power levels for use in detection equipment.

Most light obscuration smoke detectors are of the beam type and are used to protect large, open areas such as atriums. They are installed with the light source at one end of the area to be protected and the photosensitive device at the other. In some applications, mirrors determine the area of coverage by directing the beam over the desired path. For each mirror used, however, the rated beam length of the device must be progressively reduced by one-third, as illustrated in Table 5-2. Projected beam detectors are generally installed within 20 in. (51 cm) of, and parallel to, the ceiling, although angled beams are permitted to avoid obstructions.

Light scattering involves light being reflected or refracted by smoke particles. Light scattering smoke detectors are usually of the spot type and contain a light source and a photosensitive device arranged so the light normally does not fall onto the photosensitive device. When smoke particles enter the light path, light strikes the particles and is scattered onto the photosensitive device, causing the detector to respond (see Figure 5-21). A photodiode or phototransistor is usually the photosensitive device used in light scattering detectors.

Cloud Chamber Smoke Detectors: In a cloud chamber-type smoke detector, an air sample from a protected space containing submicrometer-size smoke particles is drawn through a humidifier, where it is brought to 100 percent relative humidity. The sample then passes to an expansion chamber, where the pressure is reduced with a vacuum pump, causing condensation of water on the particles. The droplets quickly grow to a visible size. A light beam passing through the

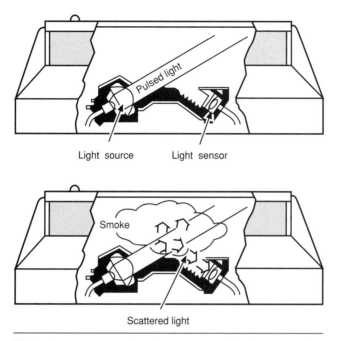

Figure 5-21. Principle of operation of a photoelectric scattering smoke detector; cross-section view of the detector (top, clean-air; bottom, with smoke).

cloud of water droplets is used to measure the concentration of droplets formed. The output voltage from the light receiver is directly proportional to the number of droplets (i.e., the number of condensation nuclei present).

The cloud chamber system uses a mechanical valve and switching arrangement to allow sampling from up to four detection zones, with as many as 10 sampling heads per zone (see Figure 5-22). Each zone is sampled once per second for 15 seconds; all four zones are sampled each minute.

Air-Sampling Smoke Detectors: Optical air monitoring equipment has been configured to provide another technique for early-warning fire detection. These particle detectors sample the air from a protected area. These detectors are capable of protecting large areas because of their inherent sensitivity. In addition, the detectors can be used in areas having high air change rates where dilute smoke concentrations or laminar airflows interfere with the proper operation of other types of smoke detectors. These air-sampling detectors can draw air through a piping network to the detector unit by an air-aspirating fan in the detector assembly. Air samples are illuminated with a high-intensity strobe light,

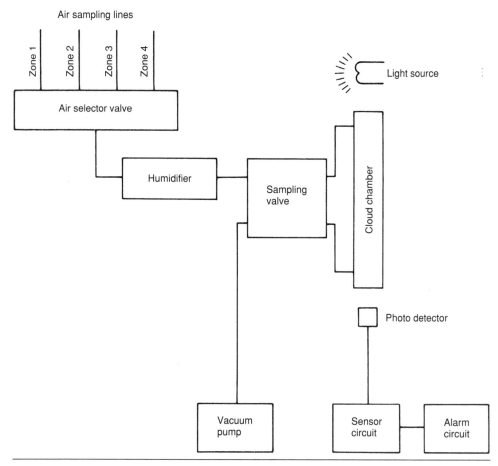

Figure 5-22. Cloud chamber smoke detector.

which causes smoke particles to reflect light to a solid-state photo receiver. An analog signal is generated from the detector to the control unit, which displays the smoke obscuration sensitivity. The detector system provides independent programmable levels of alarms to indicate different levels of fire conditions. The two main advantages of this detector are the use of sensitivity settings for incipient fire detection and the fact that one detector apparatus can cover a relatively large area, ranging up to 20,000 sq ft (1800 m²), by using perforated piping for air sampling in the protected area.

Cloud chamber and air-sampling smoke detectors can be set to much higher sensitivities than other smoke detectors because they are unaffected by air velocity, temperature, or humidity conditions in the protected area. Although they are expensive, they sample from multiple points (up to 40 for current systems) arranged in a grid pattern throughout the protected area(s) or even from within equipment or special hazards. Thus they may be less expensive than multiple spot-type detectors. A disadvantage is that they cannot identify the specific sampling port that the smoke entered—the finest resolution is by sampling zone.

Air Duct Smoke Detectors: Air duct smoke detectors are provided to prevent the recirculation of smoke from fires within the HVAC system or the spread of smoke that might be drawn into the system from a room fire. Air duct smoke detection has definite limitations. It is not a substitute for an area smoke detector or early-warning detection, and it is not a replacement for a building's regular fire detection system. Duct smoke detectors normally sample great volumes of air from large areas of coverage and cannot be expected to match the detection ability of area detectors. Dirt-contaminated air filters can restrict air flow, causing a reduction in the operating effectiveness of the duct smoke detectors. Area smoke detectors are the preferred means of controlling smoke spread because duct smoke detectors can only detect smoke when smoke-laden air is circulating in the ductwork. Fans may not be running at all times, for example, during cyclical operation or during temporary power failure.

Air duct detectors must be listed for the purpose and for the air velocities expected where they are used and securely installed in such a way as to obtain a representative sample of the air stream. Secure installation can be accomplished by:

1. Rigid mounting within the duct (see Figure 5-23)
2. Rigid mounting to the interior wall of the duct, with the sensing element protruding into the duct

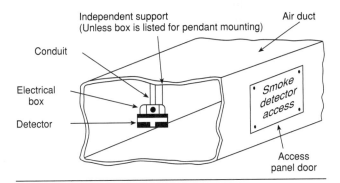

Figure 5-23. Air duct detector mounted within the duct.

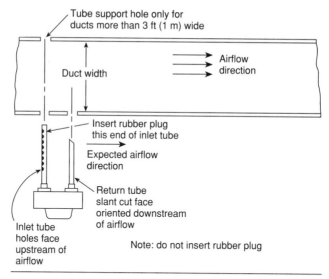

Figure 5-24. Procedure for rigid mounting of an air duct detector with sampling tubes.

3. Placement outside the duct, with rigidly mounted sampling tubes protruding into the duct (see Figure 5-24)
4. Use of a projected light beam through the duct

Design of an air duct smoke detector is based on one of the following operation principles: ionization, photoelectricity, or cloud chamber.

Gas-Sensing Fire Detectors: Many changes occur in the gas content of the environment during a fire, in particular, the addition of gases that are not usually present in such concentrations. For example, higher than usual levels of gases such as H_2O, CO, CO_2, HCl, HCN, HF, H_2S, NH_3, and various oxides of nitrogen can be present in a fire. With the exception of water vapor, CO, and CO_2, most of the evolved gases in a fire are fuel-specific and not associated with a sufficiently large number of fuels to be usable for general-purpose detection. In large-scale fire testing, where carbon monoxide and carbon dioxide measurements are generally taken, detectable levels of these gases occur between the occurrence of detectable particulate levels and detectable heat levels.

A gas-sensing fire detector can operate with simple sensors, such as a semiconductor or a catalytic element, or more complex apparatus, such as nondispersive infrared absorption or electrochemical cells. Fire-gas detectors with a semiconductor element respond to either oxidizing or reducing gases by creating electric changes in the semiconductor. The subsequent change in conductivity of the semiconductor causes detector actuation (see Figure 5-25).

In a detector with a catalytic element, the detector contains a material that remains unchanged, but accelerates the oxidation of combustible gases. The resulting temperature rise of the element causes a change in the element's resistance, which then initiates the alarm.

A metal oxide semiconductor-type element is the operating element found in one type of gas-sensing fire detector. The element consists of an *N*-type semiconductor crystal in which two heater coils are helically wound and imbedded in opposite edges. The crystal is then coated with a metal (typically tin oxide) and supported by its heater wires within a metal flame-arresting mesh. One of the two heater wires is unpowered and serves as a single electrode. The other heater wire is operated at 5 V dc to maintain the crystal at a temperature of about 662°F (350°C). The crystal must be operated at this temperature both to increase the number of free carriers available and to burn off any contaminating gases that may come in contact with the crystal. Semiconductor sensors that operate at lower temperatures or even at ambient temperatures have recently been developed. These have the advantage of reduced power requirements for operation. Care should be taken that these sensors do not become contaminated (the main function of heating the crystal is to burn off chemicals) over time nor respond to other chemicals, which might result in false alarms.

Under normal conditions, oxygen is adsorbed onto the surface of the crystal and gives it a certain conductivity. When any oxidizable gas comes in contact with the crystal, molecular oxygen is removed from the crystal surface and the conductivity increases. This increase is proportional to the gas concentration and triggers the alarm.

A disadvantage of many metal oxide elements is nonspecificity; that is, the element will respond to any oxidizable gas, the response being additive in nature. This makes the device fairly prone to false alarms in areas where oxidizable gases are normally present in small amounts. However, the metal oxide semiconductor is well suited to applications where the combustible material involved is known to produce pyrolysis products that contain oxidizable gases. Recently, a CO-specific sensor has been developed and is being utilized in a CO detector for nonfire applications.

The catalytic hot-wire gas detector consists of a coil of platinum wire imbedded in an alumina bead and connected to terminal posts (see Figure 5-26). In a clear air condition, a specified low resistance is established when power is applied to the sensor element. When the sensor makes contact with a gas mixture, the molecules of the mixture oxidize on the sensing element, increasing the element's temperature and resistance.

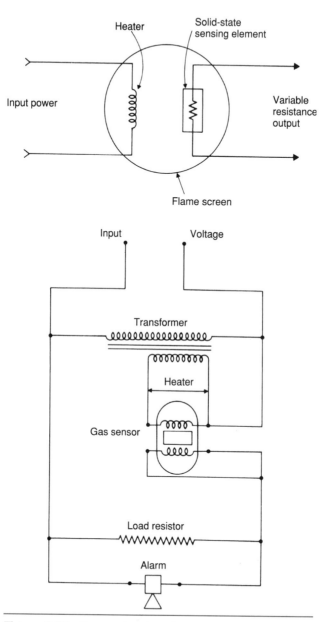

Figure 5-25. Gas-sensing fire detector with semiconductor element.

Catalytic hot-wire bridge

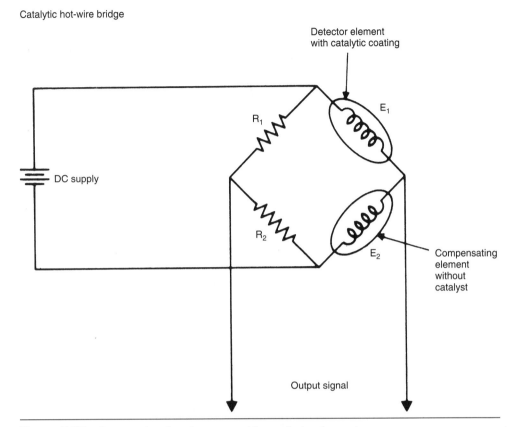

Figure 5-26. Gas-sensing fire detector with catalytic element.

To provide long-term stability, two elements are used in a Wheatstone bridge configuration. One element, the detector, has a catalytic coating. The compensating element, without a catalyst, does not respond appreciably to the presence of gas. In this configuration, power is supplied to the elements, heating them to their operating temperature. Resistors form the other side of the balanced bridge and establish a zero output signal.

When a gas mixture enters the detector housing, the mixture oxidizes on the catalytic coated element, increasing its temperature and resistance, unbalancing the bridge, and creating an output signal. A hot-wire bridge configuration compensates for changes in a detector due to temperature, aging, humidity, and power supply fluctuations. The hot-wire sensor produces a linear output with respect to gas, creating a larger relative signal shift at the operating range than a semiconductor sensor.

A catalytic hot-wire gas detector has input power fed to a regulator, then onto the sensor bridge (see Figure 5-27). The output signal of the bridge is

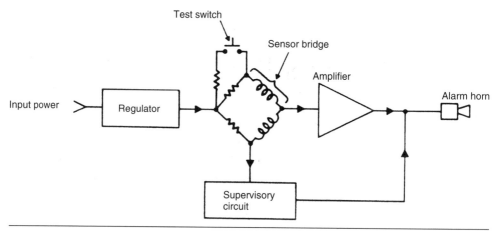

Figure 5-27. Catalytic hot-wire gas detector.

amplified, and the alarm horn is triggered and remains activated as long as the gas is present. A supervisory circuit monitors the sensing elements. Should an element fail, the bridge becomes completely unbalanced, sounding a trouble signal. An electrical test circuit is also connected to the bridge. Operation of the test switch alters the resistance of one part of the bridge, unbalancing the circuit in a way similar to the manner it is unbalanced in the presence of gas.

Another gas-sensing technique utilizes the change in color of a chemically treated material when it is exposed to a specific gas. It has been used since the 1940s for semiquantitative gas measurement; the color change is typically non-reversible, and the sensors must be changed periodically. Recently sensors have been developed for CO or CO_2 where the color reaction is completely reversible.

The Future of Gas Sensing for Fire Detection: Based on a number of research papers, especially those published in the last two AUBE conferences on automatic fire detection, one can expect that most fire detectors may one day use gas sensing exclusively. Inexpensive, low-power, highly sensitive sensors are being developed for a broad range of gases. Single sensors for 10 gases have been fabricated and sensors for 100 gases have been proposed. These utilize pattern recognition, by which specific patterns of gases present and not present indicate not only a fire, but the specific materials burning and the mode of combustion. These sensors discriminate fires from most nonfire sources, eliminating false alarms and allowing high sensitivities. They might even be able to sense prefire conditions and warn of potential failures before they occur. This is a rapidly evolving area with great potential.

Flame Detectors: Flame detectors respond to radiant energy that is either visible to the human eye (approximately 4000 to 7700 angstroms) or just outside the range of human vision. These detectors are sensitive to glowing embers, coals, or flames, which radiate energy of sufficient intensity and spectral quality to actuate the alarm.

Due to their fast detection capabilities, flame detectors are generally used only in high-hazard areas, such as fuel-loading platforms, industrial process areas, hyperbaric chambers, high-ceiling areas, and atmospheres in which explosions or very rapid fires may occur. There are three general types of flame detectors: (1) infrared (IR), (2) ultraviolet (UV), and (3) combination IR/UV. Because flame detectors must be able to "see" the fire, they must not be blocked by objects placed in front of them. (The infrared type of flame detector, however, has some capability for detecting radiation reflected from walls.)

Infrared-Type Flame Detectors: An IR detector is basically a filter and lens system that screens out unwanted wavelengths and focuses the incoming energy onto a photovoltaic or other type of cell sensitive to infrared energy (see Figure 5-28). Infrared flame detectors can respond to the total IR component of the flame alone or to a combination of IR with flame flicker in the 5- to 30-Hz frequency range as a way to minimize alarms to hot surfaces where there is no flame.

Infrared detectors that receive total IR radiation can be subject to interference from solar radiation since the intensity of solar energy can be considerably larger than that of a small fire. This problem can be partially solved by using filters that exclude all IR radiation not in the 2.5- to 2.8-micrometer and/or 4.2- to 4.5-micrometer ranges. These ranges represent absorption peaks due to the presence of CO_2 and water, respectively, in the atmosphere, and so those wavelengths do not reach the ground. When IR detectors are used in locations that are shielded from the sun, such as vaults, this filtering is not necessary. Another approach to the solar interference problem is to employ two detection circuits, often referred to as a *two-color* system. One circuit is sensitive to solar radiation in the 0.6- to 1.0-micrometer range and is used to indicate the presence of sunlight. The second circuit responds to wavelengths between 2 and 5 micrometers. A signal from the solar sensor circuit can be used to block the output from the fire-sensing cell, giving the detection unit the ability to discriminate against solar-based false alarms.

For most flame detector applications, flame flicker sensor circuits are preferred since flicker and modulation characteristics of flaming combustion are not components of either solar or manmade interference sources. Use of flame flicker circuits results in an improved signal-to-noise ratio. These detectors

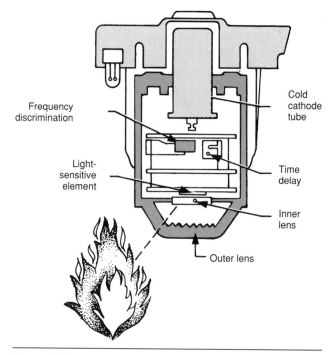

Figure 5-28. Infrared-type flame detector.

use frequency-sensitive amplifiers with inputs tuned to respond to an alternating current signal in the flame flicker range. However, false alarms have been reported from sunlight reflecting off the surface of water when waves produced the required flickering of the proper frequency.

Flame detectors are designed for volume supervision in either a fixed or a scanning mode. The fixed units continuously observe a conical volume limited by the viewing angle of the lens system and the alarm threshold. The viewing angles range from 15 to 170 degrees for typical commercial units. Some flame detectors can be fitted with lenses that restrict their view, for use in protecting specific equipment or hazards. One scanning device has a 400-ft (122-m) range and uses a mirror rotating at six revolutions per minute through 360 degrees horizontally, with a 100-degree viewing angle. The mirror stops when a signal is received. To screen out transients, the unit sounds an alarm only if the signal persists for 15 seconds.

Some flame detectors are designed to respond to passing sparks or flame fronts in ducts, such as those used in textile mills. These detectors do not contain a flame flicker circuit, but scan for glowing lint fibers in air ducting that might cause fires in the downstream filters. The detector turns on a water spray that extinguishes the glowing fiber before it reaches the filter.

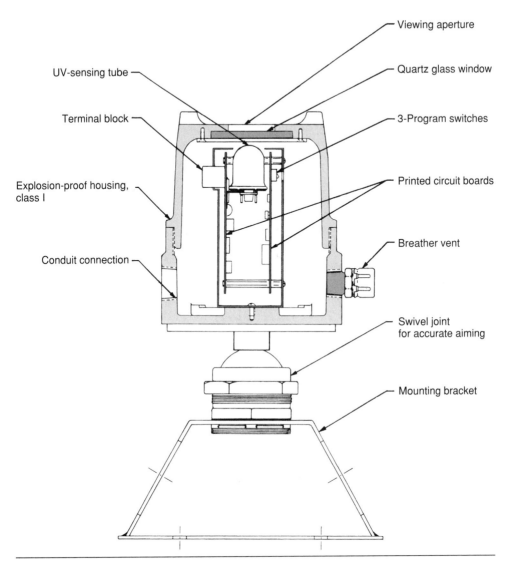

Figure 5-29. Ultraviolet-type flame detector.

Ultraviolet-Type Flame Detectors: The UV component of flame radiation is also used for fire detection (see Figure 5-29). The sensing element of UV detectors can be either a solid-state device, such as silicon carbide or aluminum nitride, or a gas-filled tube in which the gas is ionized by UV radiation and becomes conductive, activating the alarm. Ultraviolet detectors operate in the range of 0.17 to 0.30 micrometers, and in that range are essentially insensitive to both sunlight and artificial light. UV detectors are also volume detectors, with viewing angles from less than 90 to 180 degrees. Ordinary glass is opaque to UV wavelengths, so the windows through which the element looks is quartz. There have been instances where a broken lens has been replaced with a glass flashlight lens, rendering the flame detector blind.

Combination-Type Detectors: The combination of UV and IR sensing has been applied to aircraft and hyperbaric chamber fire protection. These complex devices sound an alarm when there is a preset deviation from the prescribed ambient UV–IR discrimination level in conjunction with a signal from a continuous wire overheat detector. The analysis of the deviation is performed by an on-board minicomputer.

Flame detector sensitivities are usually expressed in terms of the specific size of fires that can be detected at a specific distance, for example, a 1-foot-square pan of gasoline at 50 feet. Additionally, flame detectors are supplied with maps of their field of view along two axes to assist in their positioning.

ALARM SYSTEMS AND FIRE EXTINGUISHING SYSTEMS

Automatic fire extinguishing systems are usually required to be supervised, and this supervision is performed by the fire alarm signaling system. The fire alarm system initiates a signal on discharge of the extinguishing agent and also provides supervisory alarm service for critical components and functions of the extinguishing system. This supervisory service indicates any off-normal condition of the system that would prevent normal operation and also indicates the return of the system to a normal condition.

There are many types of extinguishing systems used with a fire alarm system; some of these are described in Table 5-3.

Functions of Alarm and Supervisory Signals

A sprinkler system with a waterflow alarm serves two functions: notification of building occupants and first responders of a fire and extinguishment or control of the fire until the first responders arrive. Immediate notification of the operation of sprinklers is important to allow for the complete extinguishment of a fire and ultimate return of the system to service. Under some conditions, the sprinklers do not immediately or completely extinguish the fire (known as the control mode); therefore, it is vital to have manual fire-fighting forces notified to complete extinguishment either by portable extinguishing devices, private hose streams, or fire department equipment.

The amount of loss or damage by water after the fire has been extinguished can be held to a minimum by closing the control valve immediately after the

TABLE 5-3. Fire Extinguishing Systems

Type	Description	Comments
1. Wet-pipe automatic sprinkler system	A permanently piped water system under pressure, using heat-actuated sprinklers; when a fire occurs, the sprinklers exposed to the high heat open and discharge water individually to control or extinguish the fire	Automatically detects and controls fire; protects structure; may cause water damage to unprotected books, manuscripts, records, paintings, specimens, or other valuable objects; not to be used in spaces subject to freezing; on–off types may limit water damage; see NFPA 13, *Standard for the Installation of Sprinkler Systems,* and NFPA 22, *Standard for Water Tanks for Private Fire Protection*
2. Preaction automatic sprinkler system	A system employing automatic sprinklers attached to a piping system containing air that may or may not be under pressure, with a supplemental fire detection system installed in the same area as the sprinklers; actuation of the fire detection system by a fire opens a valve that permits water to flow into the sprinkler system piping and to be discharged from any sprinklers that are opened by the heat from the fire	Automatically detects and controls fire; may be installed in areas subject to freezing; minimizes the accidental discharge of water due to mechanical damage to sprinkler heads or piping, and thus is useful for the protection of paintings, drawings, fabrics, manuscripts, specimens, and other valuable or irreplaceable articles that are susceptible to damage or destruction by water; see NFPA 13 and NFPA 22
3. On–off automatic sprinkler system	A system similar to the preaction system, except that the fire detector operation acts as an electric interlock, causing the control valve to open at a	In addition to the favorable feature of the automatic wet-pipe system, these systems have the ability to automatically

(Continues)

TABLE 5-3. (Continued)

Type	Description	Comments
	predetermined temperature and close when normal temperature is restored; should the fire rekindle after its initial control, the valve reopens and water again flows from the opened heads; the valve continues to open and close in accordance with the temperature sensed by the fire detectors; another type of on–off system is a standard wet-pipe system with on–off sprinkler heads; here, each individual head has incorporated in it a temperature-sensitive device that causes the head to open at a predetermined temperature and close automatically when the temperature at the head is restored to normal	stop the flow of water when no longer needed, thus eliminating unnecessary water damage; see NFPA 13
4. Dry-pipe automatic sprinkler system	Has heat-operated sprinklers attached to a piping system containing air under pressure; when a sprinkler operates, the air pressure is reduced, a dry-pipe valve is opened by water pressure, and water flows to any opened sprinklers	See No. 1; can protect areas subject to freezing; water supply must be in a heated area; see NFPA 13 and NFPA 22
5. Standpipe and hose system	A piping system in a building to which hoses are connected for emergency use by building occupants or the fire department	A desirable complement to an automatic sprinkler system; staff requires training to use hoses effectively; see NFPA 14, *Standard for the Installation of Standpipe, Private Hydrant, and Hose Systems*

TABLE 5-3. (Continued)

Type	Description	Comments
6. Clean-agent fire extinguishing system	A permanently piped system using a limited supply of an extinguishing gas under pressure, and discharge nozzles to totally flood an enclosed area; released automatically by a suitable detection system; most extinguish fires by inhibiting the chemical reaction of fuel and oxygen; most gaseous agents have a minimum concentration (by volume) for successful extinguishment that may have to be held for some time; thus the alarm system may also be used to stop ventilation fans, close dampers and doors, etc.; where agents may present a personnel hazard, evacuation signals may be provided before the agent is released	No agent damage to unprotected books, manuscripts, records, paintings, or other irreplaceable valuable objects; no agent residue; older systems use halon 1301 or 1211, but production of these agents was banned due to environmental damage; older systems may remain in use; new agents are being developed with similar properties; these systems require special precautions to avoid damage effects caused by their extremely rapid release; the high-velocity discharge from nozzles may be sufficient to dislodge substantial objects directly in the path; see NFPA 2001, *Standard on Clean Agent Fire Extinguishing Systems*
7. Carbon dioxide automatic system	Same as No. 6, except uses carbon dioxide gas; extinguishes fires by reducing oxygen content of air below combustion support point	Same as No. 6; appropriate for service and utility areas; personnel must evacuate before agent discharge to avoid suffocation; may not extinguish deep-seated fires in ordinary solid combustibles, such as paper, fabrics, etc.; but effective on surface fires in these materials; see NFPA 12, *Standard on Carbon Dioxide Extinguishing Systems*

(Continues)

TABLE 5-3. (Continued)

Type	Description	Comments
8. Dry chemical automatic system	Same as No. 6, except uses a dry chemical powder; usually released by mechanical thermal linkage; effective for surface protection	Should not be used in personnel-occupied areas; leaves powdery deposit on all exposed surfaces; requires cleanup; excellent for service facilities having kitchen range hoods and ducts; may not extinguish deep-seated fires in ordinary solid combustibles, such as paper, fabrics, etc.; but effective on surface fires in these materials; see NFPA 17, *Standard for Dry Chemical Extinguishing Systems*
9. High-expansion foam system	A fixed extinguishing system that generates a foam agent for total flooding of confined spaces and for volumetric displacement of vapor, heat, and smoke; acts on the fire by preventing free movement of air, reducing the oxygen concentration at the fire, and cooling	Should not be used in occupied areas; the discharge of large amounts of high-expansion foam may inundate personnel, blocking vision, making hearing difficult, and creating some discomfort in breathing; leaves residue and requires cleanup; high-expansion foam when used in conjunction with water sprinklers will provide more positive control and extinguishment than either extinguishment system used independently, when properly designed;

TABLE 5-3. (Continued)

Type	Description	Comments
		see NFPA 11A, *Standard for Medium- and High-Expansion Foam Systems*
10. Water mist system	Water-based system that uses high-pressure or special nozzles to produce a fine mist that is effective on some types of fires and minimizes water damage; system is activated by smoke detectors	Water mist systems must be carefully designed and applied to be successful; for example, mist discharged at the ceiling may not penetrate strong fire plumes to the fuel surface without sufficient pressure to overcome buoyancy; see NFPA 750, *Standard on Water Mist Fire Protection Systems*

need for sprinkler discharge has passed. One or two sprinklers may extinguish the fire, but water damage can be considerable unless the water is shut off as soon as it is safe to do so.

In addition to waterflow alarms, sprinkler systems are frequently equipped with devices to signal abnormal conditions that could make the system inoperative or ineffective. In general, these supervisory devices give warning of troubles with system components, for example, shut valves, low water pressure, and features requiring action by maintenance or security personnel. Waterflow devices initiate an alarm for any condition causing water to flow through the system, including frozen or broken pipes, but due to the rarity of these and the potential for water damage, a false alarm is tolerated.

Waterflow Alarm and Supervisory Signal Service

Provision should be made to indicate the flow of water in any sprinkler system, even a single head in an apartment trash room. Waterflow is indicated by an alarm signal initiated within 90 seconds after any flow of water equal to or greater than that from a single sprinkler of the smallest orifice size installed in the system. Movement of water due to waste, surges, or variable pressure need not be indicated, and this is the reason for the 90-second delay allowance. Most waterflow switches contain a pneumatic retard that provides an adjustable time delay not exceeding 90 seconds.

Waterflow alarm system design for sprinklers utilizing on–off heads should ensure that an alarm will be received in the event of a waterflow condition. On–off sprinklers open and close at predetermined temperatures. With certain types of fires, water flow can occur in a series of 10- to 30-second bursts. Waterflow detection devices with built-in time delays may not detect waterflow under these conditions. It is recommended that an excess pressure system or one that operates on pressure drop be considered to facilitate waterflow detection for on–off sprinkler systems.

Excess-pressure systems can be used with or without alarm valves. An excess-pressure system with an alarm valve consists of an excess-pressure pump with pressure switches to control pump operation. The inlet of the pump is connected to the supply side of the alarm valve and the outlet is connected to the sprinkler system. The pump control pressure switch is of the differential type, maintaining a constant sprinkler system pressure above the main pressure. Another switch monitors low sprinkler system pressure and initiates a trouble signal in the event of pump failure or other malfunction. An additional pressure switch can be used to stop pump operation in the event of a deficiency in water supply. Another pressure switch is connected to the alarm outlet of the alarm check valve to initiate a waterflow alarm signal when waterflow exists. This type of system inherently prevents false alarms due to water surges. The sprinkler alarm retard chamber should be eliminated to enhance the detection capability of the system for short-duration flows. Off-normal conditions that could prevent normal operation of the system should also be monitored in sprinkler systems.

A dry-pipe sprinkler system equipped for waterflow alarm signals provides supplementary supervision of the system air pressure to avoid false signals due to neglect in maintaining air pressure. Signals transmitted should distinctively indicate the particular malfunction, for example, valve position, temperature pressure, and so on, of the automatic sprinkler system, plus its restoration to a normal condition.

Sprinkler system waterflow alarm and supervisory initiating devices and their circuits should be so designed and installed that they cannot be readily tampered with, opened, or removed from system connection without initiating a supervisory signal. This provision specifically includes junction boxes that have been installed on the outside of buildings to facilitate access to the initiating-device circuit. No more than five waterflow switches should be connected to produce the same alarm signal, and no more than 20 supervisory switches should be connected to produce the same supervisory signal.

Monitoring of the various components of a sprinkler system supervisory signal service requires a distinctive signal indicating both the component off-normal condition and a different signal indicating return to normal condition. Supervisory signals are different and distinctive from alarm and trouble signals, both at the protected premise and at any monitoring station. Generally, the required conditions essential to the proper operation of sprinkler systems should be supervised, except for those conditions related to water mains, tanks, cisterns, reservoirs, and other containers of water controlled by a municipality or a public utility.

Sprinkler system supervisory devices and equipment described in this chapter are considered to be a part of sprinkler installations even though they may be manufactured, installed, and maintained by a central station or other supervisory service organization. Sprinkler system supervision is commonly provided for (1) water supply control valves, (2) low water level in water supply tanks, (3) low temperature in water supply tanks or ground-level reservoirs, (4) high or low water level in pressure tanks, (5) high or low air pressure in pressure tanks, (6) high or low air pressure in dry-pipe sprinkler systems, (7) failure of electric power supply to fire pumps, (8) automatic operation of electric fire pumps, and (9) fire detection devices used in conjunction with deluge and/or preaction and recycling systems. Sprinkler system devices that give waterflow alarms or supervise the condition of the installation are shown schematically in Figure 5-30.

An extinguishing system control valve should be supervised if movement of the valve from its normal position restricts or prevents the proper operation of the system. Two separate and distinct signals should be obtained, one indicating movement of the valve from its normal position and the other indicating restoration of the valve to its normal position. All supervisory parts of a system, that is, water pressure, level, and temperature, should have two separate and distinct signals as well. The off-normal signal should be obtained during the first two revolutions of the handwheel or during one-fifth of the travel distance of the valve control apparatus from its normal position.

The off-normal signal should not be restored at any valve position except normal. An attachment for supervising the position of a control valve must not interfere with the operation of the valve, obstruct the view of its indicator, or prevent access to its stuffing box. Where the signaling attachments of two or more valves utilize a common circuit, a restoration signal should be obtained only when all valves in the group are in their normal positions.

Pressure Supervision: A pressure supervisory signal attachment for a pressure tank and one for a dry-pipe sprinkler system should indicate both high- and low-pressure conditions. A signal should be obtained when the required pressure in

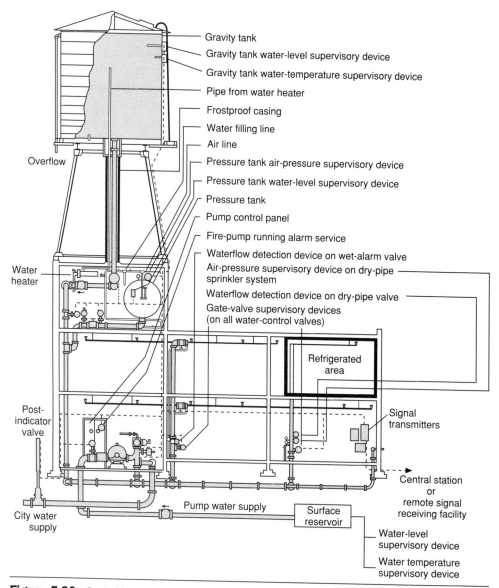

Figure 5-30. Sprinkler system waterflow alarm and supervisory devices.

the tank is increased or decreased 10 psi (70 kPa) from the required pressure value. Pressure change signals for a dry-pipe system should be used in accordance with the requirements of the authority having jurisdiction where the system is used.

A steam-pressure supervisory attachment should indicate a low-pressure condition and a signal obtained when the required pressure is reduced to a value not less than 110 percent of the minimum operating pressure of the steam-operated equipment supplied. An attachment for supervising the pressure of sources other than those previously specified should be capable of being applied and operated as required by the local authority.

Water Level Supervision: A pressure tank supervisory attachment should indicate both high- and low-level conditions in water storage. A signal should be obtained when the water level is lowered or raised 3 in. (76 mm) from the required level. A supervisory attachment for other than pressure tanks should indicate a low-level condition and a signal obtained when the water level is lowered 12 in. (300 mm) from the required level.

Water Temperature Supervision: For exposed water storage containers, one separate and distinct signal indicates that the temperature of the water has been lowered to 40°F (4.40°C) and the other indicates restoration to the required temperature.

Pump Supervision: Automatic fire pumps and special fire service pumps should be supervised as the authority having jurisdiction may require. Where supervision is applied to the electric power supplying the pump, the supervisory device should be connected on the line side of the motor starter so open fuses or open circuit breakers in the supply line to the pump will be detected promptly. All phases of operation should be supervised. All fire pump supervisory signals should be independent of, or not affected by, other signals, except that engine-drive high temperature, low oil, overspeed shutdown, battery failure, and failure to start may have a common signal.

The items that can be supervised in pump operation include, but are not limited to, electric drive (pump running, power failure, phase reversal), engine drive (pump running, controller not in "auto" condition, high engine temperature, low lubricating oil pressure, failure to start, overspeed shutdown, battery failure), and steam drive (pump running if practical, low-steam-pressure supervision). NFPA 20, *Standard for the Installation of Stationary Pumps for Fire Protection,* provides more information on pump operation.

Waterflow Sprinkler Alarms: The various types of sprinkler alarms include (1) those that operate with an actual flow of water; (2) those that activate a hydraulic or an electric alarm when a water control device such as a dry-pipe valve trips to admit water to the alarm device or mechanically operates an electric switch, whether or not water actually flows from sprinklers; and (3) those that not only signal the tripping of the control valve, but may also give supplementary warning signals either in case of damage that might impair the operation of the system or if maintenance features need attention.

Sprinkler systems are usually required to have an approved water motor gong, an electric bell, a horn, or a siren on the outside of the building. An electric bell or other audible signal device may also be located inside the building. Water-operated devices should be located near the alarm valve, dry-pipe valve, or other water control valves in order to avoid long runs of connecting pipe. All electric alarm devices, wiring, and power supplies should comply with the appropriate local codes.

Wet-Pipe Sprinkler System Alarm Devices: Waterflow alarm devices have been used to some extent ever since automatic sprinklers were first installed. They are generally located at or near the base of sprinkler risers, but may be used as floor or branch alarms. They are designed and adjusted to give an alarm if a water flow equal to the discharge of one or more automatic sprinklers occurs in the sprinkler system. The alarm signal can be given electrically, by water motor gongs, or by both. By far the most common types of waterflow alarm devices are the waterflow alarm valve and the waterflow indicator (see Figure 5-31).

The basic design of most waterflow alarm valves is that of a check valve, which lifts from its seat when water flows into a sprinkler system. The movement of the valve clapper is used in either of the following ways:

1. The valve seat ring can have a concentric groove with a pipe connection from the groove to the alarm devices. Such valves are commonly called differential-type (Figure 5-32, top) or divided-seat ring-type valves. When the clapper of the alarm valve rises to allow water to flow to sprinklers, water also enters the groove in the divided-seat ring and flows through the pipe connection to an alarm-signaling device.
2. The clapper of the alarm check may have an extension arm connected to a small auxiliary (pilot) valve having its own seat and a pipe connection to alarms (see Figure 5-32, bottom). When the auxiliary valve is lifted by movement of the main valve clapper, water is admitted to alarm devices.

Waterflow Indicators: A waterflow indicator of the paddle or vane type consists of a movable flexible vane of thin metal or plastic that is inserted through a circular

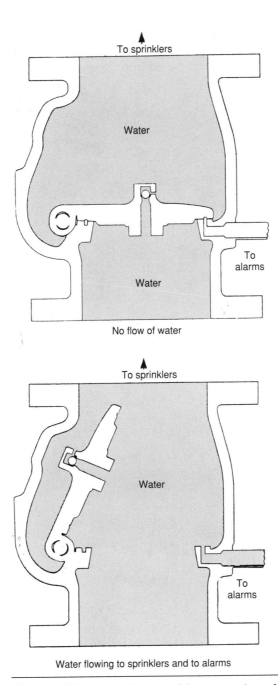

Figure 5-31. A wet-pipe sprinkler system is under water pressure at all times so water will discharge immediately when automatic sprinklers operate. The automatic alarm valve shown causes a warning signal to sound when water flows through the sprinkler piping.

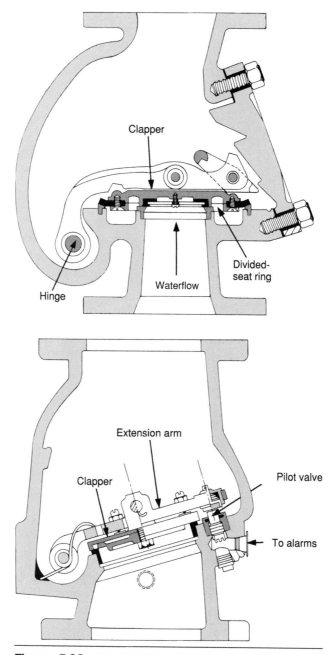

Figure 5-32. (Top) Differential-type waterflow alarm valve; (bottom) alarm check valve with auxiliary valve and pipe connection to alarms.

opening cut in the wall of a sprinkler supply pipe. The vane extends into the water-way sufficiently to be deflected by any movement of water flowing to opened sprinklers (see Figure 5-33). Motion of the vane operates an alarm-actuating electric switch or mechanically trips a signaling system transmitter. A mechanical, pneumatic, or electric time delay feature in the detector or electric circuit, or a part of the signaling system transmitter, prevents false alarms from being given by fluctuating water pressures. The retard feature is of the instantly recycling type or otherwise arranged so that the effect of a sequence of flows, each of less duration than the predetermined retard period, will not have a cumulative effect. Electrically heated thermal retards have not performed satisfactorily. Waterflow indicators have no provision for supplying water to water motor alarm gongs.

It is important that the flexible vane of a waterflow indicator be of a design and material not subject to mechanical injury or corrosion so that it cannot become detached and possibly obstruct the sprinkler piping. Devices with plastic vanes are frequently provided with markings on the paddle to trim it to the proper diameter for the pipe into which it is to be installed. These markings are

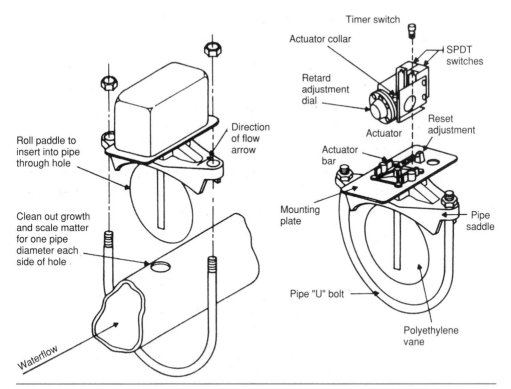

Figure 5-33. Vane-type waterflow indicator switch, single pole, double throw (SPDT).

usually for schedule 40 pipe, as this is the most common. However, for higher-pressure risers such as in high-rise buildings, thicker-walled pipe may be used. Here the outside diameter of the pipe is the same, but the inside diameter is less. Thus, if the normal trimming markings are followed, the vane may bind on the interior of the pipe, resulting in no alarm and the potential of a restricted water supply. Waterflow indicators of the vane type cannot be used in dry-pipe systems, deluge systems, or preaction systems because the vane and the mechanism are likely to be damaged by the sudden rush of water when the control valve opens.

Waterflow indicators are commonly used on new sprinkler systems. Situations where their use is most prevalent are where (1) ease of installation and economy are important in installing a waterflow alarm, (2) subdivision into several alarm areas is desired on a large sprinkler system, and (3) a central station, proprietary system, or other remote signal service is available that can receive only electric alarm signals from the protected property. Waterflow indicators are supplied both by manufacturers of sprinkler equipment and manufacturers of signaling system equipment.

Alarm-Retarding Devices: An alarm check valve that is subjected to fluctuating water supply pressures needs an alarm-retarding device in order to prevent false alarms when the check valve clapper is lifted from its seat by a transient surge of increasing pressure.

Retarding Chambers: One type of device, usually designated as a retarding chamber, is essentially a chamber inserted in the water line from the alarm check valve to the water motor gong and electric circuit closer (see Figure 5-34).

Flows of short duration from the alarm check valve to the alarm device first accumulate in the retarding chamber, which must become filled before water passes to the alarm device. The size of the chamber and the water inlet are predetermined by the manufacturer to give a delay needed to ensure continuous water flow before an alarm is given. The air displaced from the alarm piping in advance of water escapes at low pressure and does not operate the alarm devices. The retarding chamber is self-draining, so that, unless surges follow in close succession, the full retard interval is restored between surges.

Pressure-Actuated Alarm Switches

Electric switches, frequently called circuit closers, have contacts arranged to open or close an electric circuit when subjected to increased or reduced pressure. These switches are used in combination with dry-pipe valves, alarm check valves, and other types of water control valves to initiate an electric waterflow

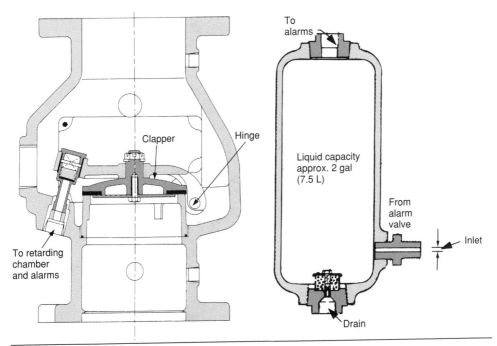

Figure 5-34. (Left) Waterflow alarm valve; (right) retarding chamber.

alarm signal when a flow of water to sprinklers occurs. Electric switches can also give a supervisory signal if pressures increase or decrease beyond established limits.

Manufacturers of dry-pipe valves, alarm check valves, and special types of water control valves regularly furnish approved pressure-operated switches to operate local electric waterflow alarms. In most cases, the motion to actuate a switch is obtained from a diaphragm exposed to the pressure on one side and opposed by a fixed, adjustable spring on the other side. A typical waterflow alarm switch is shown in Figure 5-35.

Mercury switches and other types of electric contacts can be used. Commercial pressure-actuated switches can be adapted to waterflow alarm service.

OTHER SUPERVISORY DEVICES

General requirements for supervisory signaling systems include electric supervision of circuits to indicate conditions that could prevent the required operation of the sprinkler system. The most common condition is a shut control valve. Electric switches for supervision of sprinkler system water supply control valves

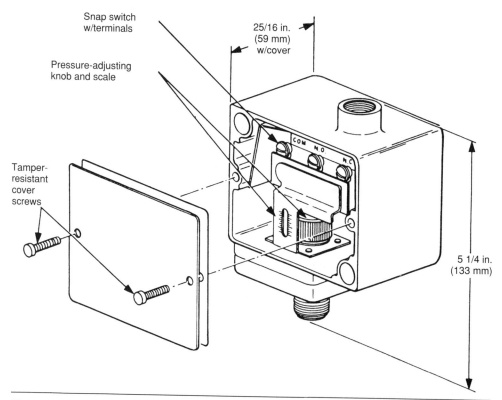

Figure 5-35. Isometric view of a waterflow alarm switch. Water pressure from the system acts upon a diaphragm in the housing at the base of the switch enclosure, causing movable contacts to close with stationary contacts. The switch can be used for normally closed or open circuits, and sensitivity to pressure differentials can be adjusted from 5 to 15 psi (34.0–103.0 kPa).

are of different mechanical designs for the different types of control valves. Electric switches can be for open- or closed-circuit supervisory systems. Supervision can also be by means of signaling transmitters mechanically tripped by operation of the gate valve.

The signal to indicate valve operation is given within two turns of the valve wheel from the fully open position. The restoration signal is given when the valve is restored to its fully open position. Figures 5-36 and 5-37 show representative approved devices for supervision of sprinkler valves.

The temperature of the water in fire service tanks exposed to cold weather is usually shown by a thermometer in the cold water return to a circulating-type heater or near the bottom of a large riser. Supervisory equipment is available for detecting dangerously low temperature near the surface of the water at the tank

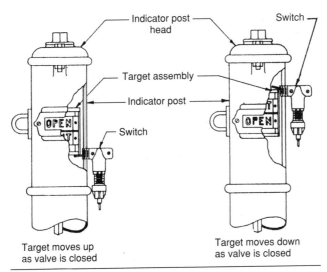

Target moves up
as valve is closed

Target moves down
as valve is closed

Figure 5-36. A post indicator valve supervisory switch. The switch operating stem is held against the movable target assembly by springs within the switch housing. The switch end of the operating stem carries an insulator that separates contact springs and opens the electric circuit when the target assembly is moved in the valve-closing direction by approximately two turns of the valve stem.

shell, where freezing is most likely to start, as well as to check manual supervision or automatic heat.

Devices for supervising the water level in pressure tanks differ from those used in water supply tanks, in that level-sensing elements are continuously subject to high pressure. The supervisory signal is given if the water level reaches 3 in. (76 mm) above or below the proper point, usually by means of an electric connection to a separate signal transmitter. Figure 5-38 shows a water-level supervisory switch for gravity tanks.

GUARD'S TOUR SUPERVISORY SERVICE

Guard's tour (or watchman's) service dates back to the 19th century (before fire alarm systems) and was especially prevalent in the textile mills of New England. Guards (or watchmen) walked an hourly tour of a building in the overnight hours (usually 6 p.m. to 6 a.m.) and every 2 hours on weekends and holidays, looking for fires or other problems.

To ensure the watchman actually made his rounds, a recording system was used, consisting of either stationary clocks at selected locations used to

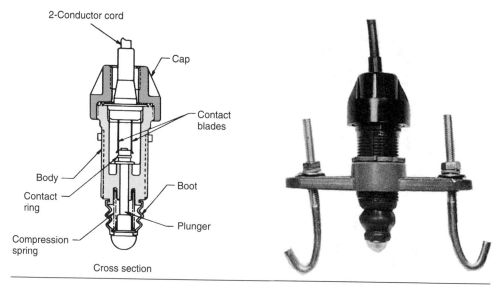

Figure 5-37. A gate valve supervisory switch. Hook bolts (right) hold the switch to the two sides of the valve yoke. When the valve is fully open, the plunger tip enters a 1/8-in. (3-mm)-deep depression drilled in the valve stem. The switch is adjusted so that, in the open-valve position, the contact ring closes the electric circuit between the contact blades. When the valve is closed not over two turns, the plunger tip rides out of the depression in the valve stem, opening the electric circuit. If the switch is tampered with or removed, the plunger takes the position shown in the sectional view (left) and opens the electric circuit.

mark a card or a portable clock carried by the guard with keys at selected locations. The keys were inserted into the portable clock to make a mark on an internal chart. If a fire was found, the alarm would be sounded either by a municipal box located within the building or a steam whistle or siren on the roof.

The idea behind guard's tour service is obvious. During normal business hours, most areas of most facilities are occupied. By their presence, the occupants provide fire protection surveillance if they are in the area when a fire occurs. A carelessly discarded cigarette, for example, may ignite the contents of a wastebasket; that fire may spread until the entire building is involved. However, someone who discovers the fire while it is still in its incipient stage may be able to extinguish it with a portable fire extinguisher. In some cases, the occupants can also detect conditions such as a malfunctioning machine that might lead to a fire. Furthermore, most companies have come to realize that increased security surveillance is vital in guarding against fires of incendiary origin.

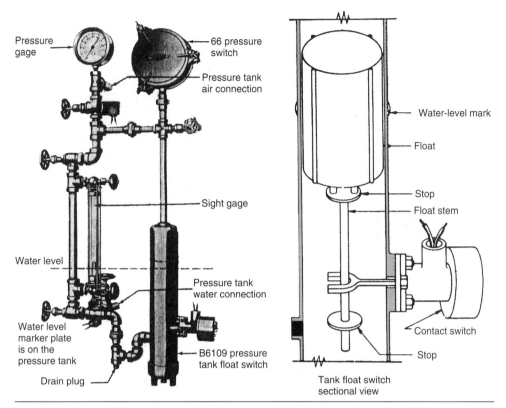

Figure 5-38. A supervisory switch for pressure water tanks. If the water level rises or falls about 3 in. (76 mm), one of the stops on the float stem in the tank float switch comes in contact with the forked lever that operates the contact switch. The switch is usually connected into the circuit of a supervisory signal transmitter.

Guard services generally serve three purposes in protecting a property against fire loss: (1) protecting the property at times when occupants are not present; (2) facilitating and controlling the movement of persons into, out of, and within the property; and (3) carrying out procedures for the orderly conduct of some operations on the property. Guards may be facility employees or employees of outside firms that provide these services on a contract basis. The duties of these individuals can be supplemented or replaced in part by various approved protective signaling systems.

Guard supervisory services designed to continuously report the performance of a guard are found in connection with local signaling service, central station service, and proprietary protective signaling systems. These services usually provide for supervised or compulsory tours.

For supervised tours, a series of patrol stations along the guard's intended route is successively operated by the guard with each station sounding a distinctive signal at a central headquarters. Customarily, the guard is expected to reach each of these stations at a definite time, and failure to do so within a reasonable grace period prompts the central station to investigate the guard's failure to signal. Frequently, manual fire alarm boxes that ordinarily transmit four or five rounds of signals for fire can also be actuated by a special watch key carried by the guard to transmit only a single round to the central station, thus signaling that the box has been visited.

By proper location of the stations, a fire or security guard can be compelled to take a definite route through the premises, and variations from that route would appear as misplaced signals on the recording tape. The order of station operation can be modified on occasion in the interests of security or to meet special conditions within the building.

With compulsory tours, one or more stations are wired to the central station. Preliminary mechanical stations condition the guard's key to operate the wired station only after the preliminary stations have been operated in a pre-arranged order. This compulsory tour arrangement is somewhat less flexible than supervised tours, but has the advantages of the absence of interconnected wires between the preliminary stations and the reduction of signal traffic. The usual arrangement is to have the guard transmit only start and finish signals that must be received at the central point at programmed reception times.

Specific features are required for a guard's tour system that is provided as part of a fire alarm signaling system. NFPA 601, *Standard for Security Services in Fire Loss Prevention,* should be consulted concerning the number of guard's reporting stations, their locations, and the route to be followed by the guard for operating the stations.

A permanent record indicating each time every signal-transmitting station is operated should be made at the main control unit. When intermediate stations that do not transmit a signal are employed in conjunction with signal-transmitting stations, distinctive signals should be transmitted at the beginning and end of each guard's tour and a signal-transmitting station provided at intervals not exceeding 10 stations. Intermediate stations that do not transmit a signal should be capable of operation only in a fixed order of succession.

In addition to these general provisions, the system should transmit a start signal to the main control unit. The guard should initiate this signal at the start of continuous tour rounds. A delinquency signal should be automatically transmitted within 15 minutes after a predetermined time if the guard fails to actuate tour stations as scheduled. A finish signal is transmitted on a predetermined schedule after the guard completes the last tour of the premises. A start signal should be transmitted at least once every 24 hours for periods of over 24 hours during which tours are continuously conducted. The start, delinquency, and finish signals should be recorded at the main control unit.

QUESTIONS

1. Manual fire alarm boxes are typically required:

    a. in every room
    b. on every floor
    c. at the main entrance to a building
    d. within egress stairwells

2. Which type of element would be used in an analog heat detector?

    a. Bimetal   b. eutectic   c. thermistor   d. ion chamber

3. The type of smoke detector most appropriate for use in rooms with high air flows is:

    a. air sampling   b. ionization   c. photoelectric   d. projected beam

4. The type of fire detector that can respond the fastest is:

    a. smoke detector   b. heat detector
    c. gas detector   d. flame detector

5. A supervisory signal is considered:

    a. an alarm signal   b. a trouble signal
    c. neither a nor b   d. both a and b

BIBLIOGRAPHY

NFPA Codes, Standards, Recommended Practices, and Manuals (see the latest *NFPA Catalog* for availability of current editions of the following documents):

NFPA 20, *Standard for the Installation of Stationary Pumps for Fire Protection.*

NFPA 70, *National Electrical Code*®.

NFPA 72®, *National Fire Alarm Code*®.

NFPA 110, *Standard for Emergency and Standby Power Systems.*

NFPA 601, *Standard for Security Services in Fire Loss Prevention.*

6

Signal Transmission Methods and Processing

INTRODUCTION

This chapter discusses three types of fire alarm system signaling circuits (channels): (1) wire, (2) wireless, and (3) optical fiber. The circuits connect alarm-initiating devices to the fire alarm control panel and other locations that require notification of a fire. Methods of transmission, compatibility of initiating devices, signal verification, and false alarms are also covered.

Use of copper wire is the oldest method of carrying a signal from one component of a fire alarm system to another. The physical and electrical use of this type of signaling circuit is more completely described in Article 760 of NFPA 70, *National Electrical Code*®.

Wireless transmission is used to transmit signals in proprietary, central station, remote-station, and public fire service communication systems. Wireless transmission also transmits alarm signals from initiating devices, such as smoke detectors, manual fire alarm boxes, and so on, to the fire alarm control unit.

Requirements for optical fiber signaling circuits are more fully described in Article 770 of *NEC*. For related information see Chapter 7, Fire Alarm Notification, and Chapter 8, Fire Alarm System Installation.

WIRE TRANSMISSION

Conventional Detectors

Transmitting alarm signals by wire can be as basic as increasing the current in the circuit by shunting the end-of-line resistor when a switch or relay contact closes in a manual fire alarm box, heat detector, or smoke detector (see Figure 6-1).

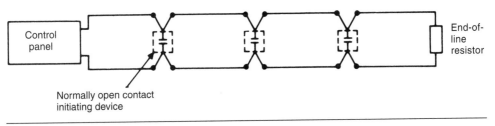

Figure 6-1. Basic wired alarm-initiating circuit.

When smoke detectors are used on the type of alarm-initiating device circuit (IDC) shown in Figure 6-1, separate conductors must be provided to power the smoke detectors (see Figure 6-2). These detectors, called four-wire detectors, can have contacts in addition to the relay alarm contacts to perform supplementary or releasing service functions. Since the smoke detector electronics and relay are not dependent on the limited power available from an alarm-initiating-device circuit, more than one smoke detector alarm relay can be energized simultaneously on the circuit.

A power supervisory relay must be installed electrically beyond the last detector on the initiating-device circuit so the detector power source is properly

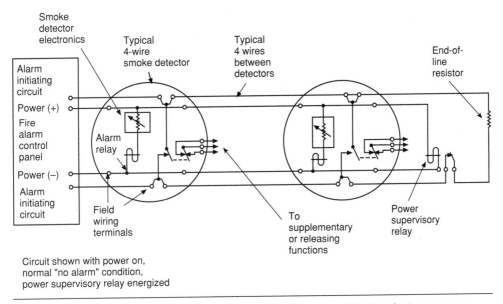

Figure 6-2. Typical circuit for a four-wire smoke detector-initiating device.

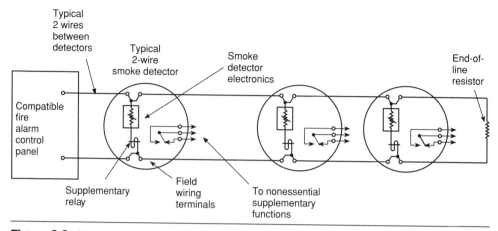

Figure 6-3. Typical circuit of a two-wire smoke detector alarm-initiating device.

supervised. Loss of power or disconnect of any field power wire causes the relay to deenergize, its contact to open, and a trouble condition to be indicated at the control panel.

The initiating-device circuit current can be increased by shunting the end-of-line resistor electronically, as in a smoke detector (a two-wire detector) that receives power and transmits an alarm signal over the same pair of wires. Field-wiring of two-wire detectors is shown in Figure 6-3.

Two-wire detectors should be used carefully. The detectors must be electronically compatible with each other and the fire alarm control panel. If relays in the detectors are intended to perform supplementary functions, most likely the first relay on that circuit would be the only one to operate unless the circuits are arranged to ensure continuous operating power to the detectors.

Relays mounted in two-wire detectors should not be used to control essential functions of the fire alarm system because operation of the supplemental relay depends on the limited power available on the initiating-device circuit. After the first smoke detector senses smoke and the relay energizes, there is usually insufficient power available to energize a relay in a second detector on that circuit. Even though it normally requires less current to maintain an energized relay than to initially energize the relay, if other initiating devices are operating and placed in parallel on the circuits, the first relay could be deprived of sufficient current, causing it to deenergize. Operation of a contact-type device, such as a heat detector or manual fire alarm box, on the same two-wire circuit would render all smoke detectors and built-in relays inoperative.

Addressable Detectors

The term *addressable* applies to various types of signal transmission between a detector and the fire alarm control panel. These devices are connected using a signaling line circuit (SLC). The following examples illustrate some of the various methods for addressable signal transmission:

1. The detector can wait to be addressed at each scanned (polled) cycle of the control panel and respond with a digital signal to indicate an alarm or normal condition. If no response is received at the control panel, a trouble (fault) condition is indicated.
2. An analog output signal that is processed at the control panel determines the moment at which the signal should be considered a fire alarm. In this example, a prealarm is indicated at the control panel when the signal reaches a certain point below the programmed alarm threshold.
3. An individual detector on a circuit can be identified by a time polling sequence in which the status of only the first detector on the circuit is shown when the panel polls the circuit. After a programmed time period, each detector switches the polling signal to each succeeding detector on the line. The individual detector can be identified at the fire alarm panel when the time slot for detector response is shown.
4. The fire alarm control panel can monitor the analog signal from a detector and determine the validity of a fire alarm signal based on the rate of change of the signal. For example, a rapid increase in the analog signals for a few scans, or a slow increase for a period of many scans, can be used to signal a fire condition.
5. A specific number of control panel scans can be counted to determine that a true alarm condition exists instead of a transient condition appearing in one or two scans.
6. A particular identification response signal can be plotted on a display at the control panel, pinpointing the location of the detector.

Addressable detectors require clear specifications that should be developed prior to purchasing a system since each method requires its own specific design features. For example, fewer wires are needed to connect addressable detectors to the control panel, or more information on the status of each detector must be obtained.

Multiplexing

Multiplexing is a signaling method characterized by simultaneous or sequential transmission or both, and reception of multiple signals in a communication channel. Multiplexing includes the means for positively identifying each signal. Multiplexing of signals has been used for central station, proprietary, and public fire service communication systems for many years.

The earliest form of signal system multiplexing used a coded system. Addressable detectors are a more recent form of multiplexing.

Some traditional coded systems are still in use; however, modern multiplexing uses either a microprocessor or a computer. There are two multiplexing methods: active and passive.

Active Multiplexing: Active multiplexing is defined as the transmission of status signals from an initiating-device circuit or an initiating device to a central receiving station upon command, enabling individual identification of signals from many locations.

In an active multiplex system, a transponder or circuit usually interfaces one or more initiating-device circuits with a signaling line circuit in a manner that permits individual identification and display of the status of each indicating circuit at a central supervisory station.

Microprocessor Control Units: Microprocessor control units in an active multiplex system consist of thousands of transistors incorporated into an integrated circuit. The microprocessor is a single device capable of performing many basic operations of a digital computer and is highly cost effective in terms of labor and wire savings. In designing very large fire alarm signaling systems, a designer needs to decide whether to incorporate microprocessor-based equipment and multiplex the information throughout the building or provide a hard-wired conventional system with a standard solid-state control panel.

Some microprocessor multiplex systems are specified with integral remote multiplexing for water and waste treatment plants, plant monitor and control systems, building automation systems, and fire alarm systems.

Microprocessor-based control equipment offers many advantages not available with standard solid-state equipment. For example, microprocessors can include system diagnostics, allowing the contractor and/or maintenance personnel to view system information and determine problems with a particular control panel.

The microprocessor control unit is the basis for an integrated fire protection system that incorporates all connected, addressable devices as well as all internal modules to ensure their proper operation. The microprocessor design allows the sensitivity of each smoke detector to be analyzed to determine alarm, normal, and trouble conditions. The interconnection line can be a single pair or noncontinuous (as in a power distribution system), that is, spider-webbed or T-tapped, respectively (see Figure 6-4).

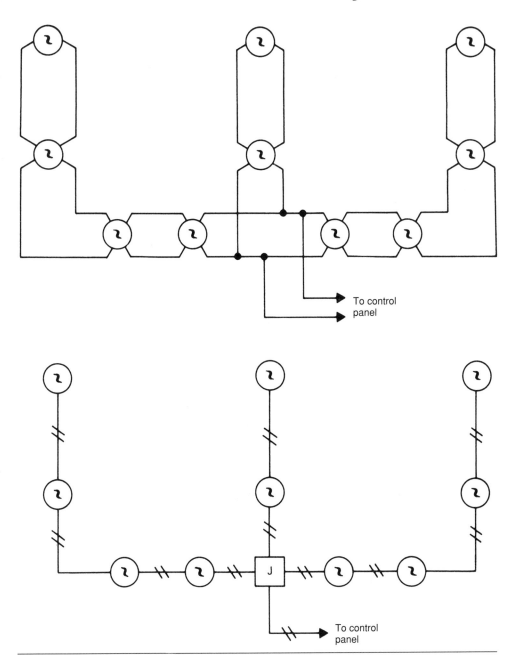

Figure 6-4. Spider-webbed/T-tapped initiating-circuit riser diagram.

Most microprocessor-based control units have been developed for use with complex multiplex systems. Systems sold today have a modular design, which allows the system capacity to match the initial design requirements and allows for system expansion. Microprocessors, however, are complex and must be serviced by qualified personnel.

The microprocessor central processing unit can automatically operate the system control points as a result of an alarm condition from any of its monitoring and sensoring points.

A microprocessor-based system will provide detailed system information, much of which may be unnecessary for the fire service, but may assist the owner to more efficiently operate the building complex or the technician to more efficiently troubleshoot the system.

The owner/operator personnel and the fire department personnel who will be responsible for fire-fighting operations in the building should all receive adequate training on the system from the manufacturer.

Multiplexing Formats for Two-Way Communication: These formats can be one of three types: (1) ripple-through, (2) sequential counting, and (3) digitally addressable.

The ripple-through format is a response to a start command and each device in turn transmits its status (see Figure 6-5). One form closes the path to the following device upon completion of transmission from the responding device. Devices are usually connected in series to the transmission pathway, which is commonly a two-wire circuit. These devices require no individual address since the address is their position in the sequential arrangement. Spider-webbed or T-tapped distribution is not possible with this format.

For the sequential counting format, station 1 reports in response to a synchronizing (reset) signal and the first transmit command. After the second transmit command, station 2 reports and this continues until all stations have responded (see Figure 6-6). Spider-webbed or T-tapped distribution is inherent in this format and end-of-line devices are not required. Emergency operation, that is, operation with a single open, is easily achievable. An individual address setting at each remote device is required, however, and interrupt capability is not possible.

For the digitally addressable format, each station has a unique digital address, making random-order communication possible (see Figure 6-7 on page 150). When the response includes the address/location of the station, interrupt capability becomes possible. This format provides maximum functional flexibility as well as interrupt capability, but also requires the largest number of characters to communicate. This slows down the polling cycle and necessitates higher speed transmission.

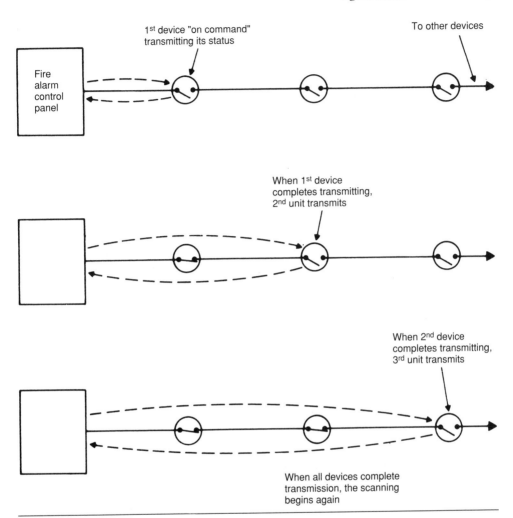

Figure 6-5. Ripple-through multiplex transmission.

The Concept of Networking: When a number of addressable detectors are connected to a multiplexed signaling line circuit loop, the central control point of the loop may in turn act as a transponder in a system's active multiplex circuit (see Figure 6-8 on page 151). The detailed information gathered from the multiplexed signaling line circuit loop can be arranged for retransmission to the system supervisory control station.

Passive Multiplexing: Passive multiplexing removes the command to transmit and permits transmitters to report at any time. The system becomes a

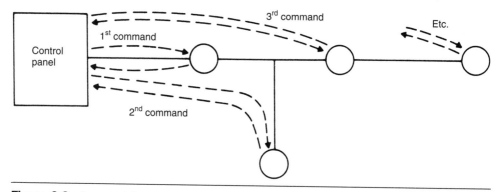

Figure 6-6. Sequential counting multiplex transmission.

one-way system with multiple transmitters sharing the same communication channel or pathway.

A simple coded fire alarm system is a passive multiplex system. More recently, battery-powered RF transmitters have been introduced in passive multiplex systems and can identify themselves and transmit this information when an alarm condition occurs.

Analog Data Transmission

Conventional spot-type detectors have two stable states: normal and alarm. The detector changes from the normal to the alarm state when the signal produced by the environment is compared in the detector to a reference level determined by the sensitivity calibration of the detector. In more sophisticated detectors, there may be multiple reference levels, allowing a detector to indicate a pre-alarm condition.

In a sense, a detector that transmits a signal indicative of the signal produced by the environment is an analog detector that simply repeats the signal from its sensor in a form compatible with the data transmission format.

Analog detectors combined with addressable devices provide a system where a detector simply reports the signal produced by its environment to the

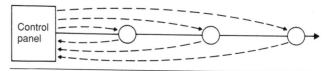

Figure 6-7. Digitally addressable random multiplex transmission.

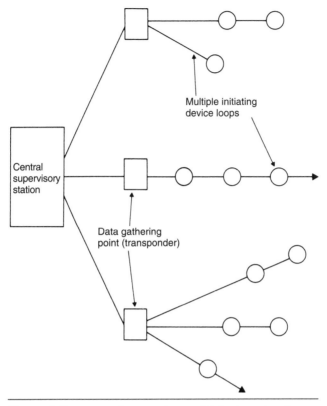

Figure 6-8. Networking, active multiplexing circuit.

control equipment, and the decision to sound an alarm or other response is produced by the control equipment.

With analog data transmission, it is possible to compensate for long-term changes in sensor response and essentially maintain a constant sensitivity. The apparent alarm threshold can be easily changed to accommodate cyclical changes (e.g., day to night) in the environment and protect against difficult hazards.

WIRELESS TRANSMISSION

A wireless initiating device is defined as any initiating device that communicates with associated control/receiving equipment by some kind of wireless transmission medium. Requirements for the use of wireless, or RF, transmission vary for each type of protective signaling system.

Central Station Systems

Wireless transmission for central station systems is accomplished by two-way RF multiplex systems. Central station equipment, transponders, transmitters, and receivers should comply with Federal Communications Commission (FCC) rules and regulations. Wiring, power supplies, and overcurrent protection requirements can be found in the *NEC.*

Radio frequency multiplex systems should be installed in accordance with the *NEC. All* external antennas should be protected in order to minimize the possibility of damage by static discharge or lightning.

If duplicate equipment for signal receiving, processing, display, and recording is not provided, the installed equipment should be so designed that any critical assembly can be replaced from on-premises spare parts and the system restored to service within 30 minutes. A malfunction in a critical assembly is one that will prevent the receipt and interpretation of signals by the central station operator. Spare modules, such as printed circuit boards, printers, and so on, should be stocked at the central station to expedite repairs if RF multiplex equipment fails.

Operation of Signal Transmission: Any status changes that occur in an initiating device or in the interconnecting communication channel from the location of the initiating device(s) to the central station should be presented in a form the operator can interpret quickly. Status change signals should provide information on type of signal (alarm, supervisory, delinquency, or trouble signal), condition (differentiate initiation of an alarm, supervisory, delinquency, or trouble signal, and restore or return to normal), and location (point of origin).

Many methods of recording and displaying or indicating change-of-status signals are possible. The following are recommended:

1. Each change-of-status signal requiring operator action should result in an audible signal, and a minimum of two independent methods should identify the type, condition, and location of the status change.
2. Each change-of-status signal should be automatically recorded with type of signal, condition, and location in addition to the time and date the signal was received.
3. Failure of an operator to acknowledge or act upon a change-of-status signal should not prevent subsequent alarm signals from being received, indicated or displayed, and recorded.
4. Change-of-status signals requiring operator action should be displayed or indicated in a manner that clearly differentiates these signals from those that have been acted upon and acknowledged.

There are maximum operating times or maximum end-to-end operating time parameters recommended for a two-way RF multiplex system. Maximum allowable time lapse from initiation of a single fire alarm signal until it is recorded at the central station should not exceed 90 seconds. Subsequent fire alarm signals should be recorded at a rate no slower than one every additional 10 seconds. The maximum allowable time lapse from the occurrence of an adverse condition in any interconnecting communication channel (until recording of the adverse condition is started) should not exceed 90 seconds.

In addition to the foregoing maximum operating time for fire alarm signals, the following are suggested: Systems with more than 500 initiating-device circuits should be able to record a minimum of 50 simultaneous status changes in 90 seconds. Systems with fewer than 500 initiating-device circuits should be able to record a minimum of 10 percent of that total number of simultaneous status changes within 90 seconds.

Central station, satellite, and repeater station radio transmitting and receiving equipment should be controlled and supervised from the central station. This can be accomplished via a supervised circuit in which the radio equipment is remotely located from the central station.

Transmitter in use (radiating), failure of ac power supplying the radio equipment, receiver malfunction, and indication of automatic switchover should be supervised at the central station. Independent deactivation of transmitters should be controlled from the central station.

RF Communication Channel: The RF multiplex communication channel terminates in a transponder at the protected premises and in a system unit at the central station. Channels can be private facilities, such as microwave, or leased facilities furnished by a communications utility company.

Occurrence of an adverse condition on the channel between a protected premises and the central station that will prevent the transmission of any change-of-status signal should be automatically indicated and recorded at the central station. This identifies the affected portions of the system so the central station operator can determine the location of the adverse condition by trunk and/or leg facility. Restoration of normal service to the affected portions of the system should be automatically recorded. The first status change of any initiating-device circuit that occurred at any of the affected premises during the service interruption should also be recorded.

Dual control should provide for redundancy between a receiving station and the central station in the form of a standby channel or similar alternative means of transmitting signals over the primary trunk portion of a communications channel. The same method of signal transmission can be used over separate routes, or different methods of signal transmission can be utilized. Public switched telephone network facilities should be used only as an alternate method of transmitting signals.

When leased telephone lines are used, that portion of the primary trunk between the central station and its serving wire center may be excepted from the separate routing requirement of the primary trunk. Where used, dual control should be supervised as follows.

Dedicated facilities that are available full-time with limited use for signaling purposes should be exercised at least once every hour. Public switched telephone network facilities should be exercised at least once every 24 hours.

Radio multiplex systems are classified based upon their ability to perform under adverse conditions that affect their communication channels. System classifications are of two types: Type 4 and Type 5.

A Type 4 system has two or more control sites. Each site has a receiver interconnected to the central monitoring point by a separate communication channel. A remote transponder should be within transmission range of at least two receiving sites. The system should contain two transmitters, either located at one site with the capability of interrogating all transponders or dispersed, with all transponders having the capability of being interrogated by two different transmitters.

The transmitter in a Type 4 system should maintain a status that permits immediate use at all times. Facilities should be provided in the central station to operate any off-line transmitter at least once every 8 hours.

Failure of any one of the receivers should not interfere with the operation of the system from the other receiver. Failure of any receiver should be annunciated. A physically separate communication channel should be provided between each transmitter/receiver site and the central station. Having more than one control site safeguards against damage caused by lightning and minimizes the effect of interference on the receipt of signals.

A Type 5 system has a single control site with a minimum of one receiving site and one transmitting site. The sites can be housed together.

The capacities of radio multiplex systems are based on the overall reliability of the signal receiving, processing, display, and recording equipment at the central station and the capability of transmitting signals during adverse conditions in the signal transmission facilities. Allowable capacities are listed in *NFPA 72*®, *National Fire Alarm Code*®. The capacity of a system unit is unlimited when the signal receiving, processing, display, and recording equipment is duplicated at the central station and a switchover can be accomplished in a maximum of 30 seconds with no loss of signals during this period.

Local Systems

Wireless transmission is used in local fire alarm systems between the initiating devices and the control panel. To facilitate this, primary (dry cell) batteries can be the sole source of power for the initiating devices; these are usually low-power devices.

Power Supplies: A primary (dry cell) battery can be used as the sole power source of a low-power wireless initiating-device transmitter under certain conditions.

Each transmitter should serve only one device and be individually identified at the receiver/control unit. The battery should be capable of operating the wireless initiating-device transmitter for at least 1 year before the battery depletion threshold is reached. A battery depletion signal should be transmitted before the battery becomes so low that it cannot support alarm transmission after 7 additional days of normal operation. This depletion signal should be distinctive from all other signals, visibly identify the affected initiating-device transmitter, and, if silenced, automatically re-sound at least once every 4 hours. A trouble signal identifying the affected initiating-device transmitter should sound at its receiver/control unit when catastrophic (open or short) battery failure occurs. If silenced, the trouble signal should automatically re-sound at least once every 4 hours. Any failure of a primary battery in an initiating-device transmitter should not affect any other initiating-device transmitter.

Supervision: Low-power wireless transmitting devices should be specifically listed for a particular transmission method and highly resistant to misinterpretation of simultaneous transmissions and interference, for example, impulse noise and adjacent channel interference.

Transmission performance can be tested using UL 985, *Household Fire Warning System Units*, and UL 1023, *Household Burglar-Alarm System Units*.

Occurrence of any single fault that disables transmission between any wireless initiating device(s) and the receiver/control unit should cause a trouble signal within 200 seconds, except when allowed otherwise by FCC regulations. A single fault on the signaling channel should not cause an alarm signal.

Normal periodic transmission from a wireless initiating device should provide additional assurance of successful alarm transmission capability by transmitting at a reduced power level or by other means.

Removal of a wireless initiating device from its installed location should cause immediate transmission of a distinctive supervisory signal, indicating its removal and identifying the individual affected device.

Reception of any unwanted (interfering) transmission by a retransmission device (repeater) or by the main control unit for a continuous period of more than 20 seconds should cause an audible and visible trouble indication at the main control unit, identifying the specific trouble condition present.

Alarm Signals: When operated, each wireless initiating-device transmitter should automatically transmit an alarm signal. (This is not intended to preclude the inclusion of verification and local test intervals prior to alarm transmission.) Each wireless initiating device in alarm should automatically repeat alarm transmissions at a maximum 60-second delay interval until the initiating device is returned to its normal condition. Fire alarm-indicating signals should have priority over all other signals.

The system should be arranged to respond with a minimum delay to an alarm signal from a wireless initiating-device transmitter. There should be a maximum 90-second allowable response delay from the first alarm transmission to receipt and display by the receiver/control unit. An alarm signal from a wireless initiating device should latch at its receiver/control unit until manually reset and should identify the particular wireless initiating device in alarm.

Remote-Station Systems and Proprietary Systems

The three retransmission methods in order of preference are (1) a dedicated circuit independent of any switching network, (2) a one-way (outgoing only) telephone at the remote station that utilizes the commercial dial network or other method acceptable to the local authority, and (3) a wireless retransmission in remote-station systems that can transmit an alarm signal from the remote station to the fire department and consists of a private radio system that uses the fire department frequency (when permitted).

Wireless transmission can also be used in proprietary systems as the signaling channel. Where a private radio is used as the signaling channel, appropriate supervised transmitting and receiving equipment should be provided at central supervising, satellite, and repeater stations.

Certain conditions are necessary for the central supervising, satellite, and repeater station radio facilities where more than five buildings or premises are protected or where 50 initiating devices or initiating device circuits are being serviced by a private radio carrier. These include:

1. Dual-supervised transmitters installed and arranged for automatic switching from one to the other in case of trouble. Where personnel manage transmitters 24 hours a day, switchboard facilities can be manually operated if the

switching can be carried out within 30 seconds. When the transmitters are located elsewhere, the circuit extending between the central supervising station and the transmitters should be a supervised circuit. Transmitters should be operated on a two-to-one time ratio basis within each 24-hour period.

2. Dual receivers installed with a means for selecting a usable output from one of the two receivers. Failure of either receiver should not interfere with the operation of the system by the other operating receiver; failure of either receiver should be annunciated.

3. Central, satellite, and repeater station radio transmitting and receiving equipment controlled and supervised for wireless transmission at the central supervising station.

Public Fire Service Communication Systems

The public fire service often uses wireless radio communication systems between coded radio fire alarm (street) boxes and the communication center. The primary use of radio communication systems is to dispatch fire companies upon receipt of an alarm at the communication center. Dispatching is done from the communication center to fire stations, mobile fire apparatus, and portable equipment. Radio devices are increasingly used to alert off-duty personnel through pagers and those on the fireground. NFPA 1221, *Standard for the Installation, Maintenance, and Use of Public Fire Services Communications Systems,* describes in detail the requirements for each use of radio equipment in public fire service communication systems.

Coded Radio Reporting Systems

Radio Box Channel Frequency: A maximum of 500 boxes (see *NFPA 72*) should be permitted on a single radio channel. All coded radio box systems should constantly monitor the frequency or frequencies in use. Both an audible and a visual indication of any sustained carrier signal (when in excess of 15 seconds) should be provided for each receiving system at the communication center.

When a box message signal to the communication center (or acknowledgment of message receipt signal from the communication center) is repeated, associated repeating facilities should meet local requirements.

Boxes: Coded radio fire alarm boxes should be designed and operated in compliance with all applicable FCC rules and regulations. Boxes should provide at least three specific and individually identifiable functions to the communication center in addition to the box number: (1) test, (2) tamper, and (3) fire. Boxes should transmit to the communication center no fewer than one round for test, one round for tamper, and three rounds for fire. The FCC requires that, "Except for test purposes, each transmission must be limited to a maximum of 2 seconds and may be automatically repeated not more than two times at spaced intervals within the following 30 seconds; thereafter, the authorized cycle may not be reactivated for 1 minute."[1]

When multifunction boxes are used to transmit request(s) for emergency service or assistance to the communication center, each such additional message function should be individually identifiable. Multifunction boxes should be so designed that the loss of supplemental or concurrently actuated messages is prevented.

Facilities and Equipment for Receipt of Box Alarms: Alarms from boxes should be automatically received and recorded at the communication center. Receipt of an alarm should be indicated by a permanent visual record and an audible signal. The permanent record should indicate the exact location from which the alarm was transmitted and the date and time of receipt of each alarm.

There are two basic types of coded radio reporting systems: Type A and Type B. For each frequency used on a Type A system, two separate receiving networks should be provided. Each should include an antenna, an audible alerting device, a receiver, a power supply, signal processing equipment, a means of providing a permanent graphic recording of the incoming message (timed and dated), and other associated equipment. Both systems and associated equipment should be in operation simultaneously; facilities should be so arranged that a failure of either receiving network will not affect the receipt of messages from boxes (see Figure 6-9).

When a polling device is incorporated into the receiving network to allow remote/selective initiation of box tests, a separate such device should be included in each of the two required receiving networks. The polling devices should be configured for automatic cycle initiation in their primary operating mode, capable of continuous self-monitoring, and integrated into the network(s) to provide automatic switchover and operational continuity in the event of failure of either device.

Test signals from boxes need not include the date as part of the permanent record, provided that the date is automatically printed on the recording tape at the beginning of each day.

For each frequency used on a Type B system, a single complete receiving network should be installed in each fire station. If alarm signals are transmitted to a fire station from the communication center using the coded radio-type receiving equipment in the fire station to receive and record the alarm message, a second receiving network should be at each fire station and employ a frequency other than that used for the receipt of box messages.

The power supplied to all required circuits and devices of the system should be supervised. Receipt of alarms depends on radio repeaters. These repeaters

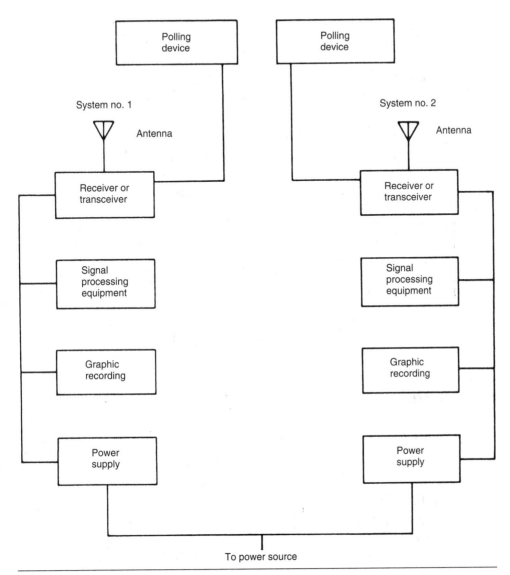

Figure 6-9. Type A coded radio reporting system. Polling is only required for transpondance (two-way) type systems.

should be provided with dual receivers and transmitters. Failure of the primary transmitter or receiver should cause an automatic switchover to the secondary receiver and transmitter. Where repeater controls are supervised by personnel on 24-hour duty, manual switchover facilities can be provided if switchover is performed within 30 seconds.

Radio Dispatch Systems

Another use of wireless transmission for public fire service communication systems is the radio dispatch system. Two separate dispatch circuits should be provided for transmitting alarms: supervised and unsupervised. A circuit terminating at a telephone instrument only is not considered as either type of required dispatch circuit.

Supervised dispatch circuits should consist of one of the following:

1. A supervised wired circuit (see Figure 6-10)
2. A radio channel with duplicate base transmitters, receivers, microphones, and antennas
3. A microwave-supervised carrier channel
4. A polling or self-interrogating radio or microwave radio system with duplicate base transmitters and equipment necessary for redundancy
5. A properly arranged and supervised telephone circuit

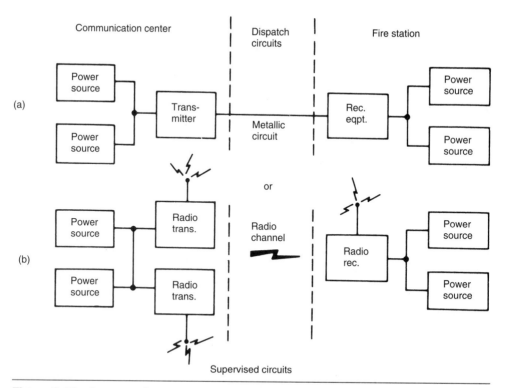

Figure 6-10. Supervised wire circuits on dispatch systems.

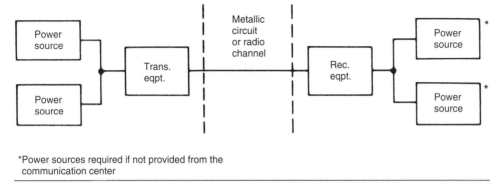

*Power sources required if not provided from the
communication center

Figure 6-11. Unsupervised circuits on a radio dispatch system.

An unsupervised dispatch circuit can be either a wired circuit or a radio channel (see Figure 6-11). Unsupervised radio does not require duplicate facilities. If both supervised and unsupervised dispatch circuits are radio, separate radio frequencies should be provided. These radio dispatch channels should be separate from channels used for routine or fireground communications. Jurisdictions that receive fewer than 600 alarms per year do not, in general, need unsupervised dispatch circuits.

Dispatching Mobile and Portable Equipment: The communication center should be equipped for radio communication with fire apparatus. All fire apparatus and other fire department emergency vehicles should be equipped with two-way radios that are FCC-type accepted or approved.

A separate frequency should be provided for fireground communications for jurisdictions or multiple jurisdictions on the same channel that receive 2500 or more alarms per year or when multiple jurisdictions share a common radio frequency.

Fire portable transceivers should be capable of multiple-frequency operation, enabling a fireground radio network to be organized independent of normal dispatch channels. Fire portable radios should be capable of continuous tone-coded squelch (CTCS) or continuous digital-coded squelch (CDCS). When data are transmitted from fire portable transceivers, the radio should transmit data without distortion, and the equipment should be so designed to ensure full data stream transmission at full power.

Where fire portable transceivers are used in a fire dispatch system, they should operate properly within the dispatch area without the use of mobile RF amplifiers.

Scanning devices, if used, should have an automatic priority feature whereby the radio automatically reverts to its primary channel whenever the channel is being used. Scanning devices should have a manual lock position solely to lock the receiver on its primary channel.

Miscellaneous Radio Devices: When radio home-alerting receivers, handheld units, pocket pagers, and similar radio devices are also used to receive fire alarms or are used on the fireground, they should meet certain requirements.

Fire portable radio equipment should be manufactured for the environment in which it will be used, water resistant, and sized for one-hand operation. Fire portable radio transceivers should not be placed into transmit mode except by the operator on a mechanically guarded switch. The operator should also be able to change channels on the multiple-frequency transceiver while wearing gloves.

The single-unit recharger for fire portable radios should be capable of recharging fully while the radio is in the receiving mode. Radio pocket pagers that are powered by replaceable batteries should indicate audibly before the battery is incapable of operating the pager.

OPTICAL FIBER TRANSMISSION

Optical fiber technology, or fiber optics, is cost effective in many different applications, including fire alarm systems. Advantages of fiber optics include its high bandwidth, low loss, small size, light weight, noise immunity, and transparency. It is also a noninterfering-type, highly secure method of transmission. On the other hand, fiber optic transmission requires active electrooptic transceivers and is an expensive method initially. There are also nonstandard products, for example, fiber, cable, and connectors, used in this new technology.

Principles of Fiber Optics

An optical fiber, in a sense, is a light pipe, where particles of light are launched into the pipe at incredible speeds (see Figure 6-12). These particles of energy, called photons, must flow (or propagate) through an intricate system of pipes without leaking out at the interconnecting fittings (connectors) and without excessive spillage at the input and output points.

Conserving flow pressure depends on the number of photons that go completely through the pipe network without getting lost. Photons can sometimes be destroyed or absorbed by contaminants in the pipe, scattered through the walls of the pipe, or lost at the fittings.

The fiber optic link in a light-wave system consists of three basic elements: (1) a transmitter, (2) a cable, and (3) a receiver (see Figure 6-13).

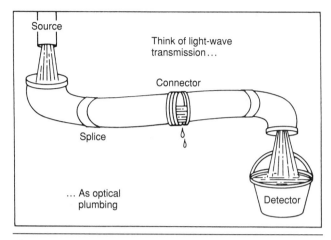

Figure 6-12. Example of light-wave transmission.

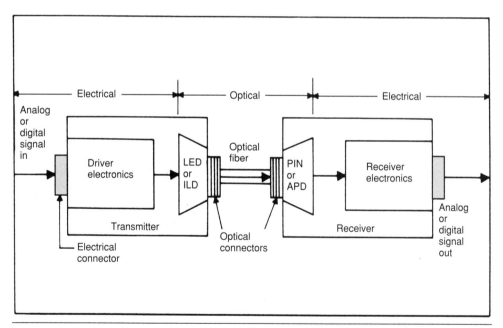

Figure 6-13. Basic light-wave system. LED = Light-emitting diode; ILD = Infrared Light Device; PIN = Pin Diode; APD = Avalanche Photo Diode.

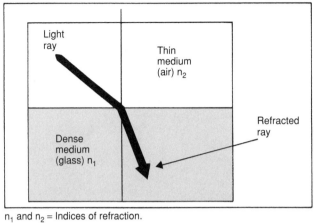

n₁ and n₂ = Indices of refraction.

Figure 6-14. Fiber optic light-ray bending.

In fiber optics, the fiber is a waveguide. Energy in the fiber is sent through the structure by total internal reflection, in contrast to electricity, which travels along the surface of the conductor. Light passing through two different but transparent media is subject to phenomena called refraction and reflection. Whenever a ray of light is incident on a boundary that separates two different media, part of the ray is reflected back into the first medium and the remainder is bent (refracted) in its path when it enters the second medium (see Figure 6-14). This creates a *critical angle*, defined as the exact angle when a light ray shifts from a state where some of its energy is refracted and some reflected to a state where all the energy is reflected. (The critical angle is a function of the ratio of the indices of refraction of the two transparent media through which the ray passes.)

External reflection occurs when a light ray in a low-index medium, for example, air, bounces off a high-index medium, for example, water or a mirror (see Figure 6-15). Internal reflection, which is more efficient, occurs when a light ray in a high-index medium, for example, water, bounces off the interface to the low-index medium, for example, air (see Figure 6-16).

A ray of light traveling in the core of the fiber is partially refracted into the clad glass unless the ray angle is less than a critical value (see Figure 6-17).

Fibers can be identified by the types of paths that the light trays (or modes) travel within the fiber core. There are two types of fibers normally used in lightwave communications: multimode and single-mode (see Figure 6-18).

There are two types of multimode fibers: step index and graded index. The step-index multimode fiber derives its name from the sharp, steplike difference

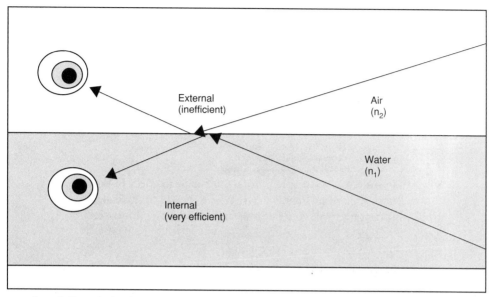

n_1 and n_2 = Indices of refraction.

Figure 6-15. Fiber optic internal/external reflection.

in the refracted index of the core and the cladding (see Figure 6-19). Step-index fibers are used primarily in image transport applications, such as medical and industrial remote viewing; multimode step-index fibers are limited in bandwidth capabilities and are mostly not used. For multimode graded-index fibers, the light waves or rays are also guided down the fiber in multiple pathways. However, unlike step-index fibers, the core contains many layers of glass, each with a lower index of refraction as one moves outward from the centerline of the fiber.

Grading causes the light waves to increase in speed as they are refracted toward the fiber centerline and travel through the outer layers. This effectively

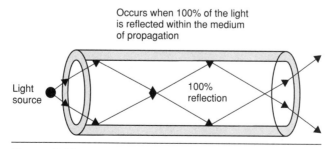

Figure 6-16. Fiber optic total internal reflection.

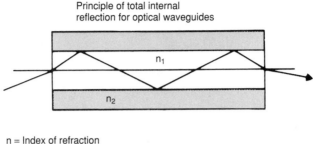

Principle of total internal
reflection for optical waveguides

n_1

n_2

n = Index of refraction
$n_1 > n_2$ gives total internal reflection

Figure 6-17. Principle of total internal reflection. In this
very efficient mechanism, whereas a typical (metallized)
mirror gives 95% reflection, total internal reflection gives
99.9999%.

matches the time of flight to those rays traveling the shorter pathways directly
down the axis of the fiber.

Fiber Cable

Fiber cable vastly multiplies the capacity of existing ducts. It is especially attractive where the expense of digging new conduit routes, such as in aerospace or
automotive facilities and/or existing road networks, is prohibitive.

A single fiber can replace any one or a combination of several copper
cables. The cable can be installed underground, aerially, or in ducts, and avoids

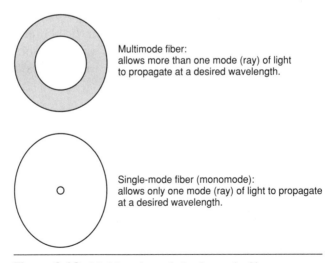

Multimode fiber:
allows more than one mode (ray) of light
to propagate at a desired wavelength.

Single-mode fiber (monomode):
allows only one mode (ray) of light to propagate
at a desired wavelength.

Figure 6-18. Multimode and single-mode fibers.

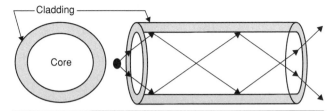

Figure 6-19. Basic multimode step-index fiber structure. The core is the central transmitting portion of the fiber, either glass to plastic. Light rays are trapped in this optically denser medium by total internal reflection. The larger the core, the more light is transmitted. The outer layer of cladding has a lower refractive index than the core and creates a containment layer for guiding light in the fiber.

transmission problems caused by lightning, ground faults, radio/television interference, and cable corrosion. Fiber cable does not need special shielding conduits or grounding because it provides extra lightning protection for computers and other equipment. Unlike copper cables, optical fibers, because of their dielectric nature, do not act as antennas or conductors if lightning strikes. Because light-wave communication is accomplished by the transmission of light energy through a glass waveguide as opposed to the movement of electrons over a conductor, spark generation is not a problem with fiber cable.

Fiber optic cables can be run almost anywhere without concern for communication interference or cross-talk factors. Fiber optic cable can run directly through hazardous areas without special expensive protection.

Installation and Use: Fiber optic cables should be installed and used in accordance with local codes (see Article 770 of the *National Electrical Code* for detailed instructions.) The cables should be tested for use in fire alarm signaling systems, preferably following requirements in UL 910, *Test Method for Fire and Smoke Characteristics of Electrical and Optical Fiber Cables*. Cables are normally specified to provide a stated loss and bandpass capability over a wide variety of operating conditions. Various standards organizations have developed test methods for simulating possible stresses to a cable. The Electronic Industry Association (EIA) in Alexandria, Virginia, has developed a comprehensive set of test methods that most manufacturers use to test their optical cables.

Meeting appropriate test requirements is not an automatic process, so the designer should work with a reliable cable supplier. The designer should understand that the cable can pass the burn test and yet not offer acceptable optical and mechanical performance.

The fiber optic system must be electrically compatible with the input-output characteristics with which it interfaces. Fiber optic systems are no more nor less sensitive to environmental problems than ordinary copper or radio-based electronics, so it is equally important to consider the environmental conditions in which the fiber optic cable will perform.

Fiber Optic Circuits

Optical fibers are sized 62.5/125 or 100/140 for use with fiber optic communication systems. The first number refers to the diameter of the fiber in micrometers and the second number is the clad diameter in micrometers. Each leg of a fiber optic link requires the use of two cables for duplex operation. These cables should be listed for use, such as in plenums, vertical risers, and general building installation, and must comply with any additional requirements of the authority having jurisdiction.

Fiber Optic Connectors

Fiber optic cables generally use ST connectors for 62.5/125 or 100/140 cable. The ST connector is a standard connector of the ferrule type, which holds and aligns the cable during installation. Some manufacturers offer fiber optic cables with factory-installed connectors. If factory-installed connectors are used, a detailed review of conduits, raceways, and cable trays must be made to determine cable length and verify that the factory-installed connectors will not be a problem during cable installation.

During field installation, a prime concern is the prevention of stress during connector installation. If the connectors are under a stress condition during installation, the stress can be transmitted to the cable, damaging the polished condition of the fibers and creating an intermittent problem condition. The manufacturer's installation procedures must be followed precisely for trouble-free results.

Another concern during field installation of fiber optic cables is the bend radius. The minimum bend radius varies with the manufacturer and diameter of the optical fiber. The minimum bend radius increases as the length of the cable pull is increased. This is to reduce the friction associated with the bends and the distance and limit the tension stress applied to the cable. As a general rule, a minimum bend radius of 1.5 in. (38 mm) for both 62.5/125 and 100/140 cables can be used as a guide.

Two other stress conditions, tension load stress and static stress, must be considered during installation to minimize problem conditions.

Fiber Optic Attenuation

Each fiber link module utilizing fiber optic cable has a *fiber optic attenuation* or *fiber optic budget,* which indicates the maximum light attenuation permitted in the fiber cable that will continue to sustain reliable communication. In order to verify that the maximum level has not been exceeded, calculations should be prepared for each fiber link. The total link loss should not exceed the fiber optic budget for the module.

Total fiber optic link attenuation is determined based on the following individual factors, the total of whose effects must be less than the fiber optic budget for the system to operate properly:

1. The fiber optic cable loss per unit distance
2. The length of the fiber optic cable
3. The number of splices in the link (as a general rule, each splice has a 2-dBA attenuation factor)
4. An estimate of the number of future splices in the link

SIGNAL TRANSMISSION COMPATIBILITY

Compatibility—the ability of equipment produced by one manufacturer to operate properly with that of another manufacturer—is critical to signal transmission in a fire alarm system. If the control panel is manufactured by one company, compatibility must be determined with other manufacturers' detectors, indicating appliances, annunciator panels, and so on. The key to system compatibility is the installation wiring diagram furnished with the equipment. If a control panel alarm-initiating-device circuit is compatible only with certain devices, those devices and pertinent installation instructions should be specified on the diagram. A circuit extending from a control panel that contains only an electrical rating and does not specify specific devices will be compatible with any device that has a similar electrical rating.

For example, when a four-wire smoke detector senses smoke, the smoke detector acts as a switch across the alarm-initiating circuit, while two other wires provide power to the smoke detector. The control panel initiating-device circuit does not differentiate between the signal from a smoke detector switch closure (relay contact) and a signal from a heat detector or a manual fire alarm box. Since a four-wire smoke detector does not impose any load on the alarm-initiating-device circuit of the control panel, it can be used with an electrically compatible fire alarm system.

In contrast, a two-wire smoke detector is powered by and sends signals over the same two wires of an alarm-initiating-device circuit. Compatibility between two-wire smoke detectors that receive power from the initiating-device circuit of a fire alarm system control panel and the panel depends upon the interaction between the circuit parameters, for example, voltage, current, frequency, and impedance of the detector, and the initiating-device circuit. A two-wire smoke detector must be compatible, therefore, not only with its control panel, but also with other two-wire detectors on the circuit.

Because a two-wire detector obtains power from the initiating-device circuit of a system control panel, its operation depends on the characteristics of the circuit to which it is connected, as the detector imposes a resistive and capacitive load on the circuit.

Compatibility between the detector and the control panel will ensure that sufficient power is furnished to the detector at startup and in the standby (supervisory) mode, and that the change in status of a smoke detector when it goes into alarm will be sensed at the control panel. If the smoke detector has an integral audible signal, annunciator light, or supplementary relay, the control panel must also be able to furnish the power for these functions.

To be listed, smoke detectors and a compatible control panel are tested for all of these conditions to ensure proper signal transmission when the system is placed into service.

SIGNAL VERIFICATION VERSUS FALSE ALARMS

When a transmission circuit is established, it is important to verify that a signal is truly an alarm and not a false alarm. Some methods used to reduce the number of false alarms from smoke detectors in equipment and in the field are as follows.

False Alarm Reduction in Equipment

1. Several alarm pulses can be counted before an alarm is signaled on smoke detectors with pulsing normal operation.
2. Detectors that have either an analog output to the control panel or a field-sensitivity adjustment on the detector can have the sensitivity (alarm set point) decreased within limits.
3. A delay-reset time of 5 to 30 seconds can be provided where the signal is held and the detector, then reset. If the detector still senses smoke for a period of time after the delay reset, the alarm signal is transmitted.

False Alarm Reduction in the Field

1. A thorough evaluation of the building should be performed to ensure that the smoke detectors are placed away from areas that would normally be

humid, dusty, smoky, or insect-laden. Detectors should also not be located in areas with temperature extremes, excessive RF noise, or other electrical noise on the power line or in the area. Any of these could be the source of false alarms.

2. Smoke detectors should be installed in proper relation to all supply and return air ducts.

3. Installation of the smoke detector heads must be delayed until all of the building construction and plaster dust has been removed.

4. A regular cleaning of all smoke detectors should be performed as recommended by the manufacturer.

5. The detector sensitivity should be regularly checked and recorded. Detectors that have drifted away from their normal setting should be replaced.

6. Smoke detectors can be installed away from smokers, if possible, or smoking prohibited in the area.

In addition to these methods of reducing false alarms, several industry programs help to reduce false alarms. The National Institute for Certification in Engineering Technologies (NICET) in Alexandria, Virginia, has established a national certification program for technicians in the fire alarm systems field. The program covers types of systems, job specifications, codes and standards, surveys, layout drawings, installation, wiring, system operation, and testing and maintenance of fire alarm systems.

Underwriters Laboratories Inc. has established a certificate service program for fire protective signaling systems. This program qualifies alarm companies to certify that a fire alarm system has been installed and tested in compliance with *NFPA 72.*

SIGNAL PROCESSING

Both automatic and manual signals are processed for the various types of fire alarm systems. Combined manual and automatic processing such as that found in guard's tour supervisory service, sprinkler system waterflow alarm and supervisory signal service, and trouble signals are also included. See also Chapter 4, Fire Alarm System Components and Circuits, and Chapter 7, Fire Alarm Notification, for further information. Processing is the dynamic link between signal input and system response. Signal processing is an interpretive function governing the response of a unit or an entire fire alarm system to the various inputs received by that unit or system. A control panel will typically receive numerous, varied signal inputs; outputs can vary as widely depending on the application of the alarm system.

In a simplified application, such as for a local or household fire alarm system, the control panel may dictate energizing of general evacuation output signals. With a more complex system, for example, proprietary, remote-station, or central station system, several response (output) modes may need to be activated for certain types of inputs. The interpretive processing routines to match appropriate outputs to inputs can be established through fixed electric or electronic circuitry or controlled with flexible, complex computer programs.

Types of Signals Processed

Manual Fire Alarm Signals: Requirements for manual fire alarm signals are addressed in *NFPA 72* and in the core chapters of NFPA *101®*, *Life Safety Code®*. Manual fire alarm signals in a protective system are typically initiated from manual fire alarm boxes located throughout the protected facility. Signals can also be initiated outside the protective system, for example, from a municipal fire alarm box, though these would not initiate an alarm within the facility. Processed outputs for manual signals vary according to the type of protective system.

Evacuation signals, whether audible (tonal or voice generated) or visible, are signal outputs in the local protective and emergency voice/alarm systems. Various prerecorded tone and voice messages can also be considered as emergency system outputs.

Manual initiation of an alarm signal in an auxiliary, remote-station, and public fire service system results in the transmission of signals to a remote point for subsequent fire department interpretation and action. Signals can be transmitted indirectly, or directly to the fire department from an on-premise alarm system.

Manual signals are processed in a similar manner for central station and proprietary systems, except that the signal is processed through private system equipment with possible retransmission to a public system. Unless system design dictates, no distinction is made between manual and automatic fire alarm signals.

Some of the differences in manual system responses are the following:

1. An auxiliary system does not normally use a coded alarm signal. The rationale for this is that auxiliary systems have historically been simple noncoded latching circuits, which did not transmit signals in a serial method to a public fire department.
2. Manual signal initiation is not part of household fire warning equipment. Since most fires occur while the occupants of the household are sleeping, initiation and detection are the only necessary functions of household fire warning equipment.

3. *NFPA 72* and NFPA 1221 assume that other system units or interfaces have processed manual signals prior to retransmission. Noncoded manual alarm inputs are processed in a manner similar to that for automatic alarm inputs.

Automatic Fire Alarm Signals: Automatic fire alarm signals for each system are addressed in *NFPA 72*. Automatic alarms are typically initiated by smoke and heat detectors, both of which can be installed on separate circuits or integrated into the same circuit. Smoke and heat detectors can be installed and maintained in protective signaling systems according to Chapter 5 of *NFPA 72* with testing requirements addressed in Chapter 10. *NFPA 72* refers to special connections for use of integral trouble contacts located on the automatic signal-initiating devices. These normally closed contacts should be connected electrically at the end of the initiating circuit to prevent disconnection of initiating devices.

The differences in automatic alarm signal generation and processing depend on the type of protective signaling system used. An evacuation alarm signal is required for building occupants when a local protective system is specified. Incoming automatic signals for central station and proprietary signaling systems are required to be identified and processed by the protected zone or other system subdivision within the protected building.

Automatic devices connected to initiating circuits are required by *NFPA 72,* to be latching, but nonlatching circuits are permitted for smoke detectors. Nonlatching smoke detector circuits are allowed by *NFPA 72* because of the transient nature of the residential conditions that can easily activate a smoke detector, such as cooking, humidity, and condensation.

Primary power, when restored, must also restore systems to their normal supervisory condition. Central station and proprietary systems have a large amount of information contained within a common system. A rapid, more discriminating means of identifying changes in system status is required in central station systems since identifying and interpreting signals is important to public fire department response.

Types of Processing

Guard's Tour Supervisory Service: In guard's tour service, a supervisory service signal is typically initiated through mechanical key-operated tour stations. These signals can also be generated electronically or optically for encoding and decoding. *Encoding* is defined as processing of a signal such as that initiated by a switch contact and subsequent transfer of the signal into information that can be interpreted by a machine. *Decoding* is the translation of this information back into signals understandable by human beings.

Guard's tour supervisory signals are not transmitted to public fire departments because the signals are additional communications for public fire service receiving stations.

Guard's tour supervisory signaling equipment generally tracks all stations within the tour schedule; all stations must report in this type of service. The beginning and end of a tour schedule must register at some remote point, and intermediate nonrecording signals must operate in fixed succession. A guard's tour system can have an exception-reporting package in which all stations must report; such a system also has a delinquency signal to identify any station that has not been operated.

Sprinkler System Waterflow Alarm and Supervisory Signal Service: In general, a sprinkler system waterflow alarm and supervisory signal service in any fire alarm system must be able to interpret the type of system signal and the particular element type that has operated. Additionally, the supervisory service must identify restoration of the element to its normal position.

Waterflow alarm supervisory signals consist of one or more of the following: gate valve supervision, pressure supervision, water level supervision, water temperature supervision, and pump supervision. The waterflow alarm supervisory signals are the only supervisory signals that can be transmitted and processed by auxiliary systems. If additional supervisory services are desired, another type of fire alarm system should be specified.

Trouble Signals: To varying degrees, all systems interpret factors, referred to as *trouble conditions,* that inhibit alarm transmission or annunciation. Many functions and/or circuits are supervised against loss of capability depending on the nature of the system and the complexity of the information processed by that system. In some cases, automatic or manual control of various system functions permits emergency operation under faults that could debilitate the entire system.

Fire alarm systems share certain characteristics concerning trouble signals. All systems interpret loss of primary (main) power as a trouble condition, which is usually indicated by audible and visible signals. Loss of initiating circuits is also a trouble condition, and protective systems process this signal information accordingly. Up to five styles of fault recognition can be specified for a proprietary fire alarm system.

Loss of indicating or evacuation circuits is a trouble condition in local and emergency voice/fire alarm communication systems. When central station, remote-station, or proprietary systems have an evacuation alarm signal, these integrated circuit systems should comply with *NFPA 72* concerning trouble signals for loss of both evacuation and indicating circuits.

Trouble signals for signaling line circuits are covered in *NFPA 72*. This standard requires that the central station, remote-station, and proprietary systems recognize a single-open and single-ground fault as an abnormal (trouble) condition, but not as an alarm or fire.

The most comprehensive process recognition design is found in the proprietary systems, in which prescribed responses are defined for single and multiple faults on both the initiating-device and signaling line circuits. Tables 4-1 and 4-2, in Chapter 4 show the performance and capacities of these circuits. Styles A through C in Table 4-1 are considered Class B circuits, and Styles D and E are considered Class A circuits. The table breaks down each style by specific performance since the circuits differ slightly in individual performance.

A central station system is also assigned system capability based on various responses, but to a lesser degree than for proprietary systems. *NFPA 72* specifies appropriate responses for virtually any kind of initiating device circuit or signaling line circuit. Although the various circuit styles do not necessarily represent an ascending order of complexity, the system processing/response implies that the styles with higher letters can accommodate greater distribution and concentration of connected devices. There is also an ascending order of complexity and capability for system-processing response in a central station system.

QUESTIONS

1. A circuit that connects powered four-wire smoke detectors to the fire alarm control unit is called:

   a. a signaling line circuit   b. a notification appliance circuit
   c. a power circuit   d. an initiating-device circuit

2. Addressable heat detectors are connected to each other and the fire alarm control unit using:

   a. a detector circuit   b. an initiating-device circuit
   c. a signaling line circuit   d. a notification appliance circuit

3. Central station service requires:

   a. the fire alarm system to be certificated or placarded
   b. runners to respond to alarms
   c. guards to call in hourly
   d. monthly testing

4. A local fire alarm system will provide alarm notification:
   a. at the building and fire department
   b. at the building and the central station
   c. at the building only
   d. at the building and the remote station

5. Fire alarm systems provide the following signals:
   a. alarm and trouble   b. alarm and supervisory
   c. alarm, supervisory, and trouble   d. alarm only

BIBLIOGRAPHY

NFPA Codes, Standards, Recommended Practices, and Manuals (see the latest *NFPA Catalog* for availability of current editions of the following documents):

NFPA 70, *National Electrical Code®*.

NFPA 72®, National Fire Alarm Code®.

NFPA *101®, Life Safety Code®*.

NFPA 1221, *Standard for the Installation, Maintenance, and Use of Emergency Services Communications Systems*.

REFERENCE CITED

1. FCC, *Rules and Regulations,* Vol. V, Part 90, March 1979, Washington, DC.

ADDITIONAL READING

UL 910, *Test Method for Fire and Smoke Characteristics of Electrical and Optical Fiber Cables,* Underwriters Laboratories Inc., Northbrook, IL.

UL 985, *Household Fire Warning System Units,* Underwriters Laboratories Inc., Northbrook, IL.

UL 1023, *Household Burglar-Alarm System Units,* Underwriters Laboratories Inc., Northbrook, IL.

7

Fire Alarm Notification

INTRODUCTION

Fire alarm notification, that is, notifying the occupants of a facility that a fire condition exists, is the most important life safety feature of a fire alarm system.

Various types of audible and visible signals as well as an emergency voice/alarm communication system can be used as a notification method. The fire alarm system designer must ensure that any audible and/or visible notification signal and the appliances used to sustain them are appropriate for the protected facility and its occupants.

This chapter describes notification signals, including classification of types of signals; combination systems; use in other systems; and operation, including audible and visible notification appliances. It also addresses the issue of physically challenged occupants and presents appropriate signal design parameters based on *NFPA 72*®, *National Fire Alarm Code*®, and the Americans with Disabilities Act (ADA). Notification appliance options are detailed. Emergency voice/alarm communication systems, emergency controls, and special considerations for high-rise buildings are also described. For related information see Chapter 6, Signal Transmission Methods and Processing; Chapter 4, Fire Alarm System Components and Circuits; and Chapter 8, Fire Alarm System Installation.

NOTIFICATION METHODS

Fire alarm notification is accomplished primarily through two methods: audible and visible. Audible notification appliances employed in fire alarm systems for alarm, supervisory, and trouble signals include electromechanical and electronic appliances, such as bells, horns, buzzers, chimes, and sirens; integral electronic tone-generating appliances (i.e., a tone generator, an amplifier, and a speaker in a common housing); and speakers energized by a remote amplifier source. The suitability of the type of appliance for a particular application and its location depend on the applicable local codes. If the system is a protected-premises (local) alarm system, it is recommended that at least one audible fire alarm notification appliance be located outside the building to alert persons in the vicinity, who, in turn, could summon fire-fighting assistance.

Visible notification appliances rely on two methods: direct viewing of an appliance light source and use of reflected light.

Fire Alarm Notification Signals

Fire alarm systems provided for evacuation of occupants should have one or more audible notification appliances approved for fire alarm use on each floor of the building and be so located that their operation will be heard clearly regardless of the maximum noise level from machinery or other equipment under normal occupancy conditions. (Each section of a floor divided by a fire wall can be considered as a separate floor for the purpose of this protection.) To ensure that audible notification appliance signals are clearly heard, it is required that they have a minimum sound level of at least 15 dBA above the average ambient sound level or 5 dBA above the maximum sound level having a duration of at least 60 seconds (whichever is greater) measured 5 ft (1.5 m) above the floor in the occupiable area. The average ambient sound level is the root-mean-square A-weighted sound pressure level measured over a 24-hour period, where A is a weighted scale related to the response for normal human hearing.

Distinctive Signals

Fire alarm signals should be distinctive in sound from other signals in the protected premises and this sound should be used for no other purpose. Audible notification appliances for a fire alarm system should produce signals that are distinctive from those of other notification appliances used for other purposes in the same area.

The recommended fire alarm evacuation signal is a uniform Code 3 temporal pattern, using any appropriate sound, keyed ½ to 1 second ON, ½ second OFF, ½ to 1 second ON, ½ second OFF, ½ to 1 second ON, and 2½ seconds OFF, with timing tolerances of ±25 percent, repeated for not less than 3 minutes. Manual interruption of the repetition time is permitted by some local codes.

The recommended standard fire alarm evacuation signal is intended for use only as an evacuation signal. A separate, different-sounding, signal should be used to relocate the occupants from the affected area to a safe area within the building or for their protection in place (e.g., in high-rise buildings, health-care facilities, penal institutions, etc.).

Supervisory signals should also be distinctive in sound from other signals and used only to indicate a supervisory condition. Fire alarm signals should take precedence over all other signals.

Trouble Signals

Trouble signals should be distinctive from fire alarm signals and indicated by the operation of a sounding appliance. If an intermittent signal is used, it should sound at least once every 10 seconds with a minimum time duration of 0.5 second. An audible trouble signal can be common to several supervised circuits. The trouble signal(s) should be located in an area where it is likely to be heard, as designated by the authority having jurisdiction.

Visual Zone Alarm Indication

One of the main purposes of a fire alarm system is to reduce the amount of time it takes for the fire department to respond to the location of the fire. Providing information to the fire department using an annunciator showing the zone or location of the alarm is an efficient method of providing this information. Zoning of a fire alarm system is designed to help the fire department; therefore the fire department should be consulted when determining the appropriate location and labeling of the annunciator. It is important to remember that too much information may render the remote annunciator useless. The responding fire department needs basic fire location information to provide the fastest response to that location. Avoiding confusion due to too much detail will assist the fire department in meeting their goals.

The location of an initiating device should be visually indicated by building, floor, fire zone, or other approved subdivision by annunciation, printout, or other approved means. The visual indication at the fire alarm control unit and annunciator should not be canceled when an audible alarm-silencing switch is operated.

Notification of occupants is of prime importance during a fire emergency, but the need to notify the fire department is equally important. If NFPA *101*®, *Life Safety Code*®, is used by the local authority or the local building code requires fire department notification, this signal should automatically be transmitted by an auxiliary fire alarm system, a central station fire alarm system, a remote-station fire alarm system, or a proprietary fire alarm system operator. If a fire alarm system is not used, some type of plan should be provided to notify the fire department of a fire emergency.

CLASSIFICATION OF NOTIFICATION SIGNALS

For the purpose of this book, notification signals for fire alarm systems are classified as noncoded, coded, or textual, and as public or private operating modes.

Noncoded, Coded, or Textual Signals

Noncoded signals are audible or visible signals and convey one discrete bit or unit of information; a coded signal conveys several bits or units of information. A textual signal conveys a stream of information. For example, one stroke of an impact-type appliance is a noncoded signal, as is the continuous operation of an appliance that is continuously energized (or interrupted at a continuous uniform rate). Examples of coded signals are numbered strokes of an impact-type appliance or numbered flashes of a visible appliance. A voice message is an example of an audible textual signal.

Public and Private Operating Modes

Notification signals can have two operating modes: public and private. With a public operating mode, the inhabitants of the protected area receive the audible or visible signal from the fire alarm system. A private operating mode restricts the signal to the persons concerned with initiating and carrying out emergency actions, for example, hospital staff or a private fire brigade, in the protected area.

Audible and Visible Characteristics

Audible signals intended for operation in the public mode should have a sound level of not less than 75 dBA at 10 ft (3 m) or more than 120 dBA at the minimum hearing distance from the audible appliance. The sound level range for private mode audible signals should not be less than 45 dBA at 10 ft (3 m) or more than 120 dBA at the minimum hearing distance from the audible appliance.

Table 7-1. Average Ambient Sound Levels

Location	dBA
Business Occupancies	55
Educational Occupancies	45
Industrial Occupancies	80
Institutional Occupancies	50
Mercantile Occupancies	40
Piers and Water-Surrounded Structures	40
Places of Assembly	55
Residential Occupancies	35
Storage Occupancies	30
High-Density Urban Thoroughfares	70
Medium-Density Urban Thoroughfares	55
Rural and Suburban Thoroughfares	40
Tower Occupancies	35
Underground Structures and Windowless Buildings	40
Vehicles and Vessels	50

The sound level of an installed signal should be adequate for its intended function and should meet the requirements of either the local authority or applicable standards. The average ambient sound level for various occupancies as listed in Table 7-1 can be used for design guidance.

There are two primary visible notification methods: direct and indirect. For a direct visible notification method, the message of notification of an emergency condition is conveyed by direct viewing of the illuminating element within or attached to the notification appliance. The indirect method illuminates the area surrounding the visible notification appliance.

SIGNAL NOTIFICATION IN COMBINATION SYSTEMS

Combination systems consist of equipment used for fire alarm, sprinkler supervisory, or watch service and for other signaling systems such as burglar alarm, voice paging, or music program systems or a coded paging system. Methods using circuit wiring common to these types of system are also referred to as combination systems.

Common Wiring

Equipment not used in fire alarm systems should be connected to common wiring in a combination system so that short or open circuits or grounds in this equipment or between this equipment and the fire alarm system wiring will not interfere with either the supervision of the fire alarm system or prevent alarm or supervisory signal transmission.

Use of Speakers

Speakers can provide the audible alarm signal for local fire alarm or sprinkler alarm systems. They should be used only for emergency voice/alarm communication systems and other emergency purposes unless other uses are approved by the local authority. Selective paging is usually allowed where the fire command station is constantly attended by a trained operator.

In general, speakers used in fire alarm systems for tone signals and/or emergency voice communications should have a frequency response and power rating suitable for the application. Speaker materials that are subject to moisture absorption should be suitably protected. In a combination system, the fire alarm signal should be clearly recognizable and take precedence over any other signal even if a nonfire alarm signal is initiated first.

SIGNAL NOTIFICATION IN OTHER SYSTEMS

Notification specifications differ where a fire alarm system is not a protected-premises (local) fire alarm system. Alarm notification should be located at the central supervising station to provide alarm notification for evacuation of occupants or signals directing aid to the location of an emergency. Suitability and location of notification appliances for alarm, supervisory, and trouble signals should be determined by the system designer and the authority having jurisdiction.

Signal notification of the section of the building where the signal originated should be designated at either the central supervising station or at the protected building. Notification appliances such as remote annunciators should be acceptable to the local authority.

SIGNAL NOTIFICATION OPERATION

Closing of a smoke damper in an air-handling duct by the signal initiation from a smoke detector within that duct is an automatic action; it results from an electric or mechanical extension of the fire alarm system.

Manual action, however, for example, the evacuation of a building occupant by signal initiation from a manual fire alarm box, is physiological, that is, it depends upon human response as a result of the sense of the signal and the interpretation of its meaning. Limitations of human response must be addressed for practical implementation of manual actions. Of the five human senses, current notification appliances only consider audibility and visibility.

Because a fire alarm system must operate before and during a fire emergency, notification appliances must operate at elevated temperatures. Fire resistance of notification appliances is tested by the nationally recognized testing laboratories.

Audible Signal Characteristics

Audibility of a notification appliance must be considered for the particular place where the appliance is installed. A gong in a recessed enclosure with a grille is less audible than a surface-mounted gong. Heavy-carpeted floors, fabric walls, and acoustic ceilings are sound deadeners that must also be considered.

The audible signal notification should be approximately 30 times greater than ambient or background noise volume. Expressed in decibels (dB), a logarithmic function, this amounts to about 15 dBA above ambient.

The signal for the public mode of operation should have a sound output level of no less than 75 dBA where contributions from signal reflection are eliminated. A minimum ambient level of 85 dBA for mechanical equipment rooms is recommended. If the ambient sound level in a room is greater than 105 dBA, visible notification appliances must be used. Since private mode operation is in a controlled environment, an ambient level would probably be no greater than 35 dBA.

Design of Audible Appliances: Basic criteria for the design and placement of audible fire alarm notification appliances for a fire alarm system can be found in *NFPA 72*. However, design methodology is not addressed in *NFPA 72*. One method that provides a good estimate for the effective positioning of devices for various fire alerting systems can be found in the *SFPE Handbook of Fire Protection Engineering*.[1] This method is based upon a 1981 British study.[2]

The four main parameters required for the location of audible fire alarm devices are the following:

1. Sound characteristics of the floor, walls, and ceiling
2. Sound pressure level of the device (based upon manufacturer's data)

3. Number of directions that the sound can propagate
4. Distance from the source device

The devices must be designed and located to meet the worst-case condition, that is, achieving the minimum required sound pressure level in the most remote room.

Audible appliances for fire alarm systems should be listed and protected against effects of corrosion, temperature, humidity, and physical damage.

Location of Audible Notification Appliances: Audible notification appliances operated in the public mode should be located so the sound pressure level is not less than 15 dBA above ambient. The sound pressure level at the minimum hearing distance from the appliance is determined by the threshold of pain (120 dBA), that is, the level where sound may cause physical damage to human hearing. For private mode operation, the appliance should produce no less than 10 dBA above ambient over the private mode area, which is usually a building control room.

Where ceiling height permits, wall-mounted appliances should have their tops at heights above the finished floors of not less than 90 in. (2.30 m) and below the finished ceilings of not less than 6 in. (0.15 m). This does not preclude ceiling-mounted or recessed appliances.

Visible Signal Characteristics

There are two types of visible signal notification: when the sole means of notification is by direct viewing of its appliance light source or by reflected light within the room. Since direct appliances can be used under a variety of conditions, the boundary limitations of the signal should be determined so as to permit maximum versatility in equipment design and application. The signal can be defined in terms of its effective intensity I_{eff} by using the following equation:

$$I_{eff} = \frac{\int_{t_2}^{t_1} I(t)dt}{t_2 - t_1 + 0.2}$$

where
$I(t)$ = Instantaneous intensity (candelas)
t = Time (seconds)
t_1 and t_2 = Times chosen to maximize the effective intensity (seconds)

The boundary conditions of the signal are $A \leq I_{eff} \leq B$, where the range limits A and B can be defined as follows:

1. Where the average illuminance is less than 5 lumens/sq ft (0.093 m^2), the effective intensity is between 1.5 and 15 candelas.
2. Where the average illuminance (without motion present) is greater than 5 lumens/sq ft (0.093 m^2), the effective intensity is between 15 and 150 candelas.
3. Where the average illuminance (with motion present) is greater than 20 lumens/sq ft (0.093 m^2), the effective intensity is between 100 and 1000 candelas.

The direct signal appliance is dependent on the viewer's line of sight and the environment. The signal within the defined visible range should have the following directional characteristics:

1. Through ±85 degrees from the nominal signal direction and not less than 10 percent from 85 to 90 degrees
2. For viewing only below the horizontal, through +10 degrees to −70 degrees and not less than 10 percent from −70 degrees to −85 degrees
3. For viewing above and below the horizontal, through +45 degrees to −70 degrees and not less than 10 percent from +45 degrees to −85 degrees

Once the ambient illuminance has been determined, the average effective illuminance E_{eff} can be determined by using the physical parameters of the environment. For the standard nominal work plane 30 in. (0.76 m) above the floor,

$$E_{eff} = \frac{\int_{t_2}^{t_1} E(t)dt}{t_2 - t_1 + 0.2}$$

where
$E(t)$ = Instantaneous illuminance (lumens per square foot)
t = Time (seconds)
t_1 and t_2 = Times chosen to maximize the effective illuminance (seconds)

When E_a is the steady-state spatially averaged work plane illuminance in the room, then

$$0.02E_a \leq E_{eff} \leq 0.1E_a$$

Indirect signal appliances should not be used for environments with an average illuminance of less than 1.0 lumen/sq ft (0.093 m^2). The direct signal appliance should be used in dark environments, such as those found in theatres and cocktail lounges.

Pulse Requirements: The repetition and duration of light pulses are important design features in visible signal notification appliances.

Pulse repetition needs to be limited to a small range, and the upper limit of the range should not appear to be a flickering light. In the lower limit of the range, the visible signal should be recognized as a warning signal. Since the indirect visible signal tends to modulate the environmental light, a higher upper limit can be tolerated for the direct visible signal. Consistent pulse repetition is desirable since this feature principally identifies the warning. The range for pulse repetition should be between a flash every 3 seconds (for an indirect signal) to a maximum of 3 flashes per second (for a direct signal).

Pulse duration is the time interval between the initial and final points of 10 percent of the maximum signal time. The duty cycle is the ratio of the pulse duration time to the time between the initial points of two consecutive pulses. Optimally, a principle pulse can be made from a string of pulses, provided the repetition of this string exceeds 30 pulses per second. This allows a system to use a string of pulses for practical control reasons. Maximum pulse duration and duty cycle should be 0.2 second and 40 percent, respectively.

Color Requirements: The visible signal source should be white. The white spectral limit is defined on a chromaticity diagram (see Figure 7-1). Available light sources best suited for use as visible signals are wide-band radiators, such as xenon discharge lamps. Although other colors can be formed by using filters, this adds unnecessarily to power needs. The pulsing nature of the visible signal is the primary form of identification, not its color. Local jurisdictions may identify yellow (amber) or red as colors to be used in notification of a fire emergency. If this is the case, these colors must be identified in protected areas as only serving that purpose.

Visible Signal Location: The location of a direct signal appliance should be determined by the needs of large enclosures that are not rectangular, multistory (such as atria), or contain spatial obstructions such as manufacturing areas and libraries. Direct signal appliances should be located in these large enclosures as described herein.

The possibility of either temporary or permanent blockage of the appliance by portable objects should be minimized. This can be accomplished with a light source opening that is no less than 6.5 ft (2 m) above the highest floor level in the room.

The power level of the appliance should be specified for the particular area covered. An appliance designed to provide notification for a large area could be

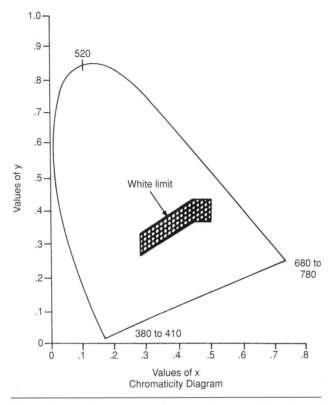

Figure 7-1. Chromaticity diagram for the definition of the white spectral limit.

objectionable to occupants of a smaller enclosed space. There should be a minimum of one additional appliance for each 7500 sq ft (697 m²) of floor area in excess of 10,000 sq ft (929 m²). At least one appliance should be directly visible for 95 percent of the occupied floor area as measured in the protection plan. It is not practical to provide a notification appliance that is within lines of sight of all building occupants. Multiple appliances can be located so that at least two appliances are separated by a minimum of 135 degrees in plain view for 75 percent of the occupied floor area. An appliance should be within +60 degrees and −30 degrees (vertical angles) over 75 percent of the floor area that the appliance covers. Considering the mobility of the lines of sight, reflections from objects in the room, and interaction between occupants, it is not necessary for a direct signal appliance to be within the fields of view at all times.

The location of the indirect signal appliance should be determined from the average effective illuminance using the space and the visible characteristics of

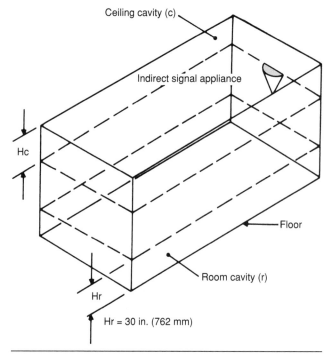

Figure 7-2. Indirect signal appliance space.

the appliance. The location parameters can be calculated using space shape and dimensions, surface reflectances, and appliance height and output (see Figure 7-2). The mathematical procedure for locating an indirect appliance solution can be approximated as follows:

1. Divide the space into a room cavity (r) between the working plane [30 in. (0.762 m) above the floor] and the height above the floor for the appliances and a ceiling cavity (c) between the height above the floor for the appliances and the ceiling. The average ceiling height should be used when the ceiling is not horizontal.
2. Determine the shape factors M_r and M_c, which are equal to the height of the cavity multiplied by the perimeter of the cavity divided by the area of the cavity. This is for any space without a reentrant perimeter (small alcoves, etc., should be disregarded). L-shaped spaces may be divided into two rectangular spaces. The shape factors simplify, for a rectangular room, to twice the height of the cavity multiplied by the sum of the width and length, divided by the product of the width and length.

3. The effective reflectance of the ceiling cavity at its lower boundary is determined by calculating

$$\frac{\text{ceiling reflectance}}{1 + 0.6M_c}$$

When the ceiling does not have uniform reflectance, the ceiling reflectance should be its spatially weighted average.

4. The flux transfer factor from the appliance to the work plane is determined by calculating

$$\frac{2 \times \text{ceiling cavity reflectance}}{1 + 0.6M_r}$$

5. The average effective illuminance is the effective luminous flux emitted by the appliances, multiplied by the flux transfer factor, divided by the floor area. The effective luminous flux emitted by the appliances ϕ_{eff} is a parameter determined by the manufacturer in accordance with the equation

$$\phi_{\text{eff}} = \frac{\int_{t_2}^{t_1} \phi(t)dt}{t_2 - t_1 + 0.2}$$

where
$\phi(t)$ = Total instantaneous spatially luminous flux emitted by the appliances (lumens)
t = Time (seconds)
t_1 and t_2 = Times chosen to maximize the luminous flux (seconds)

6. These relations can be used to determine the number and ratings of the appliances needed or to evaluate appliances already in place. They should not be used for surfaces in very dark spaces such as ceilings with less than 25 percent reflectance or walls with less than 20 percent reflectance. Spaces with very dark surfaces should be provided with direct notification appliances.

The means used for locating the indirect signal appliance indicates that ceiling reflectance has a much greater effect on the flux transfer than does wall reflectance. Generally, the ceiling reflectance range exceeds that for wall reflectance. The factor of wall reflectance may be eliminated from the calculations for locating the appliances. The error range introduced, as compared to the exact mathematical solution, is shown in Table 7-2.

Table 7-2. Error Ranges

Space Factors	Extreme Error Limits		Total
	Effective Ceiling Reflectance	*Flux Transfer Factor*	
Common Range:			
$M_r \le 2$	+25% to −40%	±20%	±50%
$M_c \le 0.8$			
Extended Range:			
$M_r \le 4$	+70% to −60%	+50% to −20%	+150% to −70%
$M_c \le 2$			
Range:			
Ceiling Reflectance	90% to 25%	90% to 20%	
Wall Reflectance	80% to 20%	60% to 20%	

Extreme errors occur for the least likely combinations of ceiling reflectance, wall reflectance, and room dimensions. The errors are generally much less than the extremes. For the common range of space factors, the maximum error will not exceed a factor of 2, and for the extended range of space factors, the maximum error will not exceed a factor of 3.

The following additional points should be considered in locating an indirect signal appliance:

1. No direct viewing of the lamp should be permitted from below the horizontal plane. No optical element (either lens or reflector) should be optically active externally at angles below the horizontal plane. This permits part of an optical component within the field of view, provided it is not directing light to that region.

2. The light from the appliance should be reasonably well distributed over the ceiling and the upper walls. There should also not be an extremely bright area adjacent to the appliance. Since the appliance is likely to be ineffective in the more distant parts of the protected space, explicit design limitations on the three-dimensional intensity distribution should ensure good light distribution.

3. For wall-mounted appliances, no more than 15 percent of the output should be directly incident on the plane of the wall. No more than 25 percent of appliance output should be incident on the ceiling within 45 degrees of the

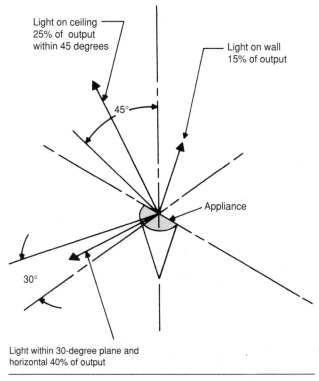

Light on ceiling
25% of output
within 45 degrees

Light on wall
15% of output

45°

Appliance

30°

Light within 30-degree plane and
horizontal 40% of output

Figure 7-3. Light output directions of an indirect notification appliance.

vertical through the appliance. A minimum of 40 percent of the output should be between the horizontal appliance plane and 30 degrees above (see Figure 7-3).

4. The vertical angle below which 10 percent of the appliance output occurs should be established. The maximum horizontal distance of coverage should not exceed four times the distance at which a line through the appliance at this angle intersects the ceiling. For example, a manufacturer might express an appliance rate for maximum coverage distance as a multiple of the appliance-to-ceiling distance (see Figure 7-4).

Visible Notification Appliances: As an alternative to audible appliances, *NFPA 72* permits visible notification appliances. These appliances can provide hearing-impaired persons with fire emergency notification equivalent to what hearing persons can receive through audible notification appliances.

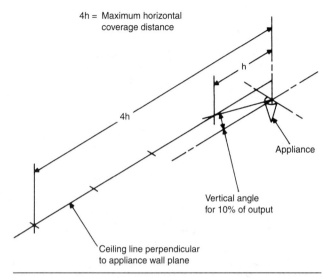

Figure 7-4. Horizontal coverage of an indirect notification appliance.

The minimum criteria for ensuring that visible notification devices can be seen by building occupants are as follows for direct primary visualization.

Average Illuminance in Protected Area	Effective Intensity Rating of Pulsing Light
<5 lumens/ft^2	0.15–15 candelas
>5 lumens/ft^2	15–150 candelas
>20 lumens/ft^2 + motion	100–1000 candelas

The minimum criteria for ensuring that the visible notification appliance can be seen by the building occupants without causing eye damage are as follows for indirect primary visualization.

1. The average effective illuminance of the pulsing signal at a standard nominal work plane 30 in. (0.76 m) above the floor should be a minimum of $\frac{1}{10}$ the steady-state spatially averaged work plane illuminance at the standard nominal plane in the room.
2. For an appliance mounted on or adjacent to a wall, it is recommended that not more than 15 percent of its output be directly incident on the plane of the wall and not more than 25 percent be incident on the ceiling within

45 degrees of the vertical through the appliance. A minimum of 40 percent of the output should be between the horizontal and +30 degrees from the horizontal.

3. Direct viewing of the lamp is not permitted from below the horizontal plane. This prevents eye damage. No optical element, either lens or reflector, should be optically active externally at angles below the horizontal. Part of the optical component within the field of view is permitted, provided it is not directing light to that region.

NFPA 72 has an exception to the above, stating that combination indirect- and direct-viewing appliances are permitted if the direct viewing component does not exceed an effective intensity of 1000 candelas.

NFPA 72 recognizes two methods of primary visible notification: direct detection and indirect detection. Direct detection methods include a visible notification method in which the message of notification of an emergency condition is conveyed by direct viewing of the illuminating element within or attached to the notification appliance. An indirect detection method is one in which the message of notification of an emergency condition is conveyed by means of illumination of the area surrounding the visible notification appliance.

It is important to remember that, when locating visible alarm notification appliances, the main objective is to notify hearing-impaired occupants of a fire emergency in a way equivalent to that provided hearing persons by audible notification appliances.

As a standard, *NFPA 72* estimates the average illuminance for various activities as listed in Table 7-3.

Table 7-3. Average Illuminances According to *NFPA 72, National Fire Alarm Code*

Type of Activity	Average Illuminance (lumens/ft²)
Public spaces with dark surroundings	2.5
Simple orientation for short, temporary visits	5–10
Working spaces where visual tasks are only performed occasionally	10–20
Performance of visual tasks of high contrast or large size	20–50
Performance of visual tasks of medium contrast or small size	50–100
Performance of visual tasks of low contrast or very small size	100–200

Since visible alarm notification appliances must be seen to be effective, it is generally necessary to use more visible appliances than audible notification appliances. Reflective techniques can be used to reduce the number of devices.

Supplementary Visible Signals: *NFPA 72* recognizes a supplementary visible signal that is utilized to augment an audible or primary visible signal. The supplementary visible signal need not meet the light intensity recommendations of primary visible signals. One of the key design parameters is that the combined intensity of the primary and the supplementary notification appliances should not exceed the upper bounds for the primary visible signal.

PHYSICALLY OR MENTALLY CHALLENGED INDIVIDUALS

The physically or mentally challenged population includes those who are hearing and sight impaired, some elderly persons, people with mental deficiencies or incapacities, and so on. This population must be considered when audible or visible notification appliances are specified for a building.

The use of an encompassing pulsing audible or visible signal must be examined to determine whether it will cause distress or confusion to some occupants. These consequences are important since pulsing audible or visible signals are used to alert persons of potentially hazardous conditions. Furthermore, pulsing audible and visible signals can have harmful effects for persons with epilepsy.

Any potential negative effect on a physically or mentally challenged occupant due to notification appliances for fire alarm systems should be examined for appropriate modifications of the signals.

Americans with Disabilities Act of 1990

Federal Register rules and regulations for the 1990 Americans with Disabilities Act (ADA) delineate specific requirements for fire alarm systems pertinent to physically or mentally challenged occupants of a building.

The ADA requires both visual and audible alarm appliances for fire alarm systems. All occupancies are included in the law except those specifically excluded (medical care facilities, federal buildings, transportation facilities, etc.). Fire alarm systems are not required by the ADA, but once a fire alarm system is planned for a facility, ADA requirements must be incorporated into the new system.

ADA requires that audible alarms have sounding levels that are 15 dBA above ambient noise levels, or levels that are 5 dBA above the maximum sound level for a 60-second duration. The regulations establish maximum alarm levels at 120 dBA for audible signals. Audible emergency signals must have an intensity and frequency that can attract the attention of individuals who have partial hearing loss. The Act indicates that people over 60 years of age have difficulty perceiving frequencies higher than 10,000 Hz. The best appliances for the partially hearing impaired are single-stroke bells, high–low alarm signals, and fast-whoop signals. The regulations suggest signals with a sound characterized by three or four clear tones without compromising "noise" in between.

The ADA parameters differ from earlier criteria established in ANSI A117.1, *Accessible and Usable Buildings and Facilities,* which is an accessibility standard used currently in the public sector. The ADA requirements are significantly different from the most recent edition of ANSI A117.1 both in terms of light intensity and strobe light distribution. Another difference is that ANSI A117.1 has a minimum lamp intensity of 185 candelas, whereas ADA has a minimum lamp intensity of 75 candelas.

Besides differences with ANSI A117.1, the ADA requirements also differ from those in *NFPA 72.* These three documents have conflicts with regard to light levels for visual alarm appliances. These conflicts concern light output levels based upon room size. The three documents also differ in appliance maximum mounting heights and flashing rates for strobe lights, which have the potential for causing epileptic seizures. Some of the differences are illustrated in Table 7-4.

Table 7-4. Minimum Required Light Output in Candelas*

	NFPA 72			ANSI A117.1	ADA
Maximum Room Size (ft)	*Maximum Ceiling Height (10 ft)*	*Maximum Ceiling Height (20 ft)*	*Maximum Ceiling Height (30 ft)*	*Mounting Height 80-96 in.*	*Mounting Height 80 in. Maximum*
20 × 20	15	30	55	15	75
30 × 30	30	45	75	30	75
40 × 40	60	80	115	60	75
50 × 50	95	115	150	95	75

*This table is a partial extraction of the criteria for light levels from *NFPA 72, National Fire Alarm Code*; ANSI A117.1, *Accessible and Usable Buildings and Facilities*, 1992; and the Americans with Disabilities Act. For SI Units: 1 in. = 25.4 mm; 1 ft = 0.30 m.

The Act also requires visual alarm notification appliances to be integrated into building or facility fire alarm systems. In essence, the features and positions of visual alarm notification appliances are described in the Act as follows:

1. The lamp must be a xenon strobe type or equivalent.
2. The color must be clear or "normal" white.
3. The maximum pulse duration must be 0.2 second, with a maximum duty cycle of 40 percent. The pulse duration is defined as the time interval between initial and final points of 10 percent of maximum signal.
4. The intensity must be a minimum of 75 candelas.
5. The flash rate must be a minimum of 1 Hz and a maximum of 3 Hz.
6. The appliance must be placed 80 in. (2 m) above the highest floor level within the space or 6 in. (150 mm) below the ceiling, whichever is lower.
7. In general, any room or space required to have a visual notification appliance must have no spot more than 50 ft (15 m) from the appliance (in the horizontal plane). In large rooms and spaces exceeding 100 ft (30 m) across and without obstruction 6 ft (1.8 m) above the finish floor, such as auditoriums, appliances can be placed around the perimeter, spaced a maximum 100 ft (30 m) apart, in lieu of suspending appliances from the ceiling.
8. No place in common corridors or hallways in which visual alarm notification appliances are required is allowed to be more than 50 ft (15 m) from the signal.

Residential Alarm Appliances for the Hearing Impaired

In sleeping areas, where visual alarm notification appliances are placed, the signal must be visible in all areas of the unit. In addition, the visual alarm notification appliances must be connected in some way to the building emergency alarm system. For visual notification appliances to be effective for the hearing impaired, the appliances must be located and oriented so that they will spread the signals and reflect the light throughout the space or raise the light level sharply in the space.

In residential-type occupancies for the hearing impaired, Underwriters Laboratories tests and studies have determined that to awaken sleepers, flashing lights with the intensity of 110 candelas are needed. This is over seven times more intense than the normal 15-candela appliance. Another way to awaken sleepers is to use bed shakers to alert hearing-impaired sleepers of fire emergencies.

NOTIFICATION APPLIANCE TYPES

Audible Alarm Notification Appliances

Bells: Bells can be used for fire alarm notification if they have a distinctive sound and will not be confused with similar audible signals used for other purposes, such as school classroom bells. Bells can be of the single-stroke or vibrating type. Single-stroke bells are used to provide audible coded signals; continuously vibrating types are used primarily for noncoded, continuously ringing applications. Continuously vibrating bells also can be used to provide coded audible signals.

Bells can be provided with 4- through 12-in. (100–300-mm) gongs in 2-in. (50-mm) increments, although the 6- and 10-in. (150- and 250-mm) sizes are the most commonly used. Bells with 4-in. (100-mm) gongs are usually reserved for use as trouble signals. Generally, the larger the diameter of the gong, the lower is the frequency and the louder is the audible signal as expressed in decibels.

Bells are usually of the underdome type and can be mounted on standard conduit boxes. Special boxes and grilles are necessary where bells must be concealed, recessed, and/or mounted flush with the wall.

Horns: Horns can be used where louder and/or more distinctive signals are needed. Horns may require more operating power than bells, and circuits should be electrically compatible where the power source supplies both types of signals.

Horns are usually of the continuously vibrating type or electronic tone type and can provide either coded or noncoded audible alarm notification. They may be of the surface-, semiflush-, or flush-mounted type. In very noisy areas, resonating, air-powered, or motor-driven horns are sometimes used because of their inherently high-decibel output. Air-powered horns using valves are controlled by ac- or dc-powered electric solenoids. Either coded or noncoded operation may be provided.

Motor-driven horns are not practical for providing coded output signals and are more widely used for continuous signals. They are particularly effective under "rolling" conditions where power is periodically applied and removed to vary the motor speed and the sound pitch.

Chimes, Buzzers, and Sirens: Chimes are soft-toned appliances that are used where panic or other undesirable actions might result from the use of loud audible alarm notification appliances. Their use is especially compatible with such areas as nurses' stations in hospitals, where only authorized personnel are alerted to a fire emergency (private mode signaling). Buzzers are generally used for trouble signals rather than fire alarm signals and are intended to provide a continuous sound. They are seldom used for coded signals. Sirens are usually motor-driven or electronic appliances and are limited to outdoor applications, but are sometimes used in extremely noisy indoor areas.

Speakers: Speakers reproduce electronic signals and can be made to sound like any mechanical notification appliance. They are capable of reproducing unique sounds that are not practical on mechanical appliances. They can also be used to provide live or recorded voice instructions. Speakers are either of the direct-radiating-cone type or the compression-driver-and-horn type, operated from audio amplifiers delivering standard output line levels of 70.7 or 25-V rms. The speakers are driven by an electronic tone generator, microphone, tape player, or voice synthesizer and electronic amplifier.

The two most widely used speakers are: the integral type, where the tone generator, amplifier, and speaker are enclosed in common housing; and the remote type, where the speaker is energized from a remotely located generator, amplifier, and so on.

Visible Notification Appliances

Visible notification appliances may be used alone or in conjunction with audible alarm notification appliances.

Visible alarm notification appliances are effective in high-noise areas, to alert a hearing-impaired occupant, or where attention directed to the light unit can provide additional information. For most applications, visible alarm notification appliances are most effective when used in a flashing mode. Flashing can be caused by periodically interrupting current to the appliance, rotating a beam of light (as is done on many emergency vehicles), using a strobe technique, or using any combination of these methods.

Examples of visible alarm notification appliances are lamp annunciators and strobe lights. Annunciators can be used with either coded or zoned noncoded systems to indicate the initiating device or zone from which the alarm originated.

Lamp Annunciators: In lamp annunciators, the zones or areas from which alarms originate are indicated by the lighting of lamps or light-emitting diodes (LEDs). Lamp annunciators can be provided with different-colored lenses (usually red for alarm, amber for trouble, and green or white for power-on indications).

Backlighted annunciators are used to light up information on a window or to illuminate a graphic annunciator. Graphic annunciators have lamps specified in a map or floor plan to provide an easier means of identifying the particular area or zone affected. The area from which the alarm is initiated can be indicated by a bull's-eye-type lamp located within the area or by backlighting the entire affected area or zone.

Liquid Crystal Displays (LCDs): In addressable and addressable analog systems, annunciation is indicated by alarm point or zone using an English-language LCD that allows a detailed description (usually limited to 40 characters) of the point in alarm.

Strobe Lights: Strobe lights operate on the energy discharge principle to produce a high-intensity, short-duration flash. These lights are very efficient, producing, for example, 8,000 candlepower for 1 W of input power. The short, bright flash is not only attention-getting, but effective with low general visibility. Strobe appliances come in a wide range of light intensities. Repetition rates are regulated by the NFPA, the UL, and the requirements of the ADA.

Combination Audible and Visible Notification Appliances

Audible and visible functions can be combined to produce both sound and light from a single appliance. For example, the audible notification appliance can be a horn, a bell, or a speaker. The visible portion of the notification appliance is a strobe with the appropriate candela rating for the application.

Permanent and Print Recorders

Permanent recorders are used on fire alarm systems to indicate the location of an alarm and to provide a permanent record of the time and date an alarm occurred. A recorder that creates a hard-copy record of the alarm location performs a similar function as an annunciator.

In many systems, the recorder can be used to record a trouble condition plus other events that are related to the receipt of an alarm. Examples of other events recorded are alarm notification appliance circuit energization, fan shutdown, door closing, or other life safety functions that may have been initiated automatically or manually.

Electronic print recorders have replaced old punch-type recorders almost entirely in modern fire alarm systems. Print recorders range from small hand-size units (similar to those found on small printing calculators) to large, sophisticated, high-speed printers found on major computer systems.

Textual Notification Appliances

Textual audible notification appliances are similar for the public and private modes and include loudspeaker, telephone, and radiofrequency appliances (radio telephone). Loudspeaker notification appliance output should be in accordance with the sound output levels previously defined for audible notification appliances. Since telephone appliances have extensive nonfire alarm system usage, existing definitions for these appliances, issued by the Federal Communications Commission and the Electronic Industries Association, are usually appropriate. *NFPA 72* specifies requirements for both loudspeaker and telephone appliances.

The textual visible notification appliance operates mostly in the private mode and consists of pictorial and alphanumeric displays that can be permanent, with a minimum retrieval time of 1 year, or temporary. Where the textual visible notification appliance operates in the public mode, for example, an annunciator in a building lobby for the fire service, additional requirements are needed to provide operation during fire conditions.

EMERGENCY VOICE/ALARM COMMUNICATIONS SYSTEMS

Fire alarm notification is increasingly found in the form of an emergency voice/alarm communications system that is added to the building fire alarm system. Emergency voice/alarm communications systems provide dedicated manual and automatic facilities for fire location and control and transmission of information and instructions concerning a fire alarm emergency to the building occupants as well as fire department personnel.

Voice/Alarm Communications Service

Voice/alarm communications service provides an automatic response upon receipt of a signal that indicates a fire emergency. Subsequent manual control capability of the transmission and audible reproduction of evacuation tone, signals, preannounce alert tone signals, and voice directions on a selective and all-call basis (as determined by the authority having jurisdiction) should also be located at the fire command station.

Automatic response is not required where the fire command station or a remote monitoring location is constantly attended by trained operators and operator acknowledgment of receipt of a fire alarm signal is received within 3 minutes.

Emergency messages should be provided in the language of the predominant building population. Where there is a probability of isolated groups that do not speak the predominant language, multilingual messages should be provided.

Multichannel Capability

The system should allow the application of an evacuation signal to one or more zones and simultaneously permit voice paging to the other zones selectively or in any combination.

Functional Sequence

After an alarm signal is initiated, the system should automatically transmit either:

1. An alert tone of 3 to 10 seconds duration followed by a message (or messages when multichannel capability is provided). The message should be given at least three times and direct the occupants of the fire initiation zone and other zones in accordance with the building's fire evacuation plan.
2. An evacuation signal in accordance with the building's fire evacuation plan.

These signals should be transmitted immediately, or after a delay allowed by the authority having jurisdiction. Failure of the message above should sound the evacuation signal automatically. Provisions for manual initiation of voice instructions or evacuation signal generation should also be provided. Live voice instructions should override any previously initiated signals.

Voice and Tone Devices

The alert tone preceding any message can be part of the voice message or transmitted automatically from a separate tone generator. When audio amplifiers and signal generators are provided, their output should be continuously monitored so an audible trouble signal results on failure.

The generating and amplifying equipment enclosed as integral parts and serving only a single listed loudspeaker need not be monitored. It is recommended to use backup amplifying and evacuation signal generating equipment with an automatic transfer feature so that, in the event of primary equipment failure, prompt restoration of service is ensured.

Voice/Alarm Communication System Components

Components of an emergency voice/alarm communication system typically include the following:

Main Fire Alarm Panel: This system panel controls more functions, operations, inputs, and outputs than a traditional fire alarm panel. It typically has the following features:

1. An automatic alert tone with a preprogrammed sequence that can be transmitted throughout the building or to any selected zone
2. A prerecorded evacuation message that is transmitted to the designated evacuation zone
3. Two independent voice communication channels to operate the paging and alert tone systems; these channels include one-way voice communication capability to all areas ("all call") or zoned operation
4. A terminal for the two-way fire department telephone

Primary Annunciator Panel: The annunciator, which is usually integrated with the control panel, can be a series of designated incandescent lights or LEDs, a graphic annunciator showing a building diagram, a CRT with a building diagram, or a video display such as a slide projector system. At a minimum, the annunciator panel should show the floor, zone, and type of alarm. If the fire command station is integrated with other building systems (e.g., security, building controls) and is remote from the fire department response point, a separate annunciator panel for the voice communication system should be provided at the primary fire department response point. Additional annunciators may be necessary in larger buildings and should be located at secondary fire department response points.

Remote Annunciators: These units generally do not have the same degree of detail as the primary annunciator panel and are provided as needed based on building size.

Fire Department Two-Way Telephone Systems: The terminal for this system is located at the fire alarm command station. Remote stations can be jacks into which portable handsets are inserted or fixed (surface- or flush-mounted telephone stations).

Remote Sounding Appliances: These consist of flush- or surface-mounted speakers, combination speaker/flashing strobe unit, or reentry horns. The fire alarm alert tone and voice messages are generated from the fire alarm control panel.

Input Devices: Manual fire alarm boxes, smoke and heat detectors, waterflow switches, and supervisory devices are typical input devices.

Peripheral Equipment and Controls: These include elevator control, smoke control, and door-unlocking control systems, which usually incorporate annunciators, override switches, and status indicators.

Voice Communication Messages and Signals

There is no standardization of emergency voice/alarm communication messages or signals. Different vendors offer different types of signals, although all evacuation tones or signals can be generated in the Temporal Code-3 pattern.

A voice message following an alert tone might be based on the following:

> May I have your attention please. There has been a fire reported in the building, do not use the elevators. Please evacuate your floor and proceed to the lobby area for further instruction.

The message tells the occupants what is happening and what they are to do next. Each voice message, whether live or prerecorded, should be clear, calm, and informative and presented so it will motivate people into action. Voice messages should differ for certain parts of the building and certain stages of occupant evacuation.

Messages to the affected areas should follow a similar pattern. When a fire is reported on any floor, several other messages need to be transmitted once the elevator message has been sent. Most importantly, the occupants of the fire floor need to be told of the situation and instructed on actions to take. The adjacent floors need to be cleared, and these occupants must also be given instructions. Finally, a message must go to the receiving floors to which these occupants will be sent.

When using voice messages to provide alarm and relocation information to the building occupants, the messages and instructions must be intelligible to be effective. Voice intelligibility should be measured in accordance with the guidelines in Annex A of IEC 60849, *Sound Systems for Emergency Purposes.*[3] The designer of an intelligible voice/alarm system should possess skills sufficient to properly design a voice/alarm system for the occupancy protected. Annex A of *NFPA 72* offers additional information on this subject.

Loudspeakers

Loudspeakers and their enclosures used in emergency voice/alarm communication service must be listed for such service.

One or more loudspeakers should be located on each floor of the building and placed so their signals will be heard clearly, regardless of the maximum noise level from machinery or other equipment under normal occupancy conditions. Loudspeakers should also be provided for each enclosed stairway in the building, with these speakers connected to separate paging zones.

Fire Command Station

A fire command station should be located near a building entrance or other suitable location. The fire command station should provide a communication center for the arriving fire department and provide for control and display of the status of detection, alarm, and communications systems. The command station may be physically combined with other building operations and security centers as permitted by the local authority. Operating controls for fire department use should be clearly marked. A fire command station should control the emergency voice/alarm communication service and the two-way telephone communication service where provided.

Monitoring Integrity of Conductors and Survivability

Because of its life safety applications, an emergency voice/alarm communication system must be in good working order at all times. Certain system conductors should be monitored and the system designed for survivability in a fire emergency.

The following interconnecting conductors should be monitored: all means of interconnecting equipment, devices, and appliances, including standby battery and two-way telephone communication service equipment when provided. The occurrence of any single ground or open in the installation conductors should be automatically signaled to and received by the building fire command station as a trouble signal within 200 seconds.

An audible trouble signal should sound when a short or open in loudspeaker installation wiring occurs. For an emergency system and where a two-way telephone communication circuit is provided, installation wires should be monitored for a short-circuit fault that could render the telephone communication circuit inoperative.

Survivability is defined as a system designed and installed so that attack by fire in an evacuation zone (causing loss of communication to this evacuation zone) should not result in loss of communication to any other evacuation zone. The system also should be so designed and installed that attack by fire causing failure of equipment or a fault on one or more installation wiring conductors of one communication path should not result in total loss of communication to any evacuation zone. Exceptions to survivability requirements can be found in *NFPA 72*.

Combination Systems

For combination systems that use common wiring or equipment for combination systems, equipment other than that for emergency voice/alarm communication systems and fire alarm systems should be connected to the common wiring of the systems so that short or open circuits or faults in this equipment or between this equipment and the common wiring do not interfere with the supervision of the fire systems or prevent alarm or supervisory signal transmission for the fire systems.

In combination systems, a fire alarm signal should be clearly recognizable and take precedence over any other signal, even if a nonfire alarm signal is initiated first. Distinctive alarm signals must be obtained between fire alarm and other functions.

Two-Way Telephone Communication Service

A two-way telephone communication service is often provided in buildings for use in emergencies. When provided, two-way telephone communication service should be available for use by the fire service.

The service can also be used for notification and communication for a building fire warden organization, notification and communication for reporting a fire and other emergencies (e.g., voice call box service), notification and communication for guard service, and other uses. Any variations in equipment and systems operation that create additional use of the two-way telephone communication service must not adversely affect its performance when used by the fire service.

Two-way telephone communication service should be capable of permitting simultaneous operation of any five telephone stations in a common talk mode.

Off-Hook Indicator: There should be a distinctive notification signal that indicates the off-hook condition of a calling telephone circuit at the fire command station. When a selective talk telephone system is supplied, there should be a distinct visible indicator for each selectable circuit so all off-hook conditions are continuously and visibly indicated.

Fire Service Use: There should be at the minimum a common talk circuit, that is, a conference or a party line, for fire service use. The minimum requirement for fire warden use, where provided, should be a selective talking system controlled at the fire command station. Either system should be capable of operating with five telephone stations connected together. There should be a minimum of one fire service telephone station or jack per floor and a minimum of one per exit stairway and, where provided, at least one fire warden station or jack to serve each fire paging zone.

EMERGENCY CONTROLS

Emergency control or actuation of other systems in the building, for example, air-handling systems, extinguishing systems, is another form of signal notification. This function is usually required either by the local code and/or local authority.

NFPA *101* sets certain requirements for emergency controls. A fire alarm system should be arranged to automatically actuate control functions necessary to make the protected premises safer for building occupants.

The fire alarm system can actuate the following emergency controls:

1. Elevator capture and control, usually in accordance with ASME/ANSI Al7.1, *Safety Code for Elevators and Escalators*
2. Release of automatic door closers
3. Stairwell or elevator shaft pressurization
4. Smoke management or smoke control systems
5. Initiation of automatic fire-extinguishing equipment
6. Emergency lighting control
7. Unlocking of doors
8. Emergency shutoff for gas and fuel supplies that may be hazardous, provided continuation of service is not essential to life safety

Performance of emergency control functions must not impair the effective response of all required alarm notification functions. It is required by NFPA *101* that wiring to required emergency control devices be monitored for integrity (electrically supervised) to within 3 ft (0.9 m) of the device to be actuated.

Location

Operator controls, visual alarm annunciators, and manual communications capability for the emergency control function should be installed in a control center at a convenient location. Controls used solely by the fire department should be located adjacent to an entrance at a fire command station or as designated by the authority having jurisdiction.

Automatic detection should be provided for the central controls when controls are not in continuously occupied areas. A detector should be provided for each central control location.

Elevator Recall

Any smoke detectors used in elevator lobbies and to recall elevators for fire-fighter service should be connected to the building fire alarm system. An elevator lobby smoke detector should be capable of initiating an alarm when all other devices on the same alarm-initiating circuit have been manually or automatically placed in the alarm condition. (Elevator recall can also be provided by the fire alarm control equipment.)

When actuated, each elevator lobby smoke detector or smoke zone should initiate an alarm condition on the building fire alarm system. The alarm initiation circuit or zone from which the alarm originated should be indicated at the control panel and required remote annunciators. Elevator lobby smoke detectors, as with all automatic fire detectors, should be tested regularly.

Smoke detectors for elevator recall should be installed in the elevator lobbies and elevator machine room. Even if facilities are not equipped with a fire alarm system, elevator recalls are required to be controlled with a smoke detection system.

ANSI A17.1, *Safety Code for Elevators and Escalators,* which is referenced by NFPA *101,* states that smoke detectors for each group of elevators serviced by the same call button should be provided with two elevator detection circuits. These circuits extend to the elevator machine room, where the elevator controller is located. Typically, the circuits use normally closed dry contacts. In the event of an electrical failure of the circuit, the elevator recall is activated upon a loss of signal.

Detection circuits should be required in lobbies, hoistways, and machine rooms. Upon activation of one of the detectors on this circuit, the elevator responds to a designated floor that is specified by the authority having jurisdiction to be a safe egress level. An additional smoke detection circuit covers the elevator lobby designated recall level. When this smoke detector activates, the elevator responds to an alternative recall level, as determined by the authority having jurisdiction. Usually, the hoistway smoke detectors are also connected to this second detection circuit.

Activated smoke detectors recall all elevators to the designated recall level, opening all elevator doors, while all call buttons become inoperative. These elevators then become available to the fire department personnel for evacuation and fire-fighting functions.

Release of Automatic Door Closers

A door designed to be kept normally closed in a means of egress, such as a door to a stair enclosure or a horizontal exit, should be a self-closing door and should not at any time be secured in the open position.

Circumstances where alternatives to self-closing doors can be used are found in NFPA *101*, and are permitted by this code for buildings with low- or ordinary-hazard contents or when permitted by local authorities. Doors can be released by an automatic smoke detection system; an automatic sprinkler system can activate the door closing, but does not substitute for smoke detector actuation.

Activation of Mechanical Ventilation Systems

Both mechanical ventilation and pressurized stair enclosure systems can be initiated by a smoke detector installed in an approved location within 10 ft (3 m) of the entrance to the smokeproof enclosure. The system can also be activated by manual controls accessible to the fire department and waterflow signal from a complete automatic sprinkler system or a general evacuation alarm signal. Activation of the closing device in any door should activate the closing devices on all doors in the smokeproof enclosure at all levels.

Smoke Management

Each air-transfer opening or duct penetration of a required smoke barrier should have an approved damper designed to resist the passage of smoke. NFPA *101* permits exceptions to provisions for smoke dampers; they may be omitted

1. In ducts or air-transfer openings that are part of an engineered smoke control system
2. In ducts where the air continues to move and the installed air-handling system prevents recirculation of exhaust or return air under fire emergency conditions
3. Where ducts penetrate floors that serve as smoke barriers

NFPA *101* also allows smoke dampers to be omitted in other circumstances.

Smoke dampers in ducts penetrating smoke barriers should close upon detection of smoke by the facility's smoke detectors. They should also close on detection of smoke by local smoke detectors on either side of the smoke barrier door opening (when ducts penetrate smoke barriers above the smoke barrier doors) or smoke detectors located within the ducts in existing installations. Details on smoke management are given in *NFPA 72;* NFPA 90A, *Standard for the Installation of Air-Conditioning and Ventilating Systems;* and NFPA *101.*

Automatic Shutdown

In systems of more than 2000 cu ft/min (944 L/sec) capacity, listed smoke detectors should be installed to detect the presence of smoke. The detectors should automatically stop the fan(s) in the supply system downstream of the filters, in the return system on each floor at the point of entry into the common return, or in a system of smoke detectors providing total area coverage. A smoke detector in the return air stream is not essential in systems of less than 15,000 cu ft/min (7,080 L/sec) capacity.

Detectors in the supply and return air stream may also be omitted, provided (1) the system is of less than 15,000 cu ft/min (7,080 L/sec) capacity, (2) the entire system is within the space served, and (3) such space is protected by an area smoke detection system.

Smoke Dampers

Smoke dampers should be installed in systems of over 15,000 cu ft/min (7,080 L/sec) capacity to isolate the air-handling equipment (including filters) from the remainder of the system. This restricts circulation of smoke and is arranged to close automatically when the system is not operating.

The protection provided by the installation of smoke detectors and automatic shutdown is intended to prevent the dispersion of smoke through the supply air duct system and, preferably, to exhaust a significant quantity of smoke to the outside. However, these functions will not guarantee early detection of fire or the detection of smoke concentrations prior to dangerous smoke conditions if smoke moves other than through the supply air system. Where facility smoke control protection is needed, a specifically designed smoke control system should be provided that includes placement of smoke detectors so that early detection of developing smoke conditions and initiation of the smoke control measures are provided.

Ideally, actuation of any duct smoke detector should sound either a fire alarm or supervisory signal and provide for control of the ventilation systems. *NFPA 72* provides requirements for installation and connections. Activation of air duct smoke detectors installed in a building that is not equipped with a building fire alarm system should also sound an audible alarm in a normally occupied area. A trouble condition should be indicated visibly or audibly in a normally occupied area and identified as air duct detector trouble.

Automatic Fire-Extinguishing Equipment

The potential fire loss of a facility may require a certain type of automatic extinguishing system in addition to an automatic sprinkler system. Examples of other types of automatic extinguishing systems are carbon dioxide, dry chemical, foam, gaseous suppression, and water spray systems.

Emergency Lighting

Emergency lighting should be provided in the stair shaft and vestibule. Batteries or a standby generator installed for the smokeproof enclosure's mechanical ventilation equipment may be used as the power supply.

Unlocking of Doors

Stairwell doors should allow reentry from the stairwell to the interior of the building, or an automatic release should be provided to unlock all stairwell doors. Such automatic release should be actuated with the initiation of the building fire alarm system.

Special Locking Arrangements

For buildings with a supervised automatic fire detection system or sprinkler system, doors in low- and ordinary-hazard areas can be equipped with approved listed locking devices. The devices should:

1. Unlock upon actuation of a supervised automatic fire detection system or sprinkler system
2. Unlock upon loss of power controlling the lock or locking mechanism
3. Release the lock within 15 seconds whenever a maximum force of 15 lb (67 N) is continuously applied to the release device for not more than 3 seconds. Such doors should be relocked manually. An operation signal in the vicinity of the door should activate when the lock releases

A sign on the door adjacent to the release device should be provided, stating "Push until alarm sounds. Door can be opened in 15 seconds." Emergency lighting should also be provided at the door.

Emergency Shutoff

Emergency shutoff of fuel supplies is usually achieved with a solenoid-operated valve.

SIGNAL NOTIFICATION IN HIGH-RISE BUILDINGS

Delayed detection of fire is a significant problem in high-rise buildings. Early discovery of fire is important to ensure prompt evacuation, but total evacuation within reasonable time limits is not always possible in high-rise buildings. Exits are not designed to handle the total occupant load of a high-rise building at one time. A 44-in. (1.12-m) stairway will discharge about 140 persons per minute, so only 140 persons can enter the stairway per minute once the stairway has been filled with people. Most building codes do not require the width of exits to increase during descent toward the stair discharge. In analyzing egress in high-rise buildings, it becomes obvious that the total evacuation of buildings over a certain height [generally 150 ft (45 m) or 12 stories] becomes unreasonable.

Further, while occupants may be trying to evacuate the building, fire fight-ers may need to use the stairs at the same time to reach the fire floor. For this and other reasons, partial evacuation is usually a more realistic approach than complete evacuation in high-rise buildings.

There are two options that can be used to alert occupants of a fire in a high-rise building: total and selective. Total building-alerting capability should be provided if it is desired to alert all occupants. For selective evacuation, the fire alarm system designer and code officials should decide the number of occupants to be alerted and the method to be used. The fire area should also be identified in some manner, since fire fighters cannot effectively search an entire building for the fire site. The fire department and building management should be notified automatically when the alarm system is activated.

Design of an emergency voice/alarm communication system in a high-rise occupancy is based on occupant load, occupant physical characteristics, active/passive protection systems, manual fire-fighting capability, and fire depart-ment access. Any voice/alarm system for a high-rise building should be capable of

1. Notifying the fire command station of the existence and location of the fire
2. Notifying the fire department of the existence of a fire
3. Directing occupants of the fire zone to the outside or to an area of refuge
4. Delaying transmission of alarms outside the immediate fire area until an authority in charge considers it appropriate to sound alarms to other areas
5. Operating in a simple and reliable manner
6. Monitoring essential interconnecting equipment and components of the system
7. Allowing a paging zone to survive the fire loss of another paging zone

The audible fire alarm signal to the occupants of the fire zone can include continuous sounding devices, voice direction that momentarily silences the con-tinuous sounding devices, or both.

A message is typically sent to the fire area and floors above and below the fire area. Facilities should be provided at the fire command station so voice and sounding appliance alarms can be manually activated in any combination of zones. The voice system should be capable of transmitting to all zones simultaneously; toward this end, the emergency voice system should not be part of the same system as the building music/paging systems. Regardless of other signals or sounds, for example, music or announcements, on the system, the fire calls should automatically override all other system traffic and sound at maximum volume. The volume of a fire call should comfortably exceed the volume of any other signal or system sounds.

If the fire command station is integrated with other building systems (e.g., security and building controls) and is remote from the fire department response point, a separate annunciator panel for the voice/communication system should be provided at the primary fire department response point. Additional annunciators may be necessary in larger buildings at secondary fire department response points.

QUESTIONS

1. The fire alarm system designer must ensure that any audible and/or visible notification signal and the appliances used to sustain them are:

   a. all of the same type
   b. located equally in the area
   c. appropriate for the facility and the occupants protected
   d. spaced in accordance with their listing

2. Audible notification appliances operated in the public mode should be located so the sound pressure level is not less than:

   a. the ambient noise level   b. 10 dBA above ambient
   c. 45 dBA   d. 15 dBA above ambient

3. "Survivability" is required:

   a. for all wiring (cables) in a voice/alarm system
   b. to ensure that loss of the speaker circuit in one evacuation zone will not affect other evacuation zones
   c. to make sure all speakers continue to operate in a fire
   d. in place of Class A wiring

4. Two of the four main parameters required for the location of audible fire alarm devices are:

   a. sound characteristics of the floor, walls, and ceiling and the sound pressure level of the device (based upon manufacturer's data)
   b. audibility and frequency

c. size of the room and carpeting type

d. none of the above

5. Emergency voice/alarm communication systems are:

a. expensive and difficult to install

b. used exclusively in high-rise buildings

c. used whenever selective notification and relocation of occupants is a design goal

d. only used in hospitals for private mode signaling

BIBLIOGRAPHY

NFPA Codes, Standards, Recommended Practices, and Manuals (see the latest *NFPA Catalog* for availability of current editions of the following documents):

NFPA 72®, *National Fire Alarm Code®*.

NFPA 90A, *Standard for the Installation of Air-Conditioning and Ventilating Systems.*

NFPA *101®*, *Life Safety Code®*.

REFERENCES CITED

1. Schifiliti, R. P., "Design of Detection Systems," *The SFPE Handbook of Fire Protection Engineering,* 1st ed., Chapter 3-1, NFPA, Quincy, MA, 1988.
2. Butler, H., Bowyer, A., and Kew, J., *Locating Fire Alarm Sounders for Audibility,* Building Services Research and Information Association, Bracknell, UK, 1981.
3. IEC 60849, *Sound Systems for Emergency Purposes;* 2nd edition, IEC, Geneva, 1998.

ADDITIONAL READING

ANSI A17.1, *Safety Code for Elevators and Escalators,* American National Standards Institute, New York, 2000.

ANSI A117.1, *Accessible and Usable Buildings and Facilities,* American National Standards Institute, New York, 1992.

ASME/ANSI Al7.1, *Safety Code for Elevators and Escalators,* American Society for Mechanical Engineers, New York, 2000.

8

Fire Alarm System Installation

INTRODUCTION

Proper installation of a fire alarm system is obviously a prime factor in subsequent system efficiency, reliability, and correct operation. Installation also affects any supplemental applications the designer may specify for the system. Listed equipment, adequate power supplies, supervised circuits, trouble signals, and necessary audible alarms should all be considered in the installation process.

This chapter covers installation of automatic fire detectors and special applications of detectors to control smoke spread, release doors, and actuate various extinguishing systems. Circuit installation and wiring for a fire alarm system are also described, and protection of a system from transients and lightning is discussed.

For additional information, see Chapter 4, Fire Alarm Systems Components and Circuits.

INSTALLATION OF AUTOMATIC FIRE DETECTORS

Detectors should be protected where subject to mechanical damage. They should be supported in all cases independent of their attachment to the circuit conductors. Detectors should not be recessed in any way into the mounting surface unless they have been tested and listed for such recessed mounting.

Detectors should be installed in all areas where required by the appropriate standard in effect for the facility. Most building codes require only partial protection. Partial and complete smoke detection is defined in NFPA *101*®, *Life Safety Code*®, as follows:

> Complete Smoke Detection Systems. Where a complete smoke detection system is required by another section of this Code, automatic detection

of smoke in accordance with *NFPA 72®, National Fire Alarm Code®*, shall be provided in all occupiable areas, common areas, and work spaces in those environments suitable for proper smoke detector operation.

Partial Smoke Detection Systems. Where a partial smoke detection system is required by another section of this Code, automatic detection of smoke in accordance with *NFPA 72®, National Fire Alarm Code®*, shall be provided in all common areas and work spaces, such as corridors, lobbies, storage rooms, equipment rooms, and other tenantless spaces in those environments suitable for proper smoke detector operation. Selective smoke detection unique to other sections of this Code shall be provided as required by those sections.

Where total (complete) coverage is required or desired, *NFPA 72* states that detectors should be installed in all rooms, halls, storage areas, basements, attics, lofts, spaces above suspended ceilings, and other subdivisions and accessible spaces and inside all closets, elevator shafts, enclosed stairways, dumbwaiter shafts, and chutes. Inaccessible areas that contain combustible material should be made accessible and protected by detector(s).

Detectors can be omitted from combustible blind spaces under any of the following conditions:

1. When the ceiling is attached directly to the underside of the supporting beams of a combustible roof or floor deck.
2. When the concealed space is entirely filled with noncombustible insulation. In solid joist construction, the insulation need only fill the space from the ceiling to the bottom edge of the joist of the roof or floor deck.
3. When there are small concealed spaces over rooms, provided any space in question does not exceed 50 sq ft (4.6 m^2) in area.
4. In spaces formed by sets of facing studs or solid joists in walls, floors or ceilings where the distance between the facing studs or solid joists is less than 6 in. (150 mm).

Similarly, detectors can be omitted from below open-grid ceilings when (1) grid openings are ¼ in. (6.4 mm) or larger in the least dimension, (2) the thickness of the material does not exceed the least dimension, and (3) the openings constitute at least 70 percent of the area of the ceiling material.

Detectors can be required under large benches, shelves or tables and inside cupboards or other enclosures. Detectors should also be installed underneath open loading docks or platforms and their covers and for accessible underfloor spaces of buildings without basements.

Detectors can be omitted when *all* of the following conditions prevail:

1. The space is not accessible for storage purposes or entrance of unauthorized persons and is protected against accumulation of windborne debri
2. The space contains no equipment such as steam pipes, electric wiring, shafting, or conveyors
3. The floor over the space is tight
4. No flammable liquids are processed, handled, or stored on the above floor

Connections and Wiring

Duplicate terminals or leads (or the equivalent) should be provided on each automatic fire detector solely to connect into the fire alarm system and provide supervision of the connections. Such terminals or leads are necessary to allow the wire run to be broken and connections to be made to the incoming and outgoing leads or terminals. Detectors that provide the equivalent supervision are excepted (see Figures 8-1 through 8-5).

If looped wire connections or unbroken wire runs (secured in a notch) are used and a circuit wire disengages, the detector is rendered inoperative and a trouble signal will not sound (see Figure 8-2).

Figure 8-6 shows an incorrect method of wiring; if the conductor from the device to the wire nut becomes disconnected, supervision is still maintained. No indication of trouble would be indicated at the fire alarm panel, even though the device was no longer connected. When the device is wired as shown in Figure 8-7, a trouble signal will sound when any of the conductors is disconnected.

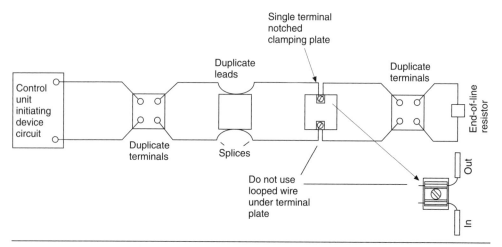

Figure 8-1. Correct wiring method for two-wire detectors and audible appliances.

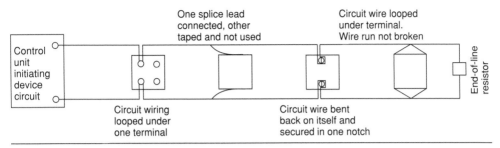

Figure 8-2. Incorrect wiring method for two-wire detectors and audible appliances.

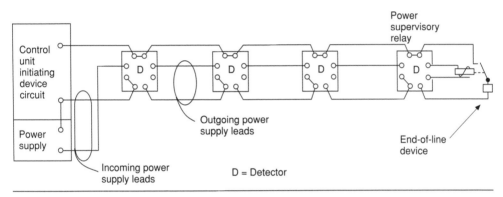

Figure 8-3. Correct wiring method for three-wire connections.

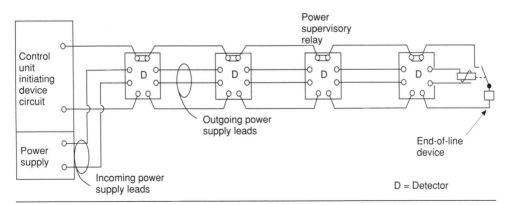

Figure 8-4. Correct wiring method for four-wire connections.

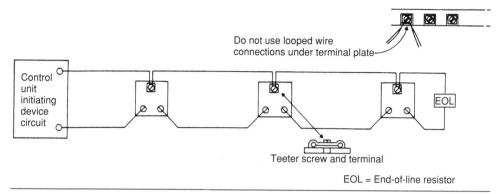

Figure 8-5. Correct wiring method for two-wire detectors.

A multiriser initiating or indicating circuit must also be installed correctly so supervision is maintained (see Figures 8-8 and 8-9).

SPECIAL APPLICATIONS FOR AUTOMATIC FIRE DETECTORS

Fire detectors (smoke detectors) should be properly located and spaced according to the local code. When automatic fire detectors are used in special applications, placement and spacing is often modified from the usual ceiling spacing. This places the detector in the path of the heat or smoke. Detector spacing for special applications is more fully described in *NFPA 72*. To minimize dust

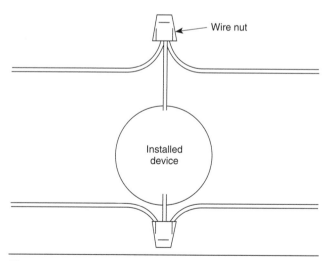

Figure 8-6. Incorrect wiring method for pigtail connections.

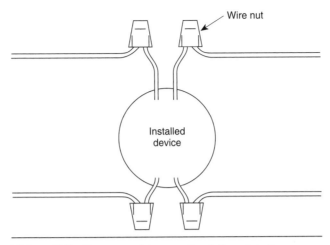

Figure 8-7. Correct wiring method for pigtail connections.

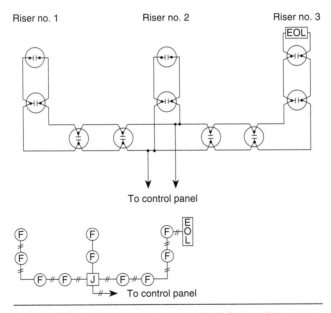

Figure 8-8. Incorrect wiring method for multiriser circuit. F, alarm-notification appliance; J, Junction box; EOL, End-of-line resistor.

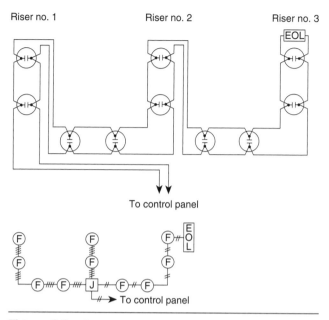

Figure 8-9. Correct wiring method for multiriser circuit. F, alarm-indicating appliance; J, Junction box; EOL, End-of-line resistor.

contamination of smoke detectors when they are installed under raised room floors and similar spaces, the detectors should be mounted only in an orientation for which they have been listed (see Figure 8-10).

High Rack Storage

Special care must be taken when installing detection systems in high rack storage areas since the potential fire load makes early detection of fire essential. Detection systems are often installed in addition to suppression systems in high rack storage. Where smoke detectors are installed for early warning in high rack storage areas, the designer should consider installing detectors at several levels in the racks to ensure quicker response to smoke. Detectors can be installed to actuate a suppression system (NFPA 231C, *Standard for Rack Storage of Materials*, contains more information on detector actuation of a suppression system in rack storage).

For the most effective detection of fire in high rack storage areas, detectors should be located on the ceiling above each aisle and at intermediate levels in the racks. This is necessary to detect smoke that may be trapped in the racks at an early stage of fire development (when insufficient thermal energy is released

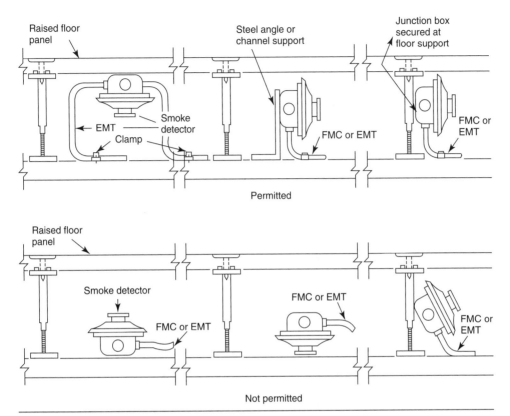

Figure 8-10. Mounting installations for smoke detectors under raised floors. EMT, Electrical Metallic Tubing; FMC, Flexible Metallic Cable.

to carry the smoke to the ceiling). Earliest detection of smoke can be achieved by locating the intermediate-level detectors adjacent to alternate pallet sections as shown in Figures 8-11 and 8-12.

Detector placement in closed rack storage protects solid storage in which transverse and longitudinal flue spaces are irregular or nonexistent (as for slatted or solid shelved storage). Detectors in open rack storage protect palletized storage or no shelved storage in which regular transverse and longitudinal flue spaces are maintained. Placement of detectors should be according to detector manufacturer's instructions.

An alternative to spot-type detection in racks is the use of active air-sampling smoke detectors (sometimes called aspirating-type detectors). Instead of a row of detectors, a single pipe can be installed at various levels of the rack. This installation greatly reduces the maintenance of the system.

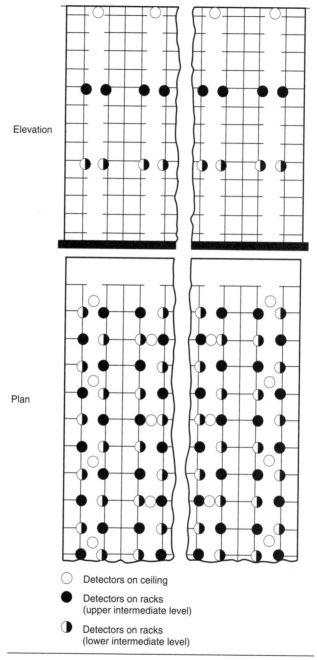

Elevation

Plan

○ Detectors on ceiling

● Detectors on racks
   (upper intermediate level)

◑ Detectors on racks
   (lower intermediate level)

Figure 8-11. Spot-type detector placement in closed rack storage.

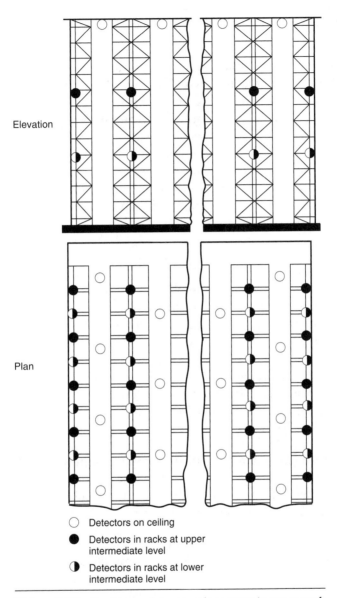

Elevation

Plan

○   Detectors on ceiling

●   Detectors in racks at upper
    intermediate level

◑   Detectors in racks at lower
    intermediate level

Figure 8-12. Spot-type detector placement in open rack
storage.

Smoke Detectors for Control of Smoke Spread

Smoke detectors can be used to prevent smoke spread by initiating control of fans, dampers, doors, and other equipment. Detectors for this use can be classified in two ways: area detectors, which are installed in the related smoke compartments; and detectors that are installed in the air duct systems.

Smoke-activated devices are also used to automatically close smoke doors in buildings to limit the spread of smoke in case of fire. Separate corridor ceiling-mounted smoke detectors can be connected to electrically operated hold-open devices on the doors. Smoke detectors can also be built into the door closure units.

Area smoke detectors that are located within a smoke compartment for open-area coverage can initiate control of smoke spread by operating doors, dampers, and other equipment.

Smoke Detectors in the Air Duct System

Air duct smoke detectors are installed either in or on return air ducts or at the return air openings of HVAC systems in buildings. This prevents recirculation of smoke from a fire through the HVAC system within the building. Upon detection of a fire, the associated control system initiates an alarm and either shuts down the circulating blowers or switches them to a smoke exhaust mode. Detectors installed in air duct systems should not be considered substitutes for open-area protection because smoke-laden air can be diluted by clean air from other parts of the building and the smoke may not be drawn into the HVAC system when the system is shut down.

Supply Air System: If detection of smoke in the supply air system is required, the detector(s) listed for the air velocity present can be located in the supply air duct downstream of both the fan and the filters. Smoke detectors should be located within all the smoke compartments served by the supply air system.

Return Air System: If the detection of smoke in the return air system is required it can be accomplished by either complete area smoke detection (the preferred method) or detector(s) listed for the air velocity present and located at every return air opening within the smoke compartment. In the second case, the detector can also be located where the air leaves each smoke compartment or in the duct system before the air enters the return air system common to more than one smoke compartment (see Figures 8-13 and 8-14).

Additional smoke detectors are not required to be installed in ducts where the air duct system passes through other smoke compartments not served by the duct (see Figure 8-15).

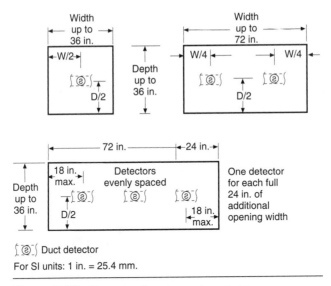

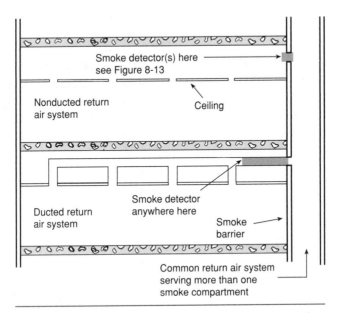

Figure 8-13. Detector in return air system.

Figure 8-14. Detector location in return air system (selective operation of equipment).

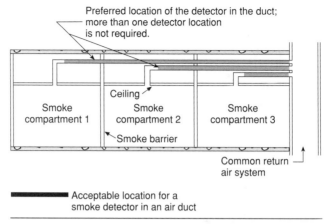

Figure 8-15. Detector location in a duct passing through compartments not served by a duct.

Detectors should be readily accessible for cleaning and mounted in accordance with the manufacturer's recommendations. Access doors or panels to the detector should be provided. Detectors mounted outside of a duct employing sampling tubes for transporting smoke from inside the duct to the detector should be arranged to permit verification of air flow from the duct to the detector.

Detectors should be suitable for proper operation over the complete range of air velocities, temperature, and humidity expected at the detector when the air handling system is operating. All penetrations of a return air duct in the vicinity of detectors installed on or in an air duct should be sealed. This prevents entrance of outside air and possible dilution or redirection of smoke within the duct.

Detectors mounted in or on return air ducts should be located at least six duct widths downstream from any duct openings, deflection plates, sharp bends, or branch connection.

If it is physically impossible to locate the detector in this manner, the detector can be positioned closer than the required six duct widths (but as far as possible from the opening, bend, or deflection plate so that smoke can still adequately be detected in the air stream).

The detectors should be installed as shown in Figure 8-13 up to 12 in. (305 mm) in front of or behind the return air opening. Detectors should be spaced according to the following dimensions:

1. Width: To 36 in. (0.9 m), one detector centered in the opening. To 72 in. (1.8 m), two detectors located at the quarter points of the opening. Over 72 in. (1.8 m), one additional detector for each full 24 in. (0.6 in) of opening.

2. Depth: The number and spacing of the detector(s) in the depth (vertical) of the opening should be the same as those given for the width (horizontal) above.
3. Orientation: Detectors should be oriented in the most favorable position for smoke entry respective to the direction of air flow. The path of a projected beam-type detector across the return air openings should be considered equivalent in coverage to a row of individual detectors.

Smoke Detectors for Door Release Service

When smoke detectors are used to close doors to prevent smoke from traveling from one side of the door to the other, they should be mounted and located as shown in Figures 8-16 and 8-17. Additional smoke detectors may be needed for more complex doorways.

ACTUATION OF EXTINGUISHING SYSTEMS

Automatic fire detectors are often used to actuate extinguishing systems, for example, preaction and deluge sprinkler systems, combined dry-pipe and pre-action sprinkler systems, and clean-agent gaseous suppression systems. Certain conditions apply when automatic detection systems are used to actuate extinguishing systems (see NFPA 13, *Standard for the Installation of Sprinkler Systems,* for more information).

The following recommendations should be followed in using automatic detection systems to actuate extinguishing systems:

1. The equipment or method used for automatic detection should be listed or approved. Such equipment or methods should detect or indicate heat, smoke, flame, combustible vapors, or any abnormal condition that could identify a fire.
2. Adequate and reliable power supplies should be used. A power supply for an electric detection system should be independent of the power supply for the protected area. If this separation is not possible, an emergency battery-powered supply with automatic switchover in case of primary power supply failure should be provided.
3. Automatic detection and actuation equipment should be supervised; preferably, indication of equipment failure should be immediately shown at a constantly attended location.
4. System operation should be indicated by audible alarms; the alarms should also alert personnel and indicate failure of a supervised device or piece of

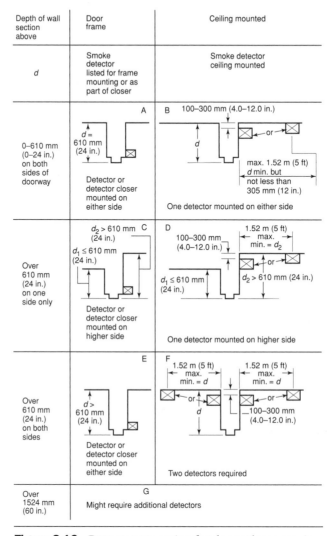

Depth of wall section above	Door frame	Ceiling mounted
d	Smoke detector listed for frame mounting or as part of closer	Smoke detector ceiling mounted
0–610 mm (0–24 in.) on both sides of doorway	**A** $d = 610$ mm (24 in.) Detector or detector closer mounted on either side	**B** 100–300 mm (4.0–12.0 in.) max. 1.52 m (5 ft) d min. but not less than 305 mm (12 in.) One detector mounted on either side
Over 610 mm (24 in.) on one side only	**C** $d_2 > 610$ mm (24 in.) $d_1 \le 610$ mm (24 in.) Detector or detector closer mounted on higher side	**D** 100–300 mm (4.0–12.0 in.) 1.52 m (5 ft) max. min. = d_2 $d_1 \le 610$ mm (24 in.) $d_2 > 610$ mm (24 in.) One detector mounted on higher side
Over 610 mm (24 in.) on both sides	**E** $d > 610$ mm (24 in.) Detector or detector closer mounted on either side	**F** 1.52 m (5 ft) max. min. = d 1.52 m (5 ft) max. min. = d d 100–300 mm (4.0–12.0 in.) Two detectors required
Over 1524 mm (60 in.)	**G** Might require additional detectors	

Figure 8-16. Detector mounting for door release service.

equipment. The number of these alarms and their locations should satisfy the local code requirements.

5. Additional alarms should be provided to show that the system has operated, to give personnel ample warning of a discharge from an extinguishing system (especially the clean-agent gaseous suppression or high-expansion-foam types), and to indicate failure of supervised devices or equipment.

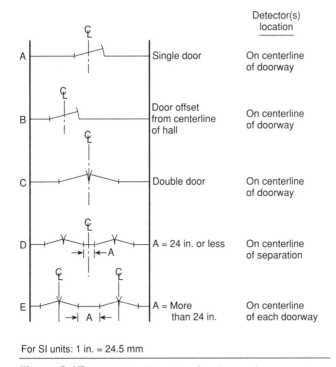

For SI units: 1 in. = 24.5 mm

Figure 8-17. Detector location for door release service.

Preaction and Deluge Extinguishing Systems

A preaction system employs automatic sprinklers attached to a piping system that contains air that may or may not be under pressure, with a supplemental fire detection system installed in the same areas as the sprinklers. Actuation of the fire detection system opens a valve that permits water to flow into the sprinkler piping system and to be discharged from any open sprinklers.

A deluge system employs open sprinklers attached to a piping system and connected to a water supply through a valve that is opened by the operation of a fire detection system installed in the same areas as the sprinklers. When this valve opens, water flows into the piping system and discharges from all sprinklers attached to the piping. Deluge systems may be needed where occupancy conditions or special hazards require quick application of large quantities of water. The fire detection devices or systems for a deluge system with more than 20 sprinklers should be automatically supervised.

Preaction and deluge systems normally do not have water in the system piping. The water supply is controlled by an automatic valve operated by means of fire detection devices and provided with manual means for operation that are independent of the sprinklers.

Many types of equipment can be included in preaction and deluge systems, for example, automatic sprinklers with sprinkler piping and fire detection devices, open sprinklers with fire detection devices, and a combination of open and automatic sprinklers with fire detection devices. The equipment may or may not be supervised.

Fire detection devices should be selected to both ensure operation and guard against premature operation of sprinklers based on normal room temperatures and draft conditions. In locations where ambient temperature at the ceiling is high (from heat sources other than fire conditions), heat-responsive devices should be selected that operate at higher than ordinary temperature and that are capable of withstanding the normal high temperature for long periods.

Fire detection devices other than automatic sprinklers should be located and spaced according to their listing by testing laboratories or to manufacturer's specifications. When automatic sprinklers are used as detectors, the distance between detectors and the area per detector should not exceed the maximum permitted for suppression sprinklers as specified by the local code.

Combined Dry-Pipe and Preaction System

A combined dry-pipe and pre-action sprinkler system employs automatic sprinklers that are attached to a piping system containing air under pressure and with a supplemental fire detection system installed in the same areas as the sprinklers. Operation of the fire detection system actuates tripping devices that open dry-pipe valves simultaneously without loss of air pressure in the system. Operation of the fire detection system also opens approved air exhaust valves at the end of the feed main, which facilitates the filling of the system with water, preceding the opening of sprinklers. The fire detection system also serves as an automatic fire alarm system.

Combined automatic dry-pipe and preaction systems should be constructed so that failure of the fire detection system does not prevent the system from functioning as a conventional automatic dry-pipe system. Also, failure of the dry-pipe system of automatic sprinklers should not prevent the fire detection system from properly functioning as an automatic fire alarm system.

Each dry-pipe valve should have an approved tripping device actuated by the fire detection system. Dry-pipe valves should be cross connected through a 1-in. (25.4-mm) pipe connection to permit simultaneous tripping of the dry-pipe valves. The connection should be equipped with a gate valve so that either dry-pipe valve can be shut off or serviced while the other remains in service (see Figure 8-18).

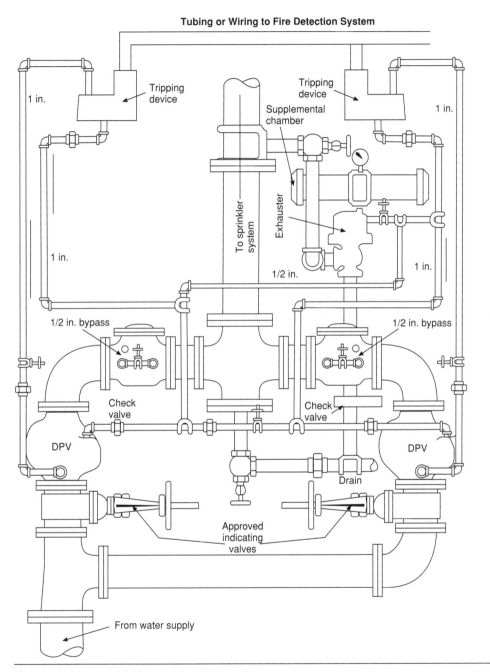

Figure 8-18. Header for combined dry-pipe and preaction sprinkler system (standard trimmings not shown). DPV, Dry-pipe valve.

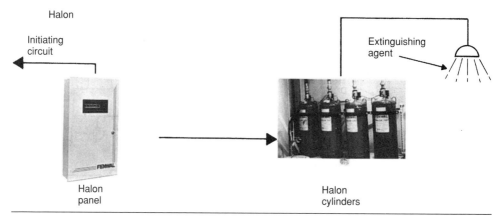

Figure 8-19. Gas discharge in an automatic extinguishing system.

Clean-Agent Gaseous Suppression Total-Flooding Extinguishing Systems

A gaseous total-flooding extinguishing system can be activated by automatic fire detectors. A gaseous total-flooding system is a permanently piped system that uses a limited stored supply of a gas under pressure and discharge nozzles to totally flood an enclosed area. The gas releases automatically via a suitable detection system and extinguishes fires by inhibiting the chemical reaction between fuel and oxygen (see Figure 8-19).

This type of extinguishing system is often installed in computer rooms, libraries, archives, or other facility where water would seriously damage contents or equipment. Gaseous suppression systems used in sprinklered or unsprinklered computer rooms can protect data in process, reduce equipment damage, and facilitate return to service. Halon was used as a fire-suppressant gas almost exclusively in gaseous suppression systems until its adverse effect on the ozone layer was found. As halon systems are being phased out, new clean agents are being used as a replacement.

Gaseous suppression systems as defined in NFPA 2001, *Standard on Clean Agent Fire Extinguishing Systems,* should be automatically actuated by an approved method of detection meeting the requirements of *NFPA 72.* To ensure detection, particular attention should be given to the choice of actuation means, the air flows usually involved in such air-handling systems, and the small heat release under fire conditions.

Where operation of the air conditioning system would exhaust the agent supply, the gaseous suppression system should be interlocked to shut down the air conditioning when the gaseous suppression system is actuated (computer equipment does not need to be deenergized prior to discharge). Alarms should warn area occupants of a discharge or pending discharge.

An automatic detection system should be installed in the space below the raised floor of a computer facility. This additional automatic detection system should meet the above requirements and sound an audible and visual alarm. Air space below the raised floor should be subdivided into areas not exceeding 10,000 sq ft (929 m²) by tight, noncombustible bulkheads.

Cross-Zone Detection for Gaseous Suppression Systems: Although gaseous suppression systems generally have operation requirements similar to systems using other types of agents, the use of automatically actuated systems is strongly recommended. This limits the size and severity of fire with which the system must deal, minimizing decomposition of the agent during extinguishment. In a Class A fire, automatic actuation coupled with sensitive detectors can often prevent the fire from becoming deep-seated.

The detector must be sensitive enough to the fuel in question to rapidly respond to the fire at an early stage, but not so sensitive as to produce false actuation of the extinguishing system. Sensitivity and reliability in a gaseous suppression system can be combined by utilizing multiple detectors connected in a double circuit or *cross-zoned* mode of operation (see Figure 8-20). Actuation

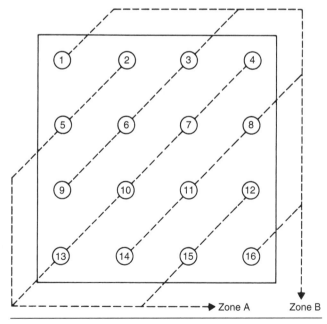

Figure 8-20. A cross-zone detector circuit.

of either zone alone activates local and remote alarms, but does not discharge the extinguishing agent. Actuation of both zones simultaneously causes the agent to discharge. This arrangement prevents a false signal by an individual detector from discharging the entire system. The two zones may contain the same or different types of detectors. Ionization and photoelectric detectors are popularly used in computer areas.

Particular attention must be given to locating detectors within a hazard. The amount and location of combustibles, the listed sensitivity spacing relationship of the detectors considered, and ventilation characteristics of the hazard are all important factors in detector location.

High-Expansion Foam Systems

There are also special considerations in automatically activating high-expansion foam extinguishing systems. This system is a fixed extinguishing system that generates a foam agent for total flooding of confined spaces and volumetric displacement of vapor, heat, and smoke. High-expansion foam extinguishes fire by preventing free movement of air, reducing the oxygen concentration at the fire, and cooling.

INSTALLATION AND WIRING FOR FIRE ALARM SYSTEM CIRCUITS

Installation wiring for a fire alarm system should be installed in accordance with the local electrical code. Often, Article 760 of NFPA 70, *National Electrical Code®*, or some variation is specified.

In wiring a fire alarm system, calculation of proper wire size is critical. The correct wire ensures that the conductor will carry the assigned load and prevents an excessive drop in voltage between the fire alarm control panel and other system components.

There are two separate and distinct circuit classifications: power-limited and non-power-limited (see Article 760 of the *NEC*). A power-limited circuit is one in which the power is limited, the circuit supervised, and the circuit durably marked where plainly visible at terminations. If the circuit is not so marked, it must be considered a non-power-limited circuit. These circuits should never be intermixed or placed in the same cable, raceway, or enclosure unless they are separated by a barrier or a mechanically defined separation of at least 2 in. (50 mm). Power-limited and non-power-limited circuits should also not share the same cable, raceway, or enclosure with electric power and lighting circuits, except that power circuits connected to the same equipment can be located in the same enclosure with power-limited conductors and in the same cable and raceway with non-power-limited circuits.

Power-limited circuits may be intermixed with Class 3 signaling circuits and Class 2 signaling circuits and communication circuits when their insulation is at least that which is required for the power-limited circuit. Non-power-limited circuits may be intermixed with Class 1 signaling circuits.

Once a circuit is designated as either power-limited or non-power-limited, it must be installed accordingly for its entire length; that is, a circuit cannot be non-power-limited for a portion of its run and power-limited for the remainder.

For example, a smoke detector with both alarm signaling contacts and additional contacts for controlling a local function (such as door release or elevator capture) should be installed so that all connected circuits are either all non-power-limited/Class 1 or all power-limited/Class 2/Class 3. This means that the classification of the controlled circuit determines the classification of the fire alarm circuit(s) and the wiring methods and materials that are permitted. Since the fire alarm circuits can terminate at the fire alarm control unit or at terminals adjacent to others for fire alarm circuits that do not have a 2-in. (50-mm) separation or adequate barrier, classifications of most of the other fire alarm circuits in the same control unit enclosure also depend upon the classification of the elevator capture or door release circuit.

Conductors

The requirements below for the number of conductors in boxes are taken from Table 314.16 of the *NEC.*

There should be boxes of sufficient size to provide free space for all conductors enclosed therein. The maximum number of conductors permitted in standard boxes is shown in Table 8-1.

Table 8-1 must be used where no fittings or devices such as fixture studs, cable clamps, hickeys, switches, or receptacles are contained in the box and where no grounding conductors are part of the wiring within the box. Where one or more of these types of fittings, such as fixture studs, cable clamps, or hickeys, are contained in the box, the number of conductors shown in the table must be reduced by one for each type of fitting; an additional deduction of one conductor must be made for each strap containing one or more devices; and a further deduction of one conductor must be made for one or more grounding conductors entering the box. Where a second set of equipment grounding conductors is present in the box, then an additional deduction of one conductor must be made. A conductor running through the box must be counted as one conductor, and each conductor originating outside of the box and terminating inside the box is counted as one conductor. Conductors no part of which leaves the box must not be counted. The volume of a wiring enclosure (box) must be the total volume of the assembled sections and, where used, the space provided

Table 8-1. Standard Metal Boxes

Box Dimensions, (in.), Shape or Type	Minimum Size (in.^{3})	Maximum Number of Conductors						
		No. 18	No. 16	No. 14	No. 12	No. 10	No. 8	No. 6
$4 \times 1\frac{1}{4}$ Round or Octagonal	12.5	8	7	6	5	5	5	2
$4 \times 1\frac{1}{2}$ Round or Octagonal	15.5	10	8	7	6	6	5	3
$4 \times 2\frac{1}{8}$ Round or Octagonal	21.5	14	12	10	9	8	7	4
$4 \times 1\frac{1}{4}$ Square	18.0	12	10	9	8	7	6	3
$4 \times 1\frac{1}{2}$ Square	21.0	14	12	10	9	8	7	4
$4 \times 2\frac{1}{8}$ Square	30.3	20	17	15	13	12	10	6
$4\frac{11}{16} \times 1\frac{1}{4}$ Square	25.5	17	14	12	11	10	8	5
$4\frac{11}{16} \times 1\frac{1}{2}$ Square	29.5	19	16	14	13	11	9	5
$4\frac{11}{16} \times 2\frac{1}{8}$ Square	42.0	28	24	21	18	16	14	8
$3 \times 2 \times 1\frac{1}{2}$ Device	7.5	5	4	3	3	3	2	1
$3 \times 2 \times 2$ Device	10.0	6	5	5	4	4	3	2
$3 \times 2 \times 2\frac{1}{4}$ Device	10.5	7	6	5	4	4	3	2
$3 \times 2 \times 2\frac{1}{2}$ Device	12.5	8	7	6	5	5	4	2
$3 \times 2 \times 2\frac{3}{4}$ Device	14.0	9	8	7	6	5	4	2
$3 \times 2 \times 3\frac{1}{2}$ Device	18.0	12	10	9	8	7	6	3
$4 \times 2\frac{1}{8} \times 1\frac{1}{2}$ Device	10.3	6	5	5	4	4	3	2
$4 \times 2\frac{1}{8} \times 1\frac{7}{8}$ Device	13.0	8	7	6	5	5	4	2
$4 \times 2\frac{1}{8} \times 2\frac{1}{8}$ Device	14.5	9	8	7	6	5	4	2
$3\frac{3}{4} \times 2 \times 2\frac{1}{2}$ Masonry Box/Gang	14.0	9	8	7	6	5	4	2
$3\frac{3}{4} \times 2 \times 3\frac{1}{2}$ Masonry Box/Gang	21.0	14	12	10	9	8	7	2
FS: Minimum Internal Depth $1\frac{3}{4}$ Single Cover/Gang	13.5	9	7	6	6	5	4	2
FD: Minimum Internal Depth $2\frac{3}{8}$ Single Cover/Gang	18.0	12	10	9	8	7	6	3
FS: Minimum Internal Depth $1\frac{3}{4}$ Multiple Cover/Gang	18.0	12	10	9	8	7	6	3
FD: Minimum Internal Depth $2\frac{3}{8}$ Multiple Cover/Gang	24.0	16	13	12	10	9	8	4

by plaster rings, domed covers, extension rings, and so on that are marked with their volume in cubic inches (cm^3) or are made from boxes the dimensions of which are listed in the table.

For combinations of conductor sizes shown in Table 8-1, the maximum number of conductors permitted must be computed using the volume per conductor listed in Table 8-2, with the deductions provided for as above, and these volume deductions must be based on the largest conductor entering the box. The maximum number and size of conductors listed in Table 8-1 must not be exceeded.

Boxes 100 cu in. (1,639 cm^3) or less other than those described in Table 8-1, conduit bodies having provision for more than two conduit entries, and nonmetallic boxes must be durably and legibly marked by the manufacturer with their cubic-inch (cm^3) capacity. The maximum number of conductors permitted must be computed using the volume required per conductor listed in Table 8-2, with the deductions provided for as above, and these volume deductions must be based on the largest conductor entering the box. Boxes described in Table 8-1 that have a larger cubic-inch (cm^3) capacity than is designated in the table must be permitted in order to have their cubic-inch (cm^3) capacity marked as required by this section and the maximum number of conductors permitted must be computed using the volume per conductor listed in Table 8-2.

Conduit bodies enclosing No. 6 conductors or smaller must have a cross-sectional area not less than twice the cross-sectional area of the largest conduit to which they are attached (see Chapter 9, Table 1, of the *NEC*).

Conduit bodies having provisions for fewer than three conduit entries must not contain splices, taps, or devices unless they comply with the provisions listed in the preceding paragraph and are supported in a rigid and secure manner.

Table 8-2. Volume Required per Conductor

Size of Conductor (AWL)	Free Space Within Box for Each Conductor (cu in.)
No. 18	1.50
No. 16	1.75
No. 14	2.00
No. 12	2.25
No. 10	2.50
No. 8	3.00
No. 6	5.00

The following information is derived from Article 760 of the *NEC* and is meant to be a general guide to circuit installation and wiring. For fire alarm system installation and wiring, a designer should consult and review the *NEC*. Circuits and equipment for fire alarm systems concerning fire spread, ducts, plenums and other air-handling spaces, hazardous (classified) locations, and corrosive, damp, or wet locations should comply with the *NEC*.

Fire alarm system circuits should be identified at terminal and junction locations in a manner that will prevent unintentional interference with the fire alarm circuit during testing and servicing. Fire alarm system line circuits that extend aerially beyond one building should meet the requirements of the *NEC*.

Fire alarm system circuits and equipment should be grounded except for dc power-limited fire alarm system circuits having a maximum current of 0.030 A. The circuit should also be electrically supervised so that a trouble signal indicates the occurrence of a single-open or a single-ground fault on any installation wiring circuit that would prevent proper alarm operation.

Interconnecting circuits of household fire warning equipment wholly within a dwelling unit do not need to be supervised. Wiring for alarm notification appliances should be similar to that for alarm-initiating devices to maintain supervision. Typical wiring circuits are shown in Figures 8-21 and 8-22.

Non-Power-Limited Fire Alarm System Circuits

The power supply of non-power-limited fire alarm system circuits should comply with the applicable code and the output voltage should not exceed 600 V nominal.

Overcurrent Protection: Conductors should be protected against overcurrent. Overcurrent protection should not exceed 7 A for No. 18 conductors and 10 A for No. 16 conductors. Other overcurrent protection may be required by

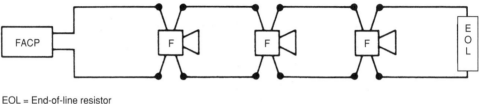

EOL = End-of-line resistor
F = Alarm notification appliance
FACP = Fire alarm control panel

Figure 8-21. Correct wiring method for typical notification appliance circuit.

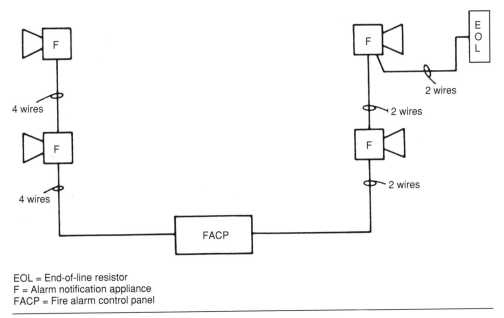

EOL = End-of-line resistor
F = Alarm notification appliance
FACP = Fire alarm control panel

Figure 8-22. Single notification appliance circuit with two risers.

the *NEC.* Overcurrent devices should be located at the point where the conductor to be protected receives its power supply.

Conductors of Different Circuits: Class I and non-power-limited fire alarm system circuits (ac or dc individual circuits) may occupy the same enclosure, cable, or raceway, provided all conductors are insulated for the maximum voltage of any conductor in the enclosure or raceway. Power supply and fire alarm system circuit conductors should be in the same enclosure, cable, or raceway only when connected to the same equipment.

Copper Conductors: Number 18 and No. 16 copper conductors can be used, provided they supply loads that do not exceed the recommended amperage and are installed in a raceway or a listed cable. Insulation on conductors should be suitable for 600 V. Conductors larger than No. 16 should comply with the *NEC* Number 16 and No. 18 conductors must be as specified in the *NEC.* Conductors with other types and thicknesses of insulation can be permitted if listed for non-power-limited fire alarm system circuit use.

Conductors should be solid or bunch-tinned (bonded) stranded copper. Wire types PTF and PAF should be permitted only for high-temperature applications between 194°F (90°C) and 482°F (250°C).

Multiconductor Cable: A multiconductor cable having two or more No. 16 or No. 18 solid or stranded copper conductors listed for this use is permitted. This type of cable can be used on fire alarm system circuits operating at 150 V or less.

Conductors in Raceways, Cable Trays, and Cables, and Derating: Where only non-power-limited fire alarm system circuits and Class 1 circuits are in a raceway, the number of conductors should be determined in accordance with the *NEC*. Derating factors should apply if such conductors carry continuous load.

Where power supply conductors and fire alarm system circuit conductors are permitted in a raceway, the number of conductors should be determined in accordance with the *NEC*. The derating factors specified in the *NEC* should apply as follows:

1. To all conductors, when the fire alarm system circuit conductors carry continuous loads and there are more than three conductors
2. To the power supply conductors only, when the fire alarm system circuit conductors do not carry continuous loads and where there are more than three power supply conductors

Fire alarm system circuit conductors installed in cable trays must comply with the *NEC*.

Power-Limited Fire Alarm System Circuits

As specified in Table 8-3 for ac circuits and Table 8-4 for dc circuits, the power for power-limited fire alarm system circuits should be either inherently limited (requiring no overcurrent protection) or limited (by a combination of a power source and overcurrent protection).

Supervision: Either a trouble or an alarm signal should indicate the occurrence of multiple ground faults or any short-circuit fault on the fire alarm system primary (main) power supply, alarm-initiating device, signaling line, or required notification appliance circuits that would prevent proper alarm operation.

The circuit should be durably marked where plainly visible at terminations to indicate that it is a power-limited fire alarm system circuit. Where overcurrent protection is required, the overcurrent protective devices should not be interchangeable with devices of higher ratings. The overcurrent should be permitted as an integral part of the power supply. Overcurrent devices, where required, should be located at the point where the conductor to be protected receives its supply.

Table 8-3. Power Limitations for AC Fire Protective Signaling Circuits*

	Value for Given Circuit Voltage V_{max}					
	Inherently Limited Power Source†			Not Inherently Limited Power Source‡		
Parameter	V_{max} = 0–20 V	V_{max} = 21–30 V	V_{max} = 31–100 V	V_{max} = 0–20 V	V_{max} = 21–100 V	V_{max} = 101–150 V
Power Limitation $(VA)_{max}$ (V-A)	—	—	—	250§	250	NA
Current Limitation I_{max} (A)	8.0	8.0	150/V_{max}	1000/V_{max}	1000/V_{max}	1.0
Maximum Overcurrent Protection (A)	—	—	—	5.0	100/V_{max}	1.0
Power Source Maximum Nameplate Ratings						
VA (V-A)	5.0 V_{max}	100	100	5.0 V_{max}	100	100
Current (A)	5.0	100/V_{max}	100/V_{max}	5.0	100/V_{max}	100/V_{max}

*V_{max}, Maximum output voltage regardless of load with rated input applied; I_{max}, maximum output current after 1 minute of operation under any noncapacitive load, including short circuit, and with overcurrent protection bypassed if used; $(VA)_{max}$, maximum volt-ampere output regardless of load and overcurrent protection bypass if used.

†Overcurrent protection not required.

‡Overcurrent protection required.

§If the power source is a transformer, $(VA)_{max} \leq 350$ when $V_{max} \leq 15$.

Table 8-4. Power Limitations for DC Fire Protective Signaling Circuits*

	Value for Given Circuit Voltage V_{max}						
	Inherently Limited Power Source†				Not Inherently Limited Power Source‡		
Parameter	$V_{max} = 0-20V$	$V_{max} = 21-30V$	$V_{max} = 31-100V$	$V_{max} = 101-250V$	$V_{max} = 0-20V$	$V_{max} = 21-100V$	$V_{max} = 101-150V$
Power Limitation $(VA)_{max}$ (V-A)	—	—	—	—	250§	250	NA
Current Limitation I_{max} (A)	8.0	8.0	150/V_{max}	0.030	1000/V_{max}	1000/V_{max}	1.0
Maximum Overcurrent Protection (A)	—	—	—	—	5.0	100/V_{max}	1.0
Power Source Maximum Nameplate Ratings							
VA (V-A)	5.0 V_{max}	100	100	0.030 V_{max}	5.0 V_{max}	100	100
Current (A)	5.0	100/V_{max}	100/V_{max}	0.030	5.0	100/V_{max}	100/V_{max}

*V_{max}: Maximum output voltage regardless of load with rated input applied; I_{max}: maximum output current after 1 minute of operation under any noncapacitive load, including short circuit, and with overcurrent protection bypassed if used; $(VA)_{max}$: maximum volt-ampere output regardless of load and overcurrent protection bypass if used.

†Overcurrent protection not required.

‡Overcurrent protection required.

§If the power source is a transformer, $(VA)_{max} \leq 350$ when $V_{max} \leq 15$.

Wiring Methods

Conductors and equipment on the supply side of overcurrent protection, transformers, or current-limiting devices should be installed in accordance with appropriate requirements. Transformers or other devices supplied from power supply conductors should be protected by an overcurrent device rated not over 20 A.

Input leads of a transformer or other power sources supplying power-limited fire alarm system circuits can be smaller than No. 14 but not smaller than No. 18 if they are not over 12 in. (305 mm) long and are insulated properly.

Circuits on the load side of overcurrent protection, transformers, and current-limiting devices should have specified wiring types and use those wiring methods and materials. Non-power-limited circuits should normally comply with Chapter 3 of the *NEC* or they can use the multiconductor cables described above. Also, conductors should be solid copper or bunch-tinned (bonded) stranded copper. Power-limited circuits can be reclassified and installed as non-power-limited circuits under certain conditions.

Power-limited circuit conductors and cables should be installed as follows:

1. In raceways or exposed on surface of ceiling and sidewalls or "fished" in concealed spaces. Where installed exposed, cable should be adequately supported and terminated in approved fittings and installed in such a way that maximum protection against physical damage is afforded by building construction such as baseboards, door frames, ledges, and so on. Where located within 7 ft (2.13 m) of the floor, cable should be securely fastened in an approved manner at intervals of not more than 18 in. (457 mm).
2. In metal raceway or rigid nonmetallic conduit when passing through a floor or wall to a height of 7 ft (2.13 m) above the floor unless adequate protection can be afforded by building construction such as detailed in Item 1 above, or unless an equivalent solid guard is provided.
3. In rigid metal conduit, intermediate metal conduit, or electrical metallic tubing when installed in hoistways.

Single- and multiconductor power-limited fire alarm system circuit cables installed as wiring within buildings should be listed Type FPL as being resistant to the spread of fire. In addition, where the cables are in a vertical run in a shaft and where the cables are installed in ducts, plenums, and other air-handling spaces, other requirements in the *NEC* may apply.

One method of defining resistance to the spread of fire is that the cables do not spread fire to the top of the tray in the vertical tray flame test in UL 1581,

Standard for Electrical Wires, Cables, and Flexible Cords, except where the cables are enclosed in a raceway or noncombustible tubing or in nonconcealed spaces where the exposed length of cable does not exceed 10 ft (3.05 m).

Single- and multiconductor power-limited fire alarm system circuit cables in a vertical run in a shaft should be of a fire-resistant type with characteristics capable of preventing the carrying of fire from floor to floor, except where the cables are encased in noncombustible tubing or are located in a fireproof shaft having firestops at each floor.

Single- and multiconductor power-limited fire alarm system circuit cables and equipment installed in ducts or plenums or other spaces used for environmental air should also be installed in accordance with other sections of the *NEC.*

Low-smoke-producing cables may be used, and can be defined by establishing an acceptable value of the smoke produced by test to a maximum peak optical density of 0.5 and a maximum average optical density of 0.15. Similarly, fire-resistant cables can be defined as having a maximum allowable flame travel distance of 5 ft (1.52 m) when tested to NFPA 262, *Standard Method of Test for Flame Travel and Smoke of Wires and Cables for Use in Air-Handling Spaces.*

Some conductors and cables on the load side of overcurrent protection, transformers, and current limiting devices should be separated.

Power-limited circuits should be separated at least 2 in. (50.8 mm) from conductors of any electric light, power, Class 1, or non-power-limited fire alarm system circuits. Separation is not required:

1. Where the electric light, power, Class 1, or non-power-limited fire alarm system circuit conductors are in raceway or in metal-sheathed, metal-clad, non-metallic-sheathed, or Type UF cables; or
2. Where the power-limited circuit conductors are permanently separated from the conductors of the other circuits by a continuous and firmly fixed nonconductor (such as porcelain tubes or flexible tubing in addition to the insulation on the wire).

Power-limited circuits should not be placed in any enclosure, raceway, cable, compartment, outlet box, or similar fitting containing conductors of electric light, power, Class 1, or non-power-limited fire alarm system circuits except:

1. Where the conductors of the different systems are separated by a partition; or
2. Where conductors in outlet boxes, junction boxes, or similar fittings or compartments where power supply conductors are introduced solely for supplying power to the power-limited fire alarm system to which the other conductors in the enclosure are connected.

Power-limited circuits should be separated by not less than 2 in. (50.8 mm) from electric light, power, Class 1, or non-power-limited fire alarm system circuit conductors run in the same shaft except:

1. Where the conductors of either the electric light, power, Class 1, the non-power-limited fire alarm system circuits, or the power-limited fire alarm system circuits are encased in noncombustible tubing; or
2. Where the electric light, power, Class 1, or the non-power-limited fire alarm system circuit conductors are in a raceway or are in metal-sheathed, metal-clad, nonmetallic-sheathed, or Type UF cables.

The requirements for separation from electric light, power, Class 1, and non-power-limited fire alarm system circuits in the *NEC* should apply even if the power-limited circuits are wired using non-power-limited circuit wiring methods.

Cables and conductors of two or more power-limited fire alarm system circuits or Class 3 circuits may be permitted in the same cable, enclosure, or raceway. Conductors of one or more Class 2 circuits may be within the same cable, enclosure, or raceway with conductors of power-limited fire alarm system circuits, provided that the insulation of the Class 2 circuit conductors in the cable, enclosure, or raceway is at least that which is required by the power-limited fire alarm system circuits.

Conductors and Cables

Conductors and cables for use with power-limited fire alarm system circuits should be listed for this use and meet or exceed the following requirements:

1. Conductors should not be smaller than No. 16 for single-conductor cable, and for multiconductor cable should be no smaller than No. 19 for two- or three-conductor cable, No. 22 for four- or five-conductor cable, and No. 24 for cables of six or more conductors.
2. Cables should be listed as being suitable for Class 3, power-limited fire alarm system, or communication circuits.
3. The cable should have a voltage rating of not less than 300 V and the jacket compound should have a high degree of abrasion resistance.
4. Coaxial cables should have a minimum No. 22 AWG copper or 30 percent minimum conductivity copper-covered steel-center conductor with an overall insulation rated at 300 V, an overall metallic shield covered by a flame-retardant nonmetallic jacket having a minimum thickness not less than 35 mils, nominal 30 mils minimum, and a high degree of abrasion resistance.

TABLE 8-5. Cable Markings*

Cable Marking	Type	NEC Reference
FPL	Power-limited fire alarm cable	760-51(f) and 760-53(c)
FPLP	Power-limited fire alarm plenum cable	760-51(d) and 760-53(a)
FPLR	Power-limited fire alarm riser cable	760-51(d) and 760-53(b)
OFC	Conductive optical fiber general-purpose cable	770-51(d) and 770-53(c)
OFCP	Conductive optical fiber plenum cable	770-51(a) and 770-53(a)
OFCR	Conductive optical fiber riser cable	770-51(b) and 770-53(b)
OFN	Nonconductive optical fiber general-purpose cable	770-51(d) and 770-53(c)
OFNP	Nonconductive optical fiber plenum cable	770-51(a) and 770-53(a)
OFNR	Nonconductive optical fiber riser cable	770-51(b) and 770-53(b)

*See the referenced NFPA 70, *National Electrical Code®* (*NEC*), sections for permitted uses.

5. Listed nonconductive and conductive optical fiber cables can also be used.
6. Listed power-limited fire alarm system cables should be marked (see Table 8-5).

Current-Carrying Continuous-Line-Type Fire Detectors

Listed continuous-line-type fire detectors, including insulated copper tubing of pneumatically operated detectors that are employed for both detection and carrying signaling currents, can be used in circuits having power-limiting characteristics.

Optical Fiber Cables

Optical fiber cables along with electric conductors can also be used for fire alarm signal transmission. Optical fiber cables transmit light for control, signaling, and communications through an optical fiber and can be grouped into three types: (1) nonconductive cables, which contain no metallic members and no other electrically conductive materials; (2) conductive cables which contain non-current-carrying conductive members, such as metallic strength members and metallic vapor barriers; and (3) hybrid cables, which contain optical fibers and current-carrying electric conductors and should be classified as electric cables in accordance with the type of electric conductors.

Optical Fibers and Electric Conductors: Optical fibers can be contained within the same hybrid cable for electric light, power, or Class 1 circuits operating at 600 V or less only where the functions of the optical fibers and the electric conductors are associated. Nonconductive optical fiber cables can occupy the same raceway or cable tray with conductors for electric light, power, or Class 1 circuits operating at 600 V or less. Conductive and hybrid optical fiber cables should not occupy the same raceway or cable tray with conductors for electric light, power, or Class 1 circuits.

Nonconductive optical fiber cables should not be permitted to occupy the same cabinet, panel, outlet box, or similar enclosure housing the electrical terminations of an electric light, power, or Class 1 circuit unless nonconductive optical fiber cable is functionally associated with the electric light, power, or Class 1 circuit.

Occupancy of the same cabinet, panel, outlet box, or similar enclosure is allowed where nonconductive optical fiber cables are installed in factory- or field-assembled control centers. Nonconductive optical fiber cables should be permitted with circuits exceeding 600 V only in industrial establishments where conditions of maintenance and supervision ensure that only qualified persons will service the installation.

Optical fibers can be in the same cable, and conductive and nonconductive optical fiber cables are permitted in the same raceway, cable tray, or enclosure with conductors of any of the following (in accordance with the *NEC*):

1. Class 2 and Class 3 remote-control, signaling, and power-limited circuits
2. Power-limited fire alarm systems
3. Communications circuits
4. Community antenna television and radio distribution systems

Non-current-carrying conductive members of optical fiber cables should be grounded.

Fire Resistance of Optical Fiber Cables: Optical fiber cables installed as wiring within buildings should be listed as being resistant to the spread of fire. Cables in a vertical run in a shaft or installed in ducts, plenums, and other air-handling spaces have other fire resistance requirements as outlined in the *NEC*. Optical fiber cables may be excepted from requirements where they are enclosed in raceway or noncombustible tubing or in nonconcealed spaces where the exposed length of cable does not exceed 10 ft (3.05 m).

Optical fiber cables in a vertical run in a shaft either should have fire-resistant characteristics capable of preventing the carrying of fire from floor to floor or be encased in noncombustible tubing or be located in a fireproof shaft having firestops at each floor.

Optical fiber cables and equipment installed in ducts or plenums or other spaces used for environmental air should also be installed in accordance with the wiring requirements of the *NEC*. Certain types of listed optical fiber cables with adequate fire resistance and that are low-smoke producing may be allowed for ducts, plenums, and other space used for environmental air.

Low-smoke-producing cables can be defined by establishing an acceptable value of the smoke produced to a maximum peak optical density of 0.5 and a maximum average optical density of 0.15. Similarly, fire-resistant cables can be defined as having a maximum allowable flame travel distance of 5 ft (1.52 m) when tested to NFPA 262.

Where exposed to contact with electric light or power conductors, the non-current-carrying metallic members of optical fiber cables entering buildings should be grounded as close to the point of entrance as practicable or interrupted as close to the point of entrance as practicable by an insulating joint or equivalent device. The point of entrance should be considered to be at the point of emergence through an exterior wall or a concrete floor slab or from a rigid metal conduit or an intermediate metal conduit that is grounded.

Listed optical fiber cables should be marked in accordance with Table 8-5.

TRANSIENT AND LIGHTNING PROTECTION

When fire alarm system equipment is submitted to a testing laboratory for listing, each type of equipment undergoes extensive transient tests. Despite this, there may be excessive electrical transients in the building that houses the fire alarm system, and installation of transient protection in the field should be considered.

Most commonly experienced high-voltage transients and interferences are of short duration and/or have low energy content. High-energy transients, although rare and usually caused by direct lightning strikes, tend to be catastrophic to electronic equipment and require elaborate protective measures.

Fire alarm systems are susceptible to many types of electrical interference, including:

1. Lightning strikes, direct or induced surges
2. Uneven power line conditions and transients

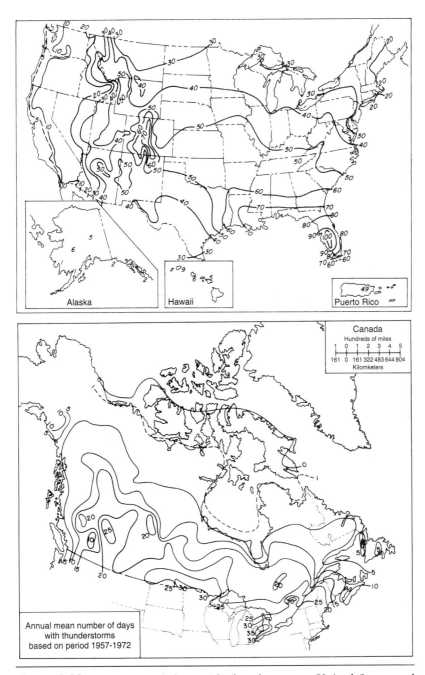

Figure 8-23. Mean annual days with thunderstorms, United States and Canada.

3. Transients generated by switching various system components, such as relays, bells, and so on
4. Interference induced by capacitive, inductive, or electromagnetic coupling to system wiring from nearby motors, neon signs, radiofrequency transmitters, and so on
5. Direct coupled transients on system wiring (other than those on power lines) that are caused by direct or secondary lightning strikes and other events (usually catastrophic to equipment)

Transient protection can be provided by transient suppressors or other components. To be effective, transient suppressors must conduct more current than the corresponding voltage rise indicates, as, for example, a resistor does not. The current increase in a resistor is directly proportional to an increase in the supply voltage to it. Nonlinear devices, such as thyristors and metal oxide varistors (MOVs), have current rises to some exponential power of the applied voltage (such as $I = KV^x$ where x, a characteristic of the device, ranges typically from 5 to 35). Extreme nonlinear devices, such as Zener diodes and spark gaps, are essentially breakover devices in that, when a certain voltage is reached, the current rises dramatically. Various suppressors can be selected for applications, depending upon cost, current-handling capability, response time, and other factors.

Suppressor selection depends upon detailed device characteristics as they apply to the intended use. Spark gaps are probably the most difficult of suppressors to apply correctly because their discharge can continue to conduct until the applied voltage is reduced below 25 V (for commonly available types). In general, protecting each conductor will provide a more reliable system, particularly in lightning-prone areas or where commercial power can destroy electronic apparatus (see Figure 8-23). Any circuit that is not sheathed in grounded conduit is highly subject to becoming an antenna, particularly if the circuits are ungrounded. Conservative design requires transient protection on each line to ground.

QUESTIONS

1. Proper installation of a fire alarm system is a prime factor in its:
   a. cost
   b. operation
   c. reliability
   d. none of the above

2. A complete smoke detection system is defined as one providing smoke detection in:

   a. all closets
   b. all offices
   c. all corridors and offices
   d. all occupiable areas, common areas, and work spaces

3. Duplicate terminals or leads are necessary to allow the wire run to be:

   a. supervised
   b. monitored for integrity
   c. broken and connections to be made to the incoming and outgoing leads or terminals
   d. continuous

4. Fire alarm system circuits must be identified at terminal and junction locations:

   a. to ensure proper use of color codes
   b. to prevent unintentional interference with the fire alarm circuit during testing and servicing
   c. to avoid mixing initiating-device circuits and notification appliance circuits
   d. to avoid mixing initiating-device circuits and control circuits

5. The correct wire size used in a fire alarm circuit ensures that the conductor will carry the assigned load and:

   a. avoid excessive fuse sizes
   b. avoid damaging the device
   c. prevent an excessive drop in voltage between the fire alarm control panel and other system components
   d. provide strength during long-distance installations

BIBLIOGRAPHY

NFPA Codes, Standards, Recommended Practices, and Manuals (see the latest *NFPA Catalog* for availability of current editions of the following documents):

NFPA 11A, *Standard for Medium- and High-Expansion Foam Systems.*

NFPA 12A, *Standard on Halon 1301 Fire Extinguishing Systems.*

NFPA 13, *Standard for the Installation of Sprinkler Systems.*

NFPA 70, *National Electrical Code®*.

NFPA 72®, National Fire Alarm Code®.

NFPA *101®, Life Safety Code®*.

NFPA 231C, *Standard for Rack Storage of Materials.*

NFPA 262, *Standard Method of Test for Flame Travel and Smoke of Wires and Cables for Use in Air Handling Spaces.*

NFPA 2001, *Standard on Clean Agent Fire Extinguishing Systems*

ADDITIONAL READING

NECA 305-2001, *NEIS Standard for Fire Alarm System Job Practices,* National Electrical Contractors Association, 2001

NFPA 75, *Standard for the Protection of Information Technology Equipment.*

NFPA 90A, *Standard for the Installation of Air-Conditioning and Ventilating Systems.*

NFPA 780, *Standard for the Installation of Lightning Protection Systems.*

UL 1581, *Standard for Electrical Wires, Cables, and Flexible Cords,* Underwriters Laboratories Inc., Northbrook, IL, 1991.

9

Fundamentals of Fire Detection System Design

INTRODUCTION

The importance of proper design for fire alarm systems cannot be overemphasized. This chapter reviews design objectives for fire alarm and detection systems and covers criteria necessary for designing a proper, effective system. Design criteria discussed in this chapter include consideration of combustibles, possible fire scenarios, and the effects of ceiling height and configuration, room ventilation, and temperature on system design.

Criteria for the selection and spacing of heat, smoke, and flame-sensing detectors for various construction types are addressed here. Further information on system design can be found in Chapter 8, Fire Alarm System Installation.

Fire protection design requires implementation of both engineering design and administrative controls. Fire hazards are controlled through fire prevention, fire protection program maintenance, and training.

Fire prevention is the preferred method for controlling fire loss in a facility. Since it is unlikely that all fires can be prevented, administrative control of fire hazards, combined with engineering design controls, can greatly reduce both the frequency and the severity of fire. Fire prevention is accomplished by controlling both combustibles and ignition sources.

Combustibles can be controlled by locating them remote from possible ignition sources, containing them within suitable enclosures, and selecting noncombustible materials. Ignition sources can be controlled by use of specialized electric equipment for hazardous locations (those involving flammable gases, vapors, combustible dusts, and ignitable fibers), minimizing static electricity buildup potential by using grounding techniques and maintaining high humidity, and designing electrical systems according to NFPA 70, *National Electrical Code*®.

An administrative control program can be implemented through operational and maintenance procedures and through the use of a permit system to control hazards.

Fire protection by design focuses on managing the impact of fire, since fire prevention cannot always be obtained. Two major categories of fire protection are passive fire protection and active fire protection. Passive fire protection centers on the control of fire through building design. For a passive system, design objectives for building design include

1. Separate hazardous areas from essential areas in the facility layout
2. Compartmentalization with fire barriers to limit fire spread
3. Use of fire-retardant materials to reduce fire growth and spread

Passive fire protection should begin in the conceptual design stage and be carried on through design and construction. Retrofitting a facility with passive fire protection is normally a very expensive and difficult process.

Active protection involves fire control with specialized fire protection systems and equipment. The general design objective for active protection is the detection and suppression of fire before unacceptable damage can occur. Key types of active fire protection are (1) fire suppression systems, (2) fire detection systems, and (3) smoke control (smoke management) systems. Common active fire suppression systems for the control and extinguishment of fire are automatic sprinkler systems, fixed-water-spray systems, clean-agent systems, CO_2 systems, and foam extinguishing systems. Heat, smoke, and flame-sensing devices are examples of fire detection systems.

Most extinguishing systems have fire detection components that automatically actuate the system.

DESIGN OBJECTIVES

The design objectives for a fire alarm and/or a fire detection system start with the basic objectives of the system itself. These objectives can include notifying building occupants of emergency conditions for evacuation purposes; alerting first responders such as a fire brigade or fire department; detecting fire at specific stages, for example, smoldering or flaming, or monitoring of development; actuating fire suppression systems; supervising fire suppression systems; and supervising processes for abnormalities that might result in a fire. These design objectives should be discussed and agreed to by all of the project stakeholders before proceeding with the design.

DESIGN CRITERIA

Design criteria provide the basis for system specifications and design drawings. Development of design criteria is essential to the fire alarm system design package and must be completed prior to preparation of the construction drawings and detailed specifications. The criteria should include an area-by-area engineering study that defines the type of combustibles in each and postulates likely fire scenarios. The construction features of the facility are of the utmost importance when considering the design of the fire alarm system. Ceiling heights and the type of ceiling construction are two features that must be considered for optimum detector placement. Other building parameters that must be analyzed are the ventilation and expected temperature range for each area. Natural and mechanical airflow patterns must be defined and quantified during the analysis stage of design.

Design criteria address system function and overall operation. Again, the design criteria must be prepared prior to the design of a fire alarm system and must include technical data defining system purpose and design objectives, design parameters, applicable design codes, standards and regulations, and quality assurance requirements.

The two essential parameters for fire detection are combustible types and possible fire scenarios. These must be evaluated through an area-by-area analysis of a facility.

1. Types of combustibles in each area. Each area of the facility where fire detection may be used should be reviewed to determine the type of fuel present, that is, solids, liquids, or gases. The type of fuel will often dictate the type of fire detection needed. For example, if the design objective is evacuation of the building occupants and the combustible material in the area is a highly smoke-producing material, the design would call for smoke detectors rather than heat detectors. If the design objective of the fire detection system in a garage facility is to actuate a deluge sprinkler system (because the garage may contain exhaust gases), the system design should consider specifying use of heat or photoelectric smoke detectors rather than ionization-type smoke detectors.

2. Fire scenarios. Fire scenarios should be based on the type of combustibles in the area and potential ignition sources as well as the characteristics of potential occupants. Fires can be categorized into smoldering or flaming types. Fire growth varies as a function of the types of combustibles, their physical arrangement, surface-to-mass ratio, and ignition source energy. Some common types of postulated fires include those involving ordinary combustibles, such

as cellulosic products and flammable/combustible liquids; electric cable insulation; plastic fuel; and combustible metals, dusts, or fibers. The types of likely fire scenarios need to be evaluated prior to selection of the actual detector. Occupant characteristics, including age and possible physical limitations, will determine the time needed for egress to a safe area. High occupant loadings may require phased evacuation or relocation to safe areas within the building. In such cases, voice evacuation systems would be required.

Other parameters that should be evaluated in the design phase of a project include ceiling height, possible stratification, ceiling configuration, room ventilation, and room temperature.

Prescriptive Requirements

Ceiling Height: Since smoke, heat, and fire-gas production in the early stages of a fire is relatively slow, the effect of ceiling height can be significant. As ceiling height increases, a larger fire is needed to actuate the same detector in the same time due to dilution by ambient air entrained into the fire plume. As fire increases in intensity and severity, the significance of the effect of ceiling height

Table 9.1 Heat Detector Spacing Reduction Based on Ceiling Height

Ceiling Height Above		Up to and Including		Multiply Listed Spacing by
m	*ft*	*m*	*ft*	
0	0	3.05	10	1.00
3.05	10	3.66	12	0.91
3.66	12	4.27	14	0.84
4.27	14	4.88	16	0.77
4.88	16	5.49	18	0.71
5.49	18	6.10	20	0.64
6.10	20	6.71	22	0.58
6.71	22	7.32	24	0.52
7.32	24	7.93	26	0.46
7.93	26	8.54	28	0.40
8.54	28	9.14	30	0.34

Source: *National Fire Alarm Code*®, NFPA, 2002, Table 5.6.5.5.1.

is reduced due to the formation of a hot upper layer. Height is the most important single dimension with regard to detector spacing when the ceiling height is over 16 ft (4.9 m). The most sensitive detectors should be used when the ceiling height is over 30 ft (9 m).

As ceiling height increases, spot-type heat detector spacing should be reduced (see Table 9-1).

Stratification: Stratification of air in a room can prevent smoke particles or gaseous combustion products from reaching ceiling-mounted smoke or fire-gas detectors.

Smoke and fire gases are carried by the fire plume to the ceiling by buoyancy: The hot material is less dense than the ambient air and rises. The rising plume entrains ambient air, which dilutes the smoke and gases and reduces the temperature. Stratification occurs when the temperature of the plume gases is the same as that of the surrounding air so there is no longer any difference in density. At that point, the plume gases will spread laterally to the walls.

In installations where detection of smoldering or small fires is desired and where the possibility of stratification exists, consideration should be given to mounting some of the detectors below the ceiling (see Figure 9-1). Specific designs for such a system should be based upon an engineering survey and analysis.

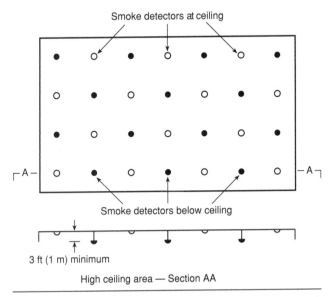

Figure 9-1. Detector location for overcoming roof/ceiling air stratification. For SI Units: 1 ft = 0.305 m.

Ceiling Configuration: Ceiling configuration greatly influences the spacing of detectors and therefore the number of detectors needed. *NFPA 72®, National Fire Alarm Code®*, provides specific instructions on detector location related to the shapes of the ceilings. Level ceilings are those that are actually level or with a slope of 1.5 in./ft (100 mm/m) or less. Sloped ceilings (of either the peaked or shed type) are defined as those with a slope greater than 1.5 in./ft (100 mm/m). Ceiling structures can take many forms, such as the following.

Beam Construction: These ceilings have solid structural or solid nonstructural members projecting down from the ceiling surface more than 4 in. (100 mm) and spaced more than 3 ft (0.9 m) on center.

Girders: For these ceilings, girders support beams or joists and run at right angles to the beams or joists. When a girder is within 4 in. (100 mm) of the ceiling, it becomes a factor in determining the number of detectors and is to be considered as a beam. When the top of the girder is more than 4 in. (100 mm) from the ceiling, it is not a factor in detector location.

Solid Joist Construction: These ceilings have solid structural or solid non-structural members projecting down from the ceiling surface a distance of more than 4 in. (100 mm) and spaced at intervals of 3 ft (0.9 m) or less on center.

Smooth Ceiling: This ceiling has a surface uninterrupted by continuous projections, such as solid joists, beams, or ducts extending more than 4 in. (100 mm) below the ceiling surface.

Note: Open truss construction is not considered to impede the flow of fire products unless the upper member is in continuous contact with the ceiling projects below the ceiling more than 4 in. (100 mm).

Room Ventilation: The effect of room ventilation on smoke detector placement is critical since air movement directly influences smoke particles and fire gases. The higher the air velocity, the more complex does the smoke detection design become. The HVAC systems found in most facilities need to be considered in smoke detection system design. Smoke detectors should be positioned near return registers, not supply registers, since detectors mounted near supply registers can be blocked from smoke by the clean air supply. Detectors located near return registers, however, will be in the path of travel of smoke and fire gases drawn toward the return. Also, air velocities near returns are lower than those near supplies. Slot diffusers common in commercial systems have some unusual flow characteristics that should be

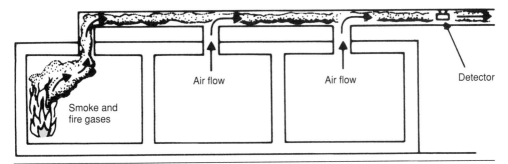

Figure 9-2. Dilution effect on the operation of an air duct detector.

considered. Guidance can be found in *NFPA 72* based on a multiyear study carried out by NIST and funded by the Fire Detection Institute through NFPA's Fire Protection Research Foundation (see Additional Reading at the end of this chapter).

Detectors installed in ducts of HVAC systems in accordance with NFPA 90A, *Standard for the Installation of Air-Conditioning and Ventilating Systems,* should not be used as a substitute for open-area protection because dilution often reduces smoke concentrations in ducts and smoke will not be drawn into the ductwork when the HVAC system is shut down (see Figures 9-2 and 9-3). Many modern HVAC systems are variable-volume systems, which vary the air flows based on conditions and demand, and at times cycle the systems off.

Other design problems associated with HVAC systems and detector placement concern perforated membrane ceilings, which evenly distribute supply air

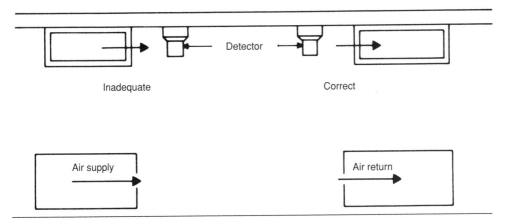

Figure 9-3. Smoke detector locations for ventilation systems.

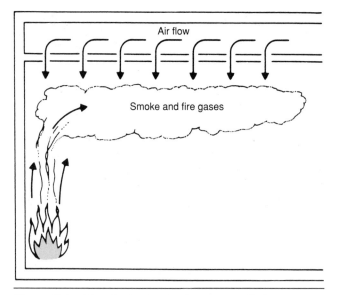

Figure 9-4. Air velocity barrier from a ventilation system.

through a suspended ceiling. In the event of fire in these rooms, smoke and fire gases would not travel to the ceiling level until the fire plume developed hot thermal gases. One technique for avoiding this problem is to use an air shield at least 3 sq ft (0.28 m^2) around the detector (see Figures 9-4 and 9-5). In clean rooms, ventilation may be distributed with supply through the ceiling and

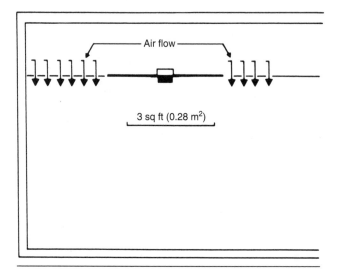

Figure 9-5. Air shield installation on a perforated membrane ceiling system.

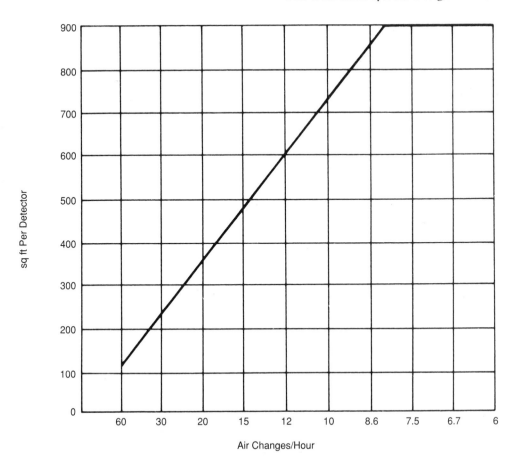

For SI units: 1 sq ft = 0.09 m^2.

Figure 9-6. Smoke detector spacing in high-air-movement areas.

return at the floor as a means of dust control. Here, smoke will not rise against the flow and other means of detection are needed.

For rooms and areas with high air movement, smoke detector spacing needs to be reduced (see Figure 9-6 and Table 9-2). The spacing coverage for high air movement should not be used for underfloor or above-ceiling spaces. Since the volume of these spaces is rather small (usually around ¹⁄₁₀ the volume of the room), reduced detector coverage is normally not required, but may be influenced by physical construction and air movement within the space.

Detectors should be placed at least 3 ft (0.9 m) from an outlet opening in a weak ventilation area and a minimum of 9 ft (2.7 m) from outlets in areas with

Table 9.2 Smoke Detector Spacing Based on Air Movement

Minutes per Air Change	Air Changes per Hour	Spacing per Detector	
		m^2	ft^2
1	60	11.61	125
2	30	23.23	250
3	20	34.84	375
4	15	46.45	500
5	12	58.06	625
6	10	69.68	750
7	8.6	81.29	875
8	7.5	83.61	900
9	6.7	83.61	900
10	6	83.61	900

Source: *National Fire Alarm Code*®, NFPA, 2002, Table 5.7.5.3.3.

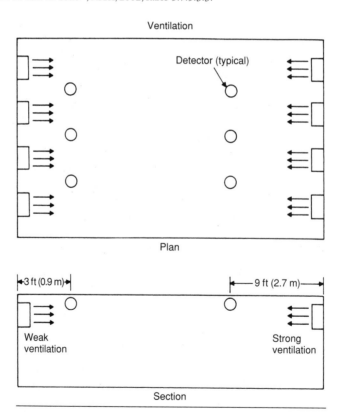

Figure 9-7. Detector placement in weak and strong ventilation areas.

strong ventilation (see Figure 9-7). Detectors must be placed so they are effective, regardless of the operating state of the ventilation system (see Figure 9-8). Rooms with exhaust ducts are best protected by a detector that is placed immediately after the last air exhaust opening (before the air enters a common duct), or dilution of the smoke could prevent early detector activation (see Figure 9-9).

Room Temperature: Room temperature is an important variable that must be determined in the design stage before detectors are selected. Many new heat

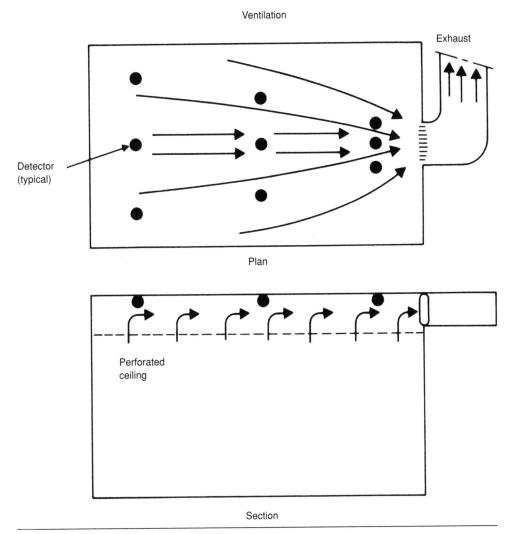

Figure 9-8. Detectors effective with HVAC system on or off.

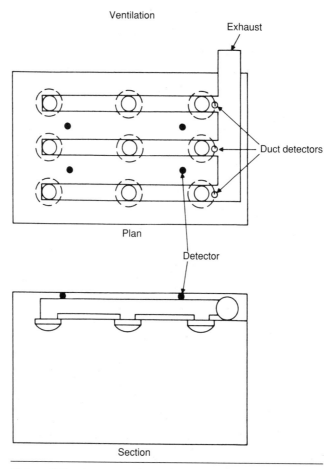

Figure 9-9. Detector placement in exhaust duct area.

detector designs have solid-state electronic components that are adversely affected by extreme temperatures.

Smoke detectors are normally listed to be installed in areas where the ambient temperature is not more than 100°F (37.8°C) or less than 32°F (0°C). Control units fall within the same temperature range because of their sensitive electronic parts. During design, local heat sources, such as ovens, boilers, and furnaces, need to be considered relative to detector placement in the room.

Fixed-temperature and rate-compensated types of heat detectors are categorized by their operational temperature (see Table 9-3).

Since most ceiling temperatures are under 100°F (37.8°C), ordinary temperature classification detectors are normally used. Local heat sources, including

Table 9.3 Temperature Classification for Heat-Sensing Fire Detectors

Temperature Classification	Temperature Rating Range		Maximum Ceiling Temperature		Color Code
	°C	°F	°C	°F	
Low*	39–57	100–134	28	80	Uncolored
Ordinary	58–79	135–174	47	115	Uncolored
Intermediate	80–121	175–249	69	155	White
High	122–162	250–324	111	230	Blue
Extra high	163–204	325–399	152	305	Red
Very extra high	205–259	400–499	194	380	Green
Ultra high	260–302	500–575	249	480	Orange

*Intended only for installation in controlled ambient areas. Units shall be marked to indicate maximum ambient installation temperature.

Source: *National Fire Alarm Code*®, NFPA, 2002, Table 5.6.2.1.1.

exhaust stacks, kilns, heaters, and lights, need to be considered in design of heat detector placement to guard against false alarms.

HEAT DETECTORS

Heat detectors should be positioned on the ceiling not less than 4 in. (100 mm) from the side wall or on the side walls between 4 and 12 in. (100 and 300 mm) from the ceiling. Heat detectors should not normally be located on the lower flange of beams or suspended from the ceiling by conduit. Where beams are less than 12 in. (300 mm) in depth and less than 8 ft (2.4 m) on center, heat detectors can be installed on the bottoms of the beams. The space where the ceiling and side wall meet is normally considered a "dead air pocket" or "void." In room fire development, the concentration of fire heat and smoke could be low in this void space, so spot-type detectors should not be located here (see Figure 9-10).

Heat detectors placed on smooth ceilings should not exceed listed spacing unless permitted by an engineering analysis. The detectors should be within a distance of one-half the listed spacing, measured at a right angle, from all walls or partitions extending to within 18 in. (0.46 m) of the ceiling. For example, a rate-compensation heat detector (UL listed) on a smooth ceiling should have 50-ft (15-m) spacing and the spacing to a wall or partition should be 25 ft (7.6 m).

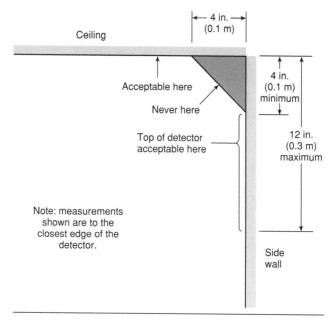

Figure 9-10. Spot-type detector locations.

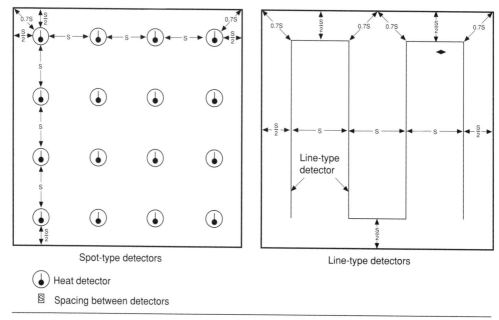

Spot-type detectors

Heat detector

S   Spacing between detectors

Figure 9-11. Heat detector spacing for a smooth ceiling.

Heat detectors on a smooth ceiling should include all points of the ceiling within a distance equal to 0.7 times the listed spacing (see Figure 9-11).

Heat detectors placed on smooth ceilings in irregularly shaped areas can have greater than listed spacing between detectors, provided the maximum spacing from a detector to the farthest point on a side wall or corner is not greater than 0.7 times the listed spacing.

Figure 9-12 shows detector placement with listed 30-ft (9.1-m) spacing. A detector listed for a 30-ft (9.1-m) spacing will cover an area of a circle with a radius of 21 ft (6.4 m). For a rectangular area, a single properly located detector will meet the spacing requirement of the diagonal of the rectangle if the diagonal does not exceed the diameter of the circle. Detector efficiency will be increased due to the reduced area of coverage.

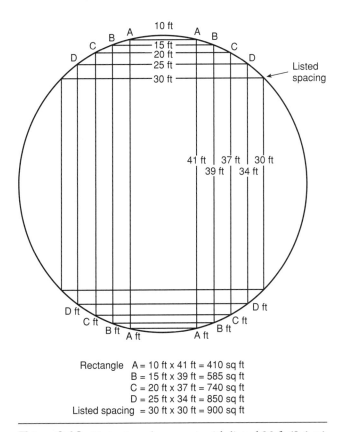

Rectangle A = 10 ft x 41 ft = 410 sq ft
B = 15 ft x 39 ft = 585 sq ft
C = 20 ft x 37 ft = 740 sq ft
D = 25 ft x 34 ft = 850 sq ft
Listed spacing = 30 ft x 30 ft = 900 sq ft

Figure 9-12. Detector placement with listed 30-ft (9.1-m) spacing. For SI Units: 1 ft = 0.30 m; 1 sq ft = 0.09 m².

In this case, 30 by 30 ft (9.1 × 9.1 m) is equal to 900 sq ft (83.6 m²), which is the maximum coverage for this detector. As an example, one detector would cover a corridor 10 by 41 ft (3 × 12.5 m) with the diagonal of the rectangle 21 ft (6.4 m) and a coverage of 410 sq ft (38 m²).

Figure 9-13 shows that spot-type heat detectors on solid joist construction should not exceed 50 percent of the smooth ceiling spacing when measured at right angles to the solid joists.

Heat detector placement on beam construction is considered to be the same as for a smooth ceiling where the beams project no more than 4 in. (100 mm) below the ceiling. If the beams project more than 4 in. (100 mm) but less than 18 in. (0.46 in), the spacing of spot-type heat detectors at right angles to the direction of beam travel should not exceed two-thirds of the smooth ceiling spacing. If the beams project more than 18 in. (0.46 m) below the ceiling and are more than 8 ft (2.4 m) on center, each bay formed by the beams is considered to be a separate area.

Heat detectors placed on peaked ceilings should be located within 3 ft (0.9 m) of the peak of the ceiling, measured horizontally (see Figure 9-14, bottom). The spacing of additional detectors should be based on the horizontal projection of the ceiling in accordance with the type of ceiling construction. Placement is similar for detectors on shed-type ceilings, except that the detector should be positioned within 3 ft (0.9 m) of the highest side of the ceiling measured horizontally and spaced in accordance with the type of construction (see Figure 9-14, top).

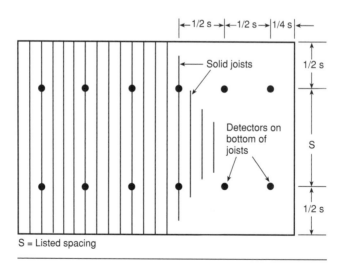

S = Listed spacing

Figure 9-13. Spot-type heat detector spacing layouts for joisted ceiling.

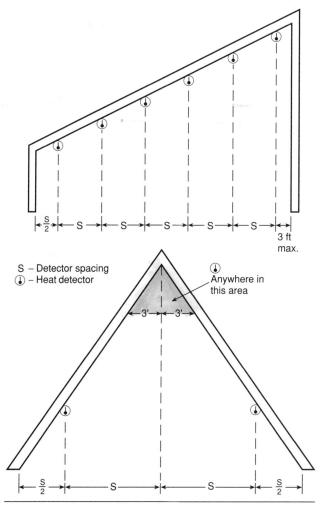

Figure 9-14. Heat and smoke detector placement for sloped ceilings (shed, top; peaked, bottom). For SI Units: 1 ft = 0.30 m.

SMOKE DETECTORS

Smoke detector spacing and location should be based on a detailed engineering study. Smoke development and travel are influenced by many parameters, including ceiling configuration, ceiling height, burning characteristics of materials, fuel arrangement, room geometry (including openings), and HVAC systems.

Since many variables affect detector placement, testing laboratories do not assign specific spacing to smoke detectors, but 30 by 30 ft (9.1 m × 9.1 m) or 900 sq ft (83.6 m²) is usually considered the maximum under optimum conditions on smooth ceilings.

Spot-type smoke detectors should be placed on ceilings not less than 4 in. (100 mm) from any side wall to the near edge, or on the side wall between 4 and 12 in. (100 and 300 mm) down from the ceiling to the top of the detector. This positioning is similar to that for spot-type heat detectors.

Exceptions to positioning spot-type smoke detectors on the ceiling or side wall include use of detectors to mitigate smoke stratification, use of solid joist construction (with which a detector can be mounted on the bottom of the joist), and use of beam construction, for which a detector can be mounted on the lower beam flange if the beams are less than 12 in. (300 mm) deep and less than 8 ft (2.4 m) on center.

For projected beam-type smoke detectors, the beam should parallel the ceiling and be within 20 in. (0.5 m) of the ceiling. The listed length of the projected beam should be reduced by 33 percent for each mirror used (see Table 9-4).

In general, a maximum smoke detector spacing of 30 ft (9.1 m) is commonly used on smooth ceilings. Smoke detector spacing for solid joist construction with joists 8 in. (200 mm) or less is considered to be the same as that for smooth ceilings. Spot-type smoke detector spacing perpendicular to the joist should be reduced if the joist is more than 8 in. (200 mm) deep.

Table 9-4. Projected Beam Using Mirrors*

Number of Mirrors		Maximum Allowable Beam Length
0	Listed Length L	
1	$(2/3) L = a + b$	
2	$(4/9) L = c + d + e$	

* Example: Maximum allowable length of beam listed for 300 ft (L) using two mirrors is (4/9) × 300 ft, or 133 ft. For SI Units: 1 ft = 0.30 m.

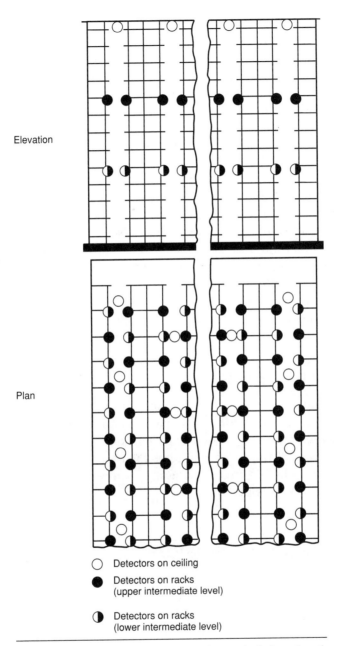

Elevation

Plan

○ Detectors on ceiling

● Detectors on racks
   (upper intermediate level)

◐ Detectors on racks
   (lower intermediate level)

Figure 9-15. Detector placement for typical closed rack storage.

Smoke detector spacing for beam construction with beams 8 in. (200 mm) or less is considered the same as that for smooth ceilings. Spacing of spot-type detectors should also be reduced perpendicular to the beams if beams exceed 8 in. (200 mm) in depth. Where beams are more than 18 in. (0.46 m) deep and more than 8 ft (2.4 m) on center, each bay should be considered a separate area containing at least one detector.

Smoke detector placement for a sloped ceiling is the same as that for heat detectors. Smoke detectors for peaked ceilings should be located within 3 ft (0.9 m) of the peak, measured horizontally. Spacing for additional detectors is based upon the horizontal projection of the ceiling (see Figure 9-14).

Smoke detectors for shed-type ceilings should be mounted within 3 ft (0.9 m) of the highest side of the ceiling measured horizontally. Spacing of additional detectors is similar to that for peaked ceilings (see Figure 9-14).

Smoke detectors for high rack storage areas provide early warning of fire. Smoke detectors should be positioned on the ceiling above each aisle and at intermediate levels in the racks (see Figure 9-15 on the preceding page). Providing detectors only at the ceiling level may hamper early detection of fire since the rack traps smoke, which lacks the thermal lift to transport itself to the ceiling detectors.

FLAME-SENSING FIRE DETECTORS

Spacing for flame-sensing fire detectors should be based on engineering judgment in accordance with their listed spacing and the manufacturer's instructions. Since flame detectors respond to either IR or UV radiation and are line-of-sight devices, some of the following items might affect the operation of flame detectors: flickering light sources, sunlight, incandescent and fluorescent lights, and cutting and welding operations.

PERFORMANCE-BASED FIRE DETECTOR DESIGN

Annex B to *NFPA 72* provides an engineering methodology for determining heat, smoke, and flame detector spacing based on the size and rate of growth of fire to be detected, various ceiling heights, and ambient temperature (see Figure 9-16). This method is applicable for flaming fires only and represents a performance-based alternative to prescriptive placement rules. It proposes that a fire detection system can be designed to detect a growing (flaming) before that fire reaches a specific heat release rate (Btu/sec). Annex B utilizes the results of fire research funded by the Fire Detection Institute.

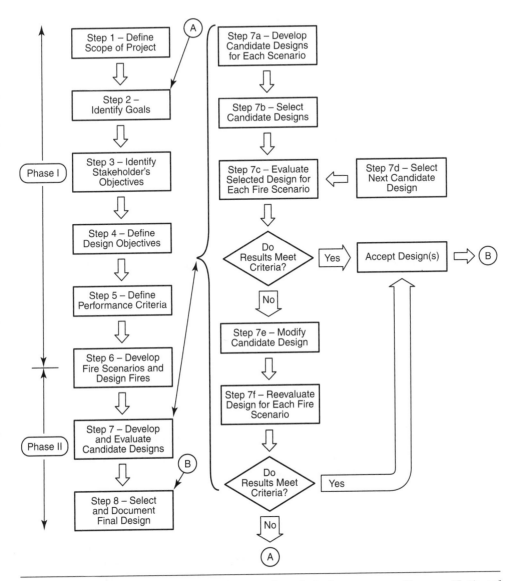

Figure 9-16. Overview of the performance-based design process. (Source: *National Fire Alarm Code*®, NFPA, 2002, Figure B.2.1.)

The data of Annex B provide a method for modifying the listed spacing of both rate-of-rise and fixed-temperature heat detectors required to achieve detector response to a geometrically growing flaming fire at a specific fire size. The data applicable to smoke detectors are limited to a theoretical analysis based on the flaming fire test data and are not intended to address the detection of smoldering fires. Flame detector design is based on theoretical considerations only.

For additional information on fire detector placement and design, see Chapter 8, Fire Alarm System Installation; and Chapter 17, Fire Warning Systems for Dwellings.

DETECTOR PLACEMENT METHODOLOGY

A method is developed in *NFPA 72* to assist fire protection design engineers in the proper placement of fire detectors based on full-scale fire test results. These test results incorporate fire plume dynamics and the rate of heat release for a variety of combustible materials. This method can be used for determining the location of fire detectors and evaluating the adequacy of existing fire detection systems.

Design engineers should refer to *NFPA 72*, Annex B, for guidance in the proper analysis of detector selection and placement. Annex B provides an analysis technique that is far superior to the outdated method of symmetrically locating detectors on electrical drawings without regard to the building variables and fire development. Electrical drawings are typically not dimensioned, so the installing electrical contractor actually locates the detection devices, not the design engineer. With the new method, use of dimensions on drawings is essential for optimum detector response, based upon the building features and detector response characteristics.

Fire research funded by the Fire Detection Institute provided the basic fire test data and analysis for the development of Annex B. The fire tests conducted for the research were full-scale tests featuring geometrically growing flaming-mode fires, so-called *t*-squared fires. The research did not include smoldering-type fire scenarios due to the complexity of predicting smoke particle movement without fire plume formation and ceiling jet flows. The understanding of smoke production, aging, agglomeration, and transport mechanisms lags behind the study of heat production. In addition, the smoke detector characteristics relating to specific fire environments are often not available to the design engineer. Therefore, the discussion of smoke detector placement in Annex B is for the detection of flaming-mode fires and not for smoldering fires.

Limitations

Annex B to *NFPA 72* has definite limitations for the placement of smoke detectors for early-warning fire notification. If the intent of the design requires signal notification for incipient or smoldering-type fires, Annex B should not be utilized. Annex B does not address the design of early-warning fire detection systems for the objective of complete evacuation of the facility occupants prior to untenable conditions.

Annex B is primarily used for property protection, such as the design of thermal detection systems that activate fire suppression systems. Facilities where Annex B should *not* be applied include: computer rooms; control rooms; clean rooms; storage vaults; electrical rooms where smoldering fires are expected, such as switchgear rooms and motor control centers; and other area types where the anticipated design fire is incipient or smoldering.

Main Variables

The primary purpose of Annex B to *NFPA 72* is to provide an engineering methodology for achieving fire detector response at a specific threshold fire size based on the fire's rate of heat release. Other key parameters affecting the detector response are the height of the ceiling (above the fire source), minimum ambient temperature, and listed spacing of the detection devices. Knowledge of these parameters and an understanding of the methodology is all that is necessary to utilize Annex B. The basic concept of Annex B is that a fire detection system can be designed and installed to detect a given fire size of a specific heat release rate.

The five main variables addressed by Annex B fire models are as follows:

1. Time of fire growth
2. Heat release rates
3. Ceiling height
4. Detector time constant (a function of the listed spacing)
5. Listed spacing

Time of fire growth and heat release rates are discussed in what follows.

Time of Fire Growth: The time of fire growth is the time from flame initiation to the time fire develops a heat release rate of 1000 Btu/sec (1055 kW/sec). Fires are categorized by their rate of growth as being slow, medium, or fast developing. A slow-developing fire is one that takes at least 400 seconds or more to reach a heat release rate of 1000 Btu/sec (1055 kW/sec). Not many fire scenarios fit the slow-developing fire classification.

A medium-developing fire is one that takes from 150 to 400 seconds to reach a heat release rate of 1000 Btu/sec (1055 kW/sec). Examples of fires that meet the medium-developing fire criteria are as follows:

1. A ⅛-in. (3.2-mm) plywood wardrobe with fire-retardant interior finish has a fire growth time of 300 seconds.
2. A metal wardrobe with 94 lb (42.6 kg) of combustible cartons has a fire growth time of 250 seconds.
3. A 36-lb (16.3-kg) metal chair frame with minimum cushion has a fire growth time of 200 seconds.

A fast-developing fire is one that takes less than 150 seconds to reach a heat release rate of 1000 Btu/sec (1055 kW/sec). Examples of fires that meet the fast-developing fire criteria are as follows:

1. Polyethylene pallets stacked 3 ft (0.9 m) high have a fire growth time of 130 seconds.
2. A single, horizontal polyurethane foam mattress has a fire growth time of 110 seconds.
3. Wood pallets stacked 5 ft (1.5 m) high with a moisture content between 6 and 12 percent have a fire growth time between 90 and 190 seconds.

Figure 9-17 illustrates time-squared (t^2)-type fire scenarios, which are characterized as slow, medium, or fast.

The 1000-Btu/sec (1055-kW/sec) heat release rate is an arbitrary figure. To put this number in perspective, the rate of heat release for listed thermal detectors that are tested with constant-burning ethyl alcohol fires is approximately 1200 Btu/sec. The listed spacing of the detectors is compared to a 160°F (71°C) sprinkler with 10-ft (3-m) spacing.

Heat Release Rates: FM Global and the National Institute of Standards and Technology performed over 100 calorimeter fire tests on furniture and warehouse materials. Results giving heat release rates, fire classifications, fire intensity coefficients, and maximum heat release rates are listed in tables in Annex B to *NFPA 72*. Typical maximum rates of heat release from the Fire Detection Institute research are as follows:

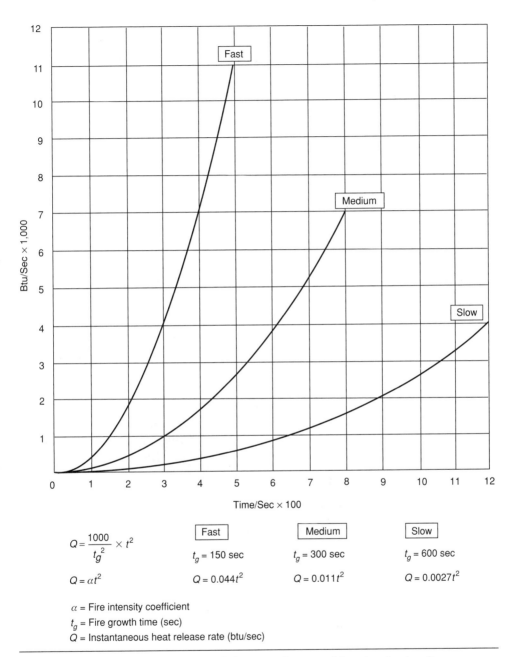

Figure 9-17. Graphic comparison of slow-, medium-, and fast-developing fires.

Substance	Heat Released
Medium wastebasket with milk cartons	100 Btu/sec
Large barrel with milk cartons	140 Btu/sec
Polyurethane foam upholstered chair	350 Btu/sec
Furnished living room (heat at door)	4000–8000 Btu/sec
Gasoline	200 Btu/sec/ft^2
Kerosene	200 Btu/sec/ft^2
Diesel oil	180 Btu/sec/ft^2
Methyl alcohol	65 Btu/sec/ft^2

Note that pool fires of flammable and combustible liquids have a steady-state heat release rate. These pool fires do not follow the t^2-type fire scenario. If Annex B is used as a guide for pool fires, time to actuation of the heat detector will be less in the pool fire than the calculated fire growth time.

Threshold Fire Size: From a design perspective, a threshold fire size Q_d, is selected by the engineer as the rate of heat release at which detection is desired. This selection needs reflect the level of desired protection. The detection system is designed and installed to detect a threshold fire size at a specific heat release rate.

For example, say a design engineer is protecting a storage tank of kerosene and wants to detect a 5-sq ft (0.45-m^2) pool fire. Since kerosene has a maximum heat release rate of 200 Btu/sec/ft^2, the rate of heat release of the liquid for the 5-sq ft (0.45-m^2) area is 1000 Btu/sec (1055 kW/sec). This is Q_d, the threshold fire size for the anticipated pool fire.

Fire Growth Model: The design fires used in Annex B to *NFPA 72* grow according to the following equation:

$$Q = (1000/t_g^2)t^2$$

where

   Q = Heat release rate (Btu/sec)
   t_g = Fire growth time (slow, medium, or fast) (sec)
   t = Time after flaming occurs (sec)
$1000/t_g^2$ = Fire intensity coefficient α

This t^2 fire formula is the mathematical model that forms the foundation of Annex B.

The equation for the rise in temperature of a heat detector is

$$\frac{\Delta T_g}{dt} = \frac{T_g - T_d}{\tau}$$

where
T_d = Temperature rating of the detector
T_g = Gas temperature at the detector
τ = Detector time constant

The detector time constant τ is a measure of the sensitivity of the particular detector. It is measured in the FM Global plunge test apparatus at a reference velocity of 5 ft/sec (1.5 m/sec), and is expressed as

$$\tau = (mc)/(hA)$$

where
τ = Detector time constant
m = Detector element mass
c = Detector element specific heat
h = Convective heat transfer coefficient
A = Surface area of the detector element

The convective heat transfer coefficient h varies as the square root of the velocity U of the gas flowing past the detector.

Response Time Index: The response time index (RTI) is equal to the detector time constant τ multiplied by the square root of the gas velocity referenced at 5 ft/sec (1.5 m/sec):

$$\text{RTI} = \tau\sqrt{u}$$

where
RTI = Response time index ($\text{sec}^{1/2} \text{ ft}^{1/2}$)
τ = Detector time constant (seconds)
U = Gas velocity set at 5 ft/sec (1.5 m/sec)

A detector having a RTI of 56 $\text{sec}^{1/2} \text{ ft}^{1/2}$ has a τ of 25 sec when measured at a gas velocity of 5 ft/sec (1.5 m/sec).

Spacing of Underwriters Laboratories Inc.-listed or FM Global-approved heat detectors is based on the detector time constant τ as established in Annex B to *NFPA 72* (see Table 9-5).

Table 9-5. Time Constant τ for Listed Heat Detectors*[,][†]

Listed Spacing (ft)	τ (sec) from UL at Given Temperature						τ from FMRC
	128°F	135°F	145°F	160°F	170°F	196°F	
10	400	330	262	195	160	97	196
15	250	190	156	110	89	45	110
20	165	135	105	70	52	17	70
25	124	100	78	48	32		48
30	95	80	61	36	22		36
40	71	57	41	18			
50	59	44	30				
70	36	24	9				

*These time constants [at a reference velocity of 5 ft/sec (1.5 m/sec)] are based on an analysis of the Underwriters Laboratories Inc. (UL) and FM Global (FMRC) listing test procedures. FMRC results are for all temperatures. Plunge test results performed on the detector to be used give a more accurate time constant. See Annex B to *NFPA 72*®, *National Fire Alarm Code*®, for a further discussion of detector time constants.
†These time constants can be converted to response time index (RTI) values by multiplying by (5 ft/sec)$^{1/2}$ [(1.5 m/sec)$^{1/2}$]. See Annex B to *NFPA 72*.
For SI Units: 1 ft = 0.30 m; °C = (5/9) (°F −32).

As an example, if the temperature rating of a heat detector is 145°F (63°C) and the UL-listed spacing is 50 ft (15 m), then τ is 30 sec. Then,

$$\text{RTI} = \tau(U)^{1/2} = 30(5)^{1/2} = 67\,\text{sec}^{1/2}\,\text{ft}^{1/2}$$

Method for Placing a Flaming-Mode Fire Detector

To decide how to place a flaming-mode fire detector:

1. Establish the goals of the fire detection system.
2. Select the threshold fire size Q_d, which is the desired rate of heat release at the time of detection.
3. Select the fire growth time t_g, which is the detection time and response time.
4. Determine the environmental conditions, including ceiling height and minimum ambient ceiling temperature.
5. Select the fire detector to be used, including type and sensitivity.
6. Use the tables from Annex B to *NFPA 72* to determine the design spacing.

7. Evaluate the design choice.
8. Redesign and calculate, if necessary.

Design of Smoke Detector Systems

The design of smoke detector systems is more complex than for heat detectors. There is no listed spacing to use as a guide to an individual model's sensitivity. Smoke detectors operating on different principles exhibit different sensitivity to different smokes and even to the same smoke at different distances. Smoke detectors are more affected by ambient conditions, especially air velocity, humidity, and the presence of particles from nonfire sources. With care, however, activation times as a function of fire characteristics and detector placement can be estimated.

Predicting Smoke Detector Activation Times

There are three parts to the prediction of activation times for smoke detectors: determining the production rate of smoke and heat (to drive the smoke) by the fire, the transport of the smoke to the detector location, and the response of the specific detector model to that smoke. To obtain a response time estimate, each can be evaluated in sequence, taking into account any factors specific to the installation that might affect the process. Alternatively, computer fire models can be used that account for these factors more rigorously than can be done by hand calculations.

Production of Smoke and Heat

This aspect is similar to that used for heat detectors, except that an additional calculation is needed to estimate the smoke release rate. Whereas the heat release rate is the fuel mass loss rate multiplied by the heat of combustion (an inherent property of the fuel), the smoke release rate is the fuel mass loss rate times the *soot yield fraction* (also an inherent property of the fuel). Soot yield fractions are measured in various test methods, often at the same time as the heat release rate, and are expressed in the units of grams of soot produced per gram of fuel burned. In some test methods a smoke parameter called *specific extinction area* (kg/m^2) is reported. Specific extinction area can be converted to soot yield fraction by dividing by 8 kg/m^2.

This allows the mass of smoke released by the fuel per unit time to be calculated. This smoke mass is distributed into the upper-layer volume and results in a smoke mass concentration (kg/m^3) that can be converted to an extinction coefficient (percent/m) by multiplying by *Seader's coefficient*, which is 33,000 m^2/g for flaming smoke and 19,000 m^2/g for smoldering smoke. Typical alarm thresholds for smoke detectors are given in Annex B of *NFPA 72*.

Transport

Smoke particulate produced by burning fuel is transported to the detector location in the fire plume and ceiling jet. Ceiling jet velocities can be estimated by Alpert's correlation. For detectors outside the ceiling jet, horizontal velocities along the ceiling for flaming fires are in the range of 1 to 2 m/sec. The transport time to the detector can be estimated by using these results.

Detector Response

The detector response can be assumed when the local smoke density exceeds the detector threshold sensitivity. Smoke detectors are known to exhibit a delay in response referred to as *smoke entry delay*, which is related to the aerodynamic restriction of the outer housing and sensor design. This entry delay can be quantified as related to the detector's *characteristic length L*. The entry delay time is the detector's characteristic length divided by the local flow velocity.

Computer Models

The calculations described above are sufficiently complex that they cannot be done by hand with any degree of accuracy. Thus, these methods have been incorporated into several of the fire models in popular use. One, called DETACT-QS, is specifically for the estimation of the activation times of heat detectors and sprinklers. DETACT can be used for smoke detectors by applying the optical density to the temperature rise correlation, as described later.

A second model that can perform activation-time estimates is the zone-type model CFAST. This model implements the calculations discussed above and is most useful for cases with simple ceiling configurations and with no interference from HVAC systems within the protected space.

The most detailed and complex model that can be used is the computational fluid dynamics model FDS. This model can be used for complex geometries (e.g., beams, open joists, sloped ceilings) and can account for HVAC interactions. FDS also predicts sprinkler activation and the cooling effect of the water spray on the upper layer.

These models are available for downloading without charge at the NIST Web site, http://fast.nist.gov/.

Hand Calculations

Two highly simplified methods for hand calculation of smoke detector response are given in Annex B of *NFPA 72*. The first uses the correlation of optical density to temperature rise first developed by Heskestad.[1] This says that there is a direct correlation between the temperature rise and the smoke density at a detector. Thus, a smoke detector can be treated as a sensitive heat detector. In the late 1970s, when this approach was developed, smoke detectors had a temperature rise sensitivity of about 13°C. Improvements in detector design (especially the reduction of entry delays) have reduced this correlation value to about 5°C. A method for using this correlation of optical density to temperature rise to estimate smoke detector response time and the associated design fire size at response are given in Annex B of *NFPA 72*.

The second estimating method for smoke detector response time in Annex B of *NFPA 72* is based on the mass optical density at the detector and follows the considerations discussed above. This method distributes the smoke mass produced into the entire room volume (because estimating the layer volume as a function of time by hand is too difficult), which yields a conservative result (the room volume is always less than the layer volume, so the estimated mass concentration and the optical density are always less, and thus the detector response is longer).

Design of Radiant Energy-Sensing Detection Systems

Radiant energy-sensing detectors respond to light energy released by a fire at certain wavelengths. Thus, the response of a radiant energy-sensing detector can be estimated from the radiant energy output of the fire that reaches the detector and the detector's sensitivity.

Annex B of *NFPA 72* provides a method for estimating the radiated power from a fire as a function of the flame volume. Since the radiative fraction from most hydrocarbon fuels is relatively constant (about 35 percent), no fuel-specific data are needed. The method accounts for atmospheric attenuation at the wavelengths of interest and for angular displacement. A similar method for spark/ember detection systems is also provided.

DESIGN EXAMPLES

The following text and worksheets are taken from *NFPA 72®*, *National Fire Alarm Code®*, 2002 edition. All references to sections, figures, and equations are to *NFPA 72*.

Fire Detection Design and Analysis Worksheet
Design Example

1.	Determine ambient temperature (T_a) ceiling height or height above fuel (H).	$T_a =$ ___10___ °C + 273 = ___283___ K $H =$ ___4___ m
2.	Determine the fire growth characteristic (α or t_g) for the expected design fire.	$\alpha =$ ___0.047___ kW/sec^2 $t_g =$ ___150___ sec
3a.	Define the characteristics of the detectors.	$T_s =$ ___57___ °C + 273 = ___330___ K RTI = ___98___ m$^{1/2}$sec$^{1/2}$ $\dfrac{dT_d}{dt} =$ _____ °C/min $\tau_0 =$ _____ sec
3b. or	*Design* — Establish system goals (t_{CR} or Q_{CR}) and make a first estimate of the distance (r) from the fire to the detector.	$t_{CR} =$ ___146___ sec $r =$ ___3.3___ m $Q_{CR} =$ ___1000___ kW
3b.	*Analysis* — Determine spacing of existing detectors and make a first estimate of the response time or the fire size at detector response ($Q = \alpha t^2$).	$r =$ _____ *1.41 = _____ = S (m) $Q =$ _____ kW $t_d =$ _____ sec
4.	Using equation B.21, calculate the nondimensional time (t^*_{2f}) at which the initial heat front reaches the detector.	$t^*_{2f} = 0.861 \left(1 + \dfrac{r}{H}\right)$ $t^*_{2f} = 1.57$
5.	Calculate the factor A defined by the relationship for A in equation B.20.	$A = \dfrac{g}{C_p T_a \rho_o}$ $A = 0.030$
6.	Use the required response time (t_{CR}) along with the relationship for t^*_p in equation B.19 and $p = 2$ to calculate the corresponding value of t^*_2.	$t^*_2 = \dfrac{t_{CR}}{A^{-1/(3+p)}\alpha^{-1/(3+p)}H^{4/(3+p)}}$ $t^*_2 = 12.98$
7.	If $t^*_2 > t^*_{2f}$ continue to step 8. If not, try a new detector position (r) and return to step 4.	
8.	Calculate the ratio $\dfrac{u}{u^*_2}$ using the relationship for U^*_p in equation B.17.	$\dfrac{u}{u^*_2} = A^{1/(3+p)}\alpha^{1/(3+p)}H^{(p-1)/(3+p)}$ $\dfrac{u}{u^*_2} = 0.356$
9.	Calculate the ratio $\dfrac{\Delta T}{\Delta T^*_2}$ using the relationship for ΔT^*_p in equation B.18.	$\dfrac{\Delta T}{\Delta T^*_2} = A^{2/(3+p)}(T_a/g)^{2/(3+p)}\alpha^{-(5-p)/(3+p)}$ $\dfrac{\Delta T}{\Delta T^*_2} = 0.913$
10.	Use the relationship for ΔT^*_2 in equation B.23 to calculate ΔT^*_2.	$\Delta T^*_2 = \left[\dfrac{t^*_2 - t^*_{2f}}{(0.146 + 0.242r/H)}\right]^{4/3}$ $\Delta T^*_2 = 105.89$
11.	Use the relationship for $\dfrac{u^*_2}{(\Delta T^*_2)^{1/2}}$ in equation B.24 to calculate the ratio $\dfrac{u^*_2}{(\Delta T^*_2)^{1/2}}$.	$\dfrac{u^*_2}{(\Delta T^*_2)^{1/2}} = 0.59\left(\dfrac{r}{H}\right)^{-0.63}$ $\dfrac{u^*_2}{(\Delta T^*_2)^{1/2}} = 0.66$
12.	Use the relationships for Y and D in equations B.27 and B.28 to calculate Y.	$Y = \left(\dfrac{3}{4}\right)\left(\dfrac{u}{u^*_2}\right)^{1/2}\left[\dfrac{u^*_2}{(\Delta T^*_2)^{1/2}}\right]^{1/2}\left(\dfrac{\Delta T^*_2}{\text{RTI}}\right)\left(\dfrac{t}{t^*_2}\right)D$ $Y = 1.533$
13.	*Fixed Temperature HD* — Use the relationship for $T_d(t) - T_d(0)$ in equation B.25 to calculate the resulting temperature of the detector $T_d(t)$.	$T_d(t) = \left(\dfrac{\Delta T}{\Delta T^*_2}\right)\Delta T^*_2\left[1 - \dfrac{(1-e^{-Y})}{Y}\right] + T_d(0)$ $T_d(t) = 57.25$
14.	*Rate of Rise HD* — Use the relationship for $\dfrac{dT_d(t)}{dt}$ in equation B.26.	$dT_d = \left[\left(\dfrac{4}{3}\right)\left(\dfrac{\Delta T}{\Delta T^*_2}\right)(\Delta T^*_2)^{1/4}\dfrac{(1-e^{-Y})}{[(t/t^*_2)D]}\right]dt$ $dT_d =$
15.	If: 1. $T_d > T_s$ 2. $T_d < T_s$ 3. $T_d = T$	Repeat Procedure Using Design Analysis 1. a larger r 1. a larger t_r 2. a smaller r 2. a smaller t_r 3. $s = 1.41 \times r =$ ___4.7___ m 3. $t_r =$ _____ sec

Figure 9-18. Fire detection design and analysis worksheet—Design Example.

Evaluate Candidate Designs

Fixed-temperature heat detectors have been selected for installation in a warehouse with a 57°C (135°F) operating temperature and a UL-listed spacing of 9.1 m (30 ft). From Table B.3.2.5 of *NFPA 72*, the time constant is determined to be 80 seconds when referenced to a gas velocity of 1.5 m/sec (5 ft/sec). When used with equation B.14 of *NFPA 72*, the detector's RTI can be calculated as follows:

$$\text{RTI} = \tau_0 u_0^{1/2}$$

$$\text{RTI} = 98 \text{ m}^{1/2} \text{ sec}^{1/2}$$

or

$$\text{RTI} = 179 \text{ ft}^{1/2} \text{ sec}^{1/2}$$

In order to begin calculations, it will be necessary to make a first guess at the required detector spacing. For this example, a first estimate of 4.7 m (15.3 ft) is used. This correlates to a radial distance of 3.3 m (10.8 ft).

These values can then be entered into the design and analysis worksheet shown in Figure 9-18 in order to evaluate the candidate design.

After 146 seconds, when the fire has grown to 1000 kW (948 Btu/sec) and at a radial distance of 3.3 m (10.8 ft) from the center of the fire, the detector temperature is calculated to be 57°C (135°F). This is the detector actuation temperature. If the calculated temperature of the detector were higher than the actuation temperature, then the radial distance could be increased. The calculation would then be repeated until the calculated detector temperature is approximately equal to the actuation temperature.

The last step is to use the final calculated value of r with the equation relating spacing to radial distance. This will determine the maximum installed detector spacing that will result in detector response within the established goals.

$$S = 2^{1/2} r$$

$$S = 4.7 \text{ m (15.3 ft)}$$

where:
S = spacing of detectors
r = radial distance from fire plume axis (m or ft)

The following example shows how an existing heat detection system or a proposed design can be analyzed to determine the response time or fire size at response. The warehouse building has existing heat detectors. The fire, building, and detectors have the same characteristics as the previous example with the exception of spacing. The detectors are spaced evenly on the ceiling at 9.1-m (30-ft) intervals.

The following equation is used to determine the maximum radial distance from the fire axis to a detector:

$$S = 1.414r$$

or

$$r = \frac{S}{1.414}$$

$$r = 6.5 \text{ m } (21.2 \text{ ft})$$

where:
S = spacing of detectors
r = radial distance from fire plume axis (m or ft)

Next, the response time of the detector or the fire size at response is estimated. In the design above, the fire grew to 1000 kW (948 Btu/sec) in 146 seconds when the detector located at a distance of 3.3 m (10.8 ft) responded. As the radial distance in this example is larger, a slower response time and thus a larger fire size at response is expected. A first approximation at the response time is made at 3 minutes. The corresponding fire size is found using the power law fire growth equation B.16 of *NFPA 72* with $p = 2$ and α from B.3.3.6.6.1 of *NFPA 72*:

$$Q = \alpha t^{p}$$

$$Q = (0.047 \text{ kW/sec}^2)(180 \text{ sec})^2$$

$$Q = 1523 \text{ kW}$$

or

$$Q = (0.044 \text{ Btu/sec}^3)(180 \text{ sec})^2$$

$$Q = 1426 \text{ Btu/sec}$$

This data can be incorporated into the fire detection design and analysis worksheet shown in Figure 9-19 in order to carry out the remainder of the calculations.

Fire Detection Design and Analysis Worksheet
Design Analysis 2

1.	Determine ambient temperature (T_a) ceiling height or height above fuel (H).	$T_a = $ ___10___ °C + 273 = ___283___ K $H = $ ___4___ m
2.	Determine the fire growth characteristic (α or t_g) for the expected design fire.	$\alpha = $ ___0.047___ kW/sec^2 $t_g = $ ___150___ sec
3a.	Define the characteristics of the detectors.	$T_s = $ ___57___ °C + 273 = __330__ K RTI = __98__ m$^{1/2}$sec$^{1/2}$ $\dfrac{dT_d}{dt} = $ _____ °C/min $\tau_0 = $ _____ sec
3b. or	*Design* — Establish system goals (t_{CR} or Q_{CR}) and make a first estimate of the distance (r) from the fire to the detector.	$t_{CR} = $ _____ sec $r = $ _____ m $Q_{CR} = $ _____ kW
3b.	*Analysis* — Determine spacing of existing detectors and make a first estimate of the response time or the fire size at detector response ($Q = \alpha t^2$).	$r = $ ___6.5___ *1.41 = ___9.2___ = S (m) $Q = $ ___1,523___ kW $t_d = $ ___180___ sec
4.	Using equation B.21, calculate the nondimensional time (t^*_{2f}) at which the initial heat front reaches the detector.	$t^*_{2f} = 0.861 \left(1 + \dfrac{r}{H}\right)$ $t^*_{2f} = $ **2.26**
5.	Calculate the factor A defined by the relationship for A in equation B.20.	$A = \dfrac{g}{C_p T_a \rho_0}$ $A = $ **0.030**
6.	Use the required response time (t_{CR}) along with the relationship for t^*_p in equation B.19 and $p = 2$ to calculate the corresponding value of t^*_2.	$t^*_2 = \dfrac{t_{CR}}{A^{-1/(3+p)} \alpha^{-1/(3+p)} H^{4/(3+p)}}$ $t^*_2 = $ **16**
7.	If $t^*_2 > t^*_{2f}$ continue to step 8. If not, try a new detector position (r) and return to step 4.	
8.	Calculate the ratio $\dfrac{u}{u^*_2}$ using the relationship for U^*_p in equation B.18.	$\dfrac{u}{u^*_2} = A^{1/(3+p)} \alpha^{1/(3+p)} H^{(p-1)/(3+p)}$ $\dfrac{u}{u^*_2} = $ **0.356**
9.	Calculate the ratio $\dfrac{\Delta T}{\Delta T^*_2}$ using the relationship for ΔT^*_p in equation B.18.	$\dfrac{\Delta T}{\Delta T^*_2} = A^{2/(3+p)} (T_a/g) \alpha^{2/(3+p)} H^{-(5-p)/(3+p)}$ $\dfrac{\Delta T}{\Delta T^*_2} = $ **0.913**
10.	Use the relationship for ΔT^*_2 in equation B.23 to calculate ΔT^*_2.	$\Delta T^*_2 = \left[\dfrac{t^*_2 - t^*_{2f}}{(0.146 + 0.242 r/H)}\right]^{4/3}$ $\Delta T^*_2 = $ **75.01**
11.	Use the relationship for $\dfrac{u^*_2}{(\Delta T^*_2)^{1/2}}$ in equation B.24 to calculate the ratio $\dfrac{u^*_2}{(\Delta T^*_2)^{1/2}}$.	$\dfrac{u^*_2}{(\Delta T^*_2)^{1/2}} = 0.59 \left(\dfrac{r}{H}\right)^{-0.63}$ $\dfrac{u^*_2}{(\Delta T^*_2)^{1/2}} = $ **0.435**
12.	Use the relationships for Y and D in equations B.27 and B.28 to calculate Y.	$Y = \left(\dfrac{3}{4}\right)\left(\dfrac{u}{u^*_2}\right)^{1/2} \left[\dfrac{u^*_2}{(\Delta T^*_2)^{1/2}}\right]^{1/2} \left(\dfrac{\Delta T^*_2}{RTI}\right)\left(\dfrac{t}{t^*_2}\right)D$ $Y = $ **1.37**
13.	*Fixed Temperature HD* — Use the relationship for $T_d(t) - T_d(0)$ in equation B.25 to calculate the resulting temperature of the detector $T_d(t)$.	$T_d(t) = \left(\dfrac{\Delta T}{\Delta T^*}\right)\Delta T^*_2 \left[1 - \dfrac{(1 - e^{-Y})}{Y}\right] + T_d(0)$ $T_d(t) = $ **41**
14.	*Rate of Rise HD* — Use the relationship for $\dfrac{dT_d(t)}{dt}$ in equation B.26.	$dT_d = \left[\left(\dfrac{4}{3}\right)\left(\dfrac{\Delta T}{\Delta T^*}\right)(\Delta T^*_2)^{1/4} \dfrac{(1 - e^{-Y})}{[(t/t^*_2)D]}\right]dt$ $dT_d = $
15.	If: 1. $T_d > T_s$ 2. $T_d < T_s$ 3. $T_d = T$	Repeat Procedure Using Design Analysis 1. a larger r 1. a larger t_r 2. a smaller r 2. a smaller t_r 3. $s = 1.41 \times r = $ _____ m 3. $t_r = $ _____ sec

Figure 9-19. Fire detection design and analysis worksheet—Design Example.

Using a radial distance of 6.5 m (21 ft) from the axis of this fire, the temperature of the detector is calculated to be 41°C (106°F) after 3 minutes of exposure. The detector actuation temperature is 57°C (135°F). Thus, the detector response time is more than the estimated 3 minutes. If the calculated temperature were more than the actuation temperature, then a smaller t would be used. As in the previous example, calculations should be repeated varying the time to response until the calculated detector temperature is approximately equal to the actuation temperature. For this example, the response time is estimated to be 213 seconds. This corresponds to a fire size at response of 2132 kW (2022 Btu/sec).

The above examples assume that the fire continues to follow the t-squared fire growth relationship up to detector activation. These calculations do not check whether this will happen, nor do they show how the detector temperature varies once the fire stops following the power law relationship. The user should therefore determine that there will be sufficient fuel, as the above correlations do not perform this analysis. If there is not a sufficient amount of fuel, then there is the possibility that the heat release rate curve will flatten out or decline before the heat release rate needed for actuation is reached.

Rate-of-Rise Heat Detector Spacing

The procedure presented above can be used to estimate the response of rate-of-rise heat detectors for either design or analysis purposes. In this case, it is necessary to assume that the heat detector response can be modeled using a lumped mass heat transfer model.

In step 3 of Figure B.3.3.4.4 of *NFPA 72*, Figure 9-18 and Figure 9-19 the user must determine the rate of temperature rise (dT_d/dt) at which the detector will respond from the manufacturer's data. [Note that listed rate-of-rise heat detectors are designed to activate at a nominal rate of temperature rise of 8°C (15°F) per minute.] The user must use the relationship for $dT_d(t)/dt$ in equation B.26 of *NFPA 72* instead of the relationship for $T_d(t) - T_d(0)$ in equation B.25 of *NFPA 72* in order to calculate the rate of change of the detector temperature. This value is then compared to the rate of change at which the chosen detector is designed to respond.

Note: The assumption that heat transfer to a detector can be modeled as a lumped mass might not hold for rate-of-rise heat detectors. This is due to the operating principle of this type of detector, in that most rate-of-rise detectors operate when the expansion of air in a chamber expands at a rate faster than it can vent through an opening. To accurately model the response of a rate-of-rise detector would require modeling the heat transfer from the detector body to the air in the chamber, as well as the air venting through the hole.

Rate Compensation-Type Heat Detectors

Rate-compensated detectors are not specifically covered by Annex B. However, a conservative approach to predicting their performance is to use the fixed-temperature heat detector guidance contained herein.

QUESTIONS

1. Which of the following would NOT be an appropriate fire alarm system design objective?

   a. Notify building occupants of the need to take action
   b. supervise fire sprinkler systems
   c. monitor security systems
   d. detect fires at a specific stage

2. Room ventilation is NOT a factor in the design of which type of detection system?

   a. Smoke detectors   b. flame detectors
   c. heat detectors   d. fire gas detectors

3. For which type of fire is the design method of *NFPA 72*, Annex B, NOT appropriate?

   a. Smoldering fires   b. hydrocarbon fires
   c. high piled storage   d. combustible metals

4. The radiative fraction for most hydrocarbon fuels is:

   a. 10 percent   b. 20 percent
   c. 35 percent   d. 60 percent

5. Smoke detector activation times can be estimated by:

   a. temperature rise correlations   b. fire models
   c. engineering judgement   d. a, b, or c

BIBLIOGRAPHY

NFPA Codes, Standards, Recommended Practices, and Manuals (see the latest *NFPA Catalog* for availability of current editions of the following documents):

NFPA 70, *National Electrical Code®*.

NFPA 72®, *National Fire Alarm Code®*.

NFPA 90A, *Standard for the Installation of Air-Conditioning and Ventilating Systems*.

REFERENCE CITED

1. Heskestad, G., and Delichatsios, M. A., "Environments of Fire Detectors—Phase I: Effect of Fire Size, Ceiling Height, and Material," NBS-GCR77-95, The Fire Detection Institute Fire Test Report, Vol. 11—Analysis, National Institute for Standards and Technology, Gaithersburg, MD.

ADDITIONAL READING

Alpert, R., Ceiling Jets, Fire Technology (NFPA) Aug 1972.

Davis, W. D., Forney, G. P., and Bukowski, R. W., "Field Modeling: Simulating the Effect of Sloped, Beamed Ceilings on Detector and Sprinkler Response," International Fire Detection Research Project, Technical Report, Year 2, National Fire Protection Research Foundation, Quincy, MA, 1994.

DiNenno, P. J. (Editor), *SFPE Handbook of Fire Protection Engineering*, National Fire Protection Association, Quincy, MA, 2002.

Forney, G. P., Bukowski, R. W., and Davis, W. D., "Fire Modeling: Effects of Flat Beamed Ceilings on Detector and Sprinkler Response," International Fire Detection Research Project, Technical Report, Year 1, National Fire Protection Research Foundation, Quincy, MA, 1993.

Klote, J. H., Forney, G. P., Davis, W. D., and Bukowski, R. W., "Field Modeling: Simulating the Effects of HVAC Induced Air Flow from Slot Diffusers on Detector Response," Technical Report, Year 3, National Fire Protection Research Foundation, Quincy, MA, 1996.

Klote, J. H., Davis, W. D., Forney, G. P., and Bukowski, R. W., "Field Modeling: Simulating the Effects of HVAC Induced Airflow from Various Diffusers and Returns on Detector Response," Technical Report, Year 4, National Fire Protection Research Foundation, Quincy, MA, 1998.

10

Engineering Documents

INTRODUCTION

Preparation of engineering documents is a major aspect in the design and installation of a fire alarm signaling system. Engineering documents, which are generally divided into drawings and specifications, must be prepared accurately, precisely, and completely for all components of a fire alarm signaling system to operate properly. Clear drawings and carefully worded specifications ensure that the facility owner, the contractor, and the fire alarm signaling system designer work toward this end. Additional information on design criteria that affect engineering documents can be found in Chapter 9, Fundamentals of Fire Detection System Design.

Conceptual engineering documents concerning system design should include defined objectives, goals, and criteria. The design objectives outline basic functions, such as life safety, actuating suppression systems, and operating emergency building features, that the system must meet to be effective.

The design criteria describe the specific parameters of the fire detection system and become a basic document for design drawings and specifications. The design criteria evaluate each area of the facility being protected. Building features analyzed on an area-by-area basis should include type of combustibles (to postulate fire scenarios), ceiling height and configuration, stratification, room ventilation and temperature, implications for automatic fire detectors, and detector selection and spacing.

After design criteria have been developed, the principle documents prepared for the installation of fire alarm signaling systems should include drawings and specifications that contain standard clauses and technical instructions.

These documents are the construction contract, which governs the rights, duties, and liabilities of the contractor or vendor who performs the work, the person or organization for whom the work is to be executed, and the engineers who design, inspect, and test the work.

DESIGN DRAWINGS

Design drawings provide the contractor with needed information in easily understood pictorial form. Drawings must be clear and accurate; carelessly prepared or inadequate drawings can cause trouble in the performance of a contract and are a prime source of a contractor's claims for extra compensation. Engineers should provide sufficient time and personnel for preparing suitable drawings. If initial contract drawings for a fire detection system are incomplete, any features added to the system by the owner or engineer will appreciably increase system cost. Some claims by the contractor to recover these extra costs are inevitable and usually justifiable. This should be understood by the owner and the engineer if the proposal drawings are not thorough and comprehensive.

In general, engineering and architectural drawings (or plans) are called design drawings when they are prepared for bidding purposes. Once the contract is awarded, design drawings become known as contract drawings.

Basically, design drawings show the general outline and sufficient detail regarding the system to enable the contractor to bid the job. These drawings seldom are entirely adequate for every need connected with the installation of the fire detection system. Engineering drawings are often supplemented with detail drawings, shop drawings, construction drawings, or working drawings that are prepared by or for the contractor. A detailed drawing, for example, could be a schematic of the electric circuitry inside a fire alarm and detection system control unit.

Types of Drawings

Four types of drawings can generally depict any size of fire alarm signaling system design. These are:

1. Equipment location drawings. These drawings are used on large fire alarm signaling systems to graphically show the location of the main annunciator, remote annunciators, printers, control units, transmitters, transponders, and municipal fire alarm system connections at the facility.
2. Elementary wiring diagram (riser diagram). These diagrams illustrate the arrangement of fire detection devices, notification appliances, and alarm boxes

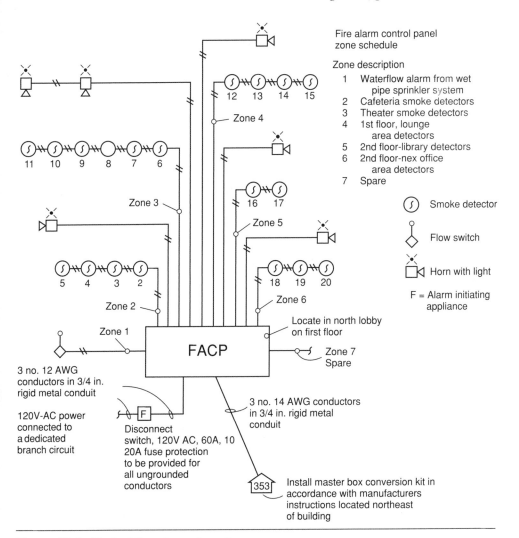

Fire alarm control panel
zone schedule

Zone description

1   Waterflow alarm from wet
    pipe sprinkler system
2   Cafeteria smoke detectors
3   Theater smoke detectors
4   1st floor, lounge
    area detectors
5   2nd floor-library detectors
6   2nd floor-nex office
    area detectors
7   Spare

Ⓢ   Smoke detector

◇   Flow switch

⬚◁ Horn with light

F = Alarm initiating
    appliance

Zone 4

12 13 14 15

11 10 9 8 7 6

Zone 3

16 17

Zone 5

5 4 3 2

Zone 2

18 19 20

Zone 6

Zone 1

FACP

Locate in north lobby
on first floor

Zone 7
Spare

3 no. 12 AWG
conductors in 3/4 in.
rigid metal conduit

120V-AC power
connected to
a dedicated
branch circuit

F

Disconnect
switch, 120V AC, 60A, 10
20A fuse protection
to be provided for
all ungrounded
conductors

3 no. 14 AWG conductors
in 3/4 in. rigid metal
conduit

353   Install master box conversion kit in
      accordance with manufacturers
      instructions located northeast
      of building

Figure 10-1. Typical fire alarm riser diagram.

related to control units. Typical data included on these diagrams consist of control panel or unit, power supply input, alarm-initiating and alarm-indicating circuits, auxiliary output function (such as suppression system actuation, HVAC system shutdown, and elevator recall systems), and so on (see Figure 10-1).

3. Fire detector location drawings. The key element of these drawings is the location dimensioning of each fire detector and alarm box. These drawings

should be prepared by engineers or system designers familiar with *NFPA 72*®, *National Fire Alarm Code*®. The following data should be available before these drawings are completed:

(a) Ceiling construction (structural and architectural drawings): smooth ceiling, beam construction, joist construction, slope construction.

(b) HVAC system (mechanical drawings): air supply and return location for each area having fire detection, HVAC system controls (see NFPA 90A, *Standard for the Installation of Air-Conditioning and Ventilating System*).

(c) Dimensional information (architectural drawings): room dimensions, sections showing ceiling height.

Without proper detector dimensions on the drawings, a detector can be easily installed in an inappropriate location (e.g., on a lower flange or truss of a beam or adjacent to HVAC equipment).

4. Conduit and wiring drawings. These drawings identify the conduit and wiring for field-routed junction boxes, control units, and detection system devices. Conduit is normally not dimensioned on drawings, but is located while the building is under construction.

Preparing Drawings

The following points should be included in the contract if the contractor is to prepare drawings:

1. The size of drawings [if the drawings are to be microfilmed, a minimum scale of ¼ in. = 1 ft (6.35 mm = 0.30 m) should be required].
2. Instructions on use of paper, mylar, or acetate for tracings; pencil or ink for the final drawings; and printed forms.
3. The general title and title block to be used for all drawings pertaining to the project.
4. The printed forms to be used on the drawings for recording revisions and approval by the engineer.
5. Statement concerning responsibility for the accuracy of drawings.
6. The need and procedure for the submission of any preliminary layouts for approval by the engineer.
7. The number of prints required and the procedure for submitting shop drawings for the approval of the engineer.

8. The number of prints required for filing purposes by the owner and the engineer, and the number of prints to be sent to the inspectors and to the engineer's field staff.

SPECIFICATIONS

Specifications are important because they are legal contractual documents. Specifications are usually divided into three parts: (1) general clauses and agreements, (2) technical instructions, and (3) acceptance of the system.

General clauses and agreements (sometimes called *boilerplate*) can also contain descriptions of a quality assurance program; shipping, handling, and storage of materials; equipment delivery; administrative procedures; basis for payment; and project work rules.

Technical instructions affect materials and workmanship, and contain the scope of the project, applicable referenced documents, design and fabrication requirements, erection/installation requirements (i.e., workmanship), and design documents.

Specifications should contain all the necessary instructions and requirements regarding equipment, performance of the system, materials used, and workmanship. Furthermore, they must be clear, concise, correct, and unambiguous (contractors have included extra charges for projects with poor specifications). An additional part of the specification is "acceptance of the system," which requires an acceptance test of the system and subsequent certification of the system (see Chapter 12, Fire Alarm System Testing).

Definitions

Typical standard clauses in specifications include such definitions as the following:

Engineer: The chief engineer, acting either personally or through duly authorized representatives, who operates within the scope of the authority vested in him or her.

Contract: The entire agreement between the parties, including the form of contract together with the specifications, the advertisement, the information for bidders, the contractor's proposal (copies of which are bound with or accompany the other data), and the contract drawings.

Owner: The corporation or individual who owns the project.

Contractor: The individual, corporation, company, firm, or other organization who or which has contracted to furnish the materials and to perform the work under the contract.

Work: All labor, plant, materials, facilities, and any other items necessary or proper for or incidental to the construction of the fire alarm signaling system.

Contract drawings: Those listed in specifications under the title "contract drawings" and any future changes and revisions of said drawings.

Inspector: Any representative designated by the engineer to act as an inspector within the scope of the engineer's authority, as delegated. The engineer, however, may review and change any decision of an inspector.

Subcontractor: Anyone other than the contractor who performs work at the construction site directly or indirectly for or in behalf of the contractor. *Subcontractor* does not include someone who furnishes personal services only.

Writing Specifications

Specifications are clear, concise, and appropriate if they are well organized, well written, and follow proper rules of grammar. The following requirements should be met:

1. A table of contents with page references and an index should be prepared.
2. A numbering system for the various sections and clauses in the documents should be established for easy future reference.
3. All parts to be covered should be outlined before the clauses are prepared. The specification document should be divided into appropriate sections.
4. All hazards and contingent information should be revealed.
5. Requirements in the specifications should be fair, reasonable, and practicable.
6. A given point should be researched if the subject matter is ambiguous or incomplete.
7. Cross-references with title and paragraph numbers should be clearly noted.
8. The written material should be coordinated with the work of persons who are making the drawings. The drafter or person who prepares documents should examine all related written materials.
9. Wording should be clear and precise. Objective review is encouraged.
10. Proper rules of grammar and punctuation must be followed. It is important to be concise: Short sentences are preferred to long, involved ones.
11. A specification is a legal document in which accuracy and clarity are critical. Complete sentences, not phrases, must be used.
12. Use of ambiguous words, words with more than one meaning, and colloquialisms should be avoided. Synonyms should be avoided and simple, clearly understood language should be used.
13. Relative and personal pronouns, such as *his* and *it,* should be avoided.

14. All implications of broad, general statements should be examined before such statements are used.
15. Omissions and irrelevant or useless repetitious information should be avoided.

There are two general types of specifications used for fire alarm signaling systems: vendor's specifications, which concern equipment, and engineering specifications, which concern system performance.

Key parameters in fire alarm signaling system specifications should define and indicate the following requirements:

1. Type of system based on applicable codes.
2. Location of fire alarm control unit(s).
3. Primary (main) and secondary (standby) power supplies required.
4. Interfaces and responsibilities for (a) waterflow switches, (b) fire suppression system(s), (c) existing facility's electrical system, and (d) fire department alarms.
5. Location of fire detectors and alarm boxes. These locations should be dimensioned on drawings prepared by the contractor.
6. Listed equipment.
7. Bill of materials.

System Records

With any fire alarm system, changes occur as the system is installed. It is important that these changes be documented as part of the permanent record. Many authorities having jurisdiction (AHJs) require record or "as-built" drawings to be submitted prior to conducting the system acceptance test to ensure that every device is tested as required by *NFPA 72*.

A Certificate of Completion must be completed and should be submitted with the as-built drawings. The Certificate of Completion is required for every fire alarm system installation, regardless of system size.

Following the completion of the acceptance test, *NFPA 72* requires the preparation of permanent records, which include as-built drawings, operation and maintenance manuals, and a written sequence of operation, often called a system narrative, approved by the AHJ and delivered to the building owner or his or her representative. The owner is then responsible for maintaining these records for the life of the system.

The operation and maintenance manuals should be written in nontechnical language and should indicate those actions that can be performed by the owner and those that should be performed under a service contract by qualified individuals. The sequence of operation should be written to instruct the owner or his or her employees who would normally be expected to operate the system on a daily basis on what they should do in response to any system signal. Of particular importance is when to silence alarm signals and whom to notify in the event of trouble or supervisory signals.

The following example specification illustrates typical components of an invitation to bid. It shows information that might be needed for a fire alarm signaling system, but is not intended to be a model specification.

TECHNICAL SPECIFICATION FOR FIRE DETECTION AND ALARM SIGNALING SYSTEM (EXAMPLE)

1.0 *Purpose*

Fire detection and alarm signaling systems are critical to ensuring life safety. They also protect valuable property, so it is imperative that these systems function properly and reliably. This specification has been prepared to ensure that all installations of these systems work properly.

2.0 *Applicability*

These requirements are applicable to new installations of fire detection and alarm signaling systems in buildings owned or leased by (company name). (These requirements are also applicable to modifications in existing systems.)

3.0 *Revisions and Circulation of Specification*

The owner is responsible for periodic review and revision of this specification. The (owner's name) may be reached at (owner's address, responsible person, and phone number).

Revisions become effective on the date of publication unless otherwise indicated. Contracts for fire detection and alarm systems are governed by the issue in force on the bid opening date.

Copies of this specification will automatically be provided to the Division of Purchase and Contract, designers of alarm systems, equipment manufacturers, and other interested parties.

4.0 *General Requirements for All Systems*

Approval of samples, cut sheets, shop drawings, and other matter submitted by the contractor must not relieve the contractor of responsibility for full compliance with the specifications, unless the attention of the engineer is called to each noncomplying feature by letter accompanying the submitted matter.

4.1 All fire alarm signaling systems must comply with (add appropriate codes), unless otherwise approved by the company.

Table 10-1. Zoning Fire Alarm Systems

Alarm-Initiating Device Circuits

Minimum of one zone for each building floor

Each zone not to exceed 20,000 sq ft (1858 m^2) or 200 (61 m) linear feet

Zone boundaries at smoke/fire partitions (if any) and/or other readily identifiable building features.

For high-rise buildings that exceed 75 ft (23 m) in height, separate zones for smoke, manual, and waterflow alarms on each floor

Notification Appliance Circuits

Maximum coverage is one floor, with total load not to exceed 75% of rated module output

4.2 The system must be nominal 24 V dc and noncoded. All equipment supplied must be listed for the purpose for which it is used and installed in accordance with any instructions included in its listing. Equipment must be new, with a minimum 1-year warranty on parts and labor from the date of final inspection and acceptance by the owner.

4.3 The fire alarm control (FAC) panel must be of modular design, for ease of future system extension or modification or both. The front of the panel must have a steady "power on" light (green color), and each zone module must have separate "alarm" (red color) and "trouble" (amber color) indicators. Zones should be configured in accordance with the contract drawings (see Table 10-1).

4.4 The FAC power supply must have a continuous rating adequate to power all zones and functions indefinitely in full alarm condition.

4.5 The system must be equipped with protective devices to prevent damage or nuisance alarms by nearby lightning strikes, stray currents, or line voltage fluctuations.

4.6 Both audible and visible alarm signals must be provided. Visible signals must be the strobe (flash discharge) type, with a minimum intensity of 8000 candlepower. When installed in corridors, visible signals must be equipped with a side-viewing lens.

4.7 The FAC panel must have an "alarm silence" switch. Subsequent alarm (alarm resound) feature is required in all buildings with two or more zones. Any annunciators/graphic displays remote from the area alarmed must also include an audible signal with subsequent alarm.

NOTE: Annunciation and graphic display requirements should be defined in this section of the specification.

4.8 The coverage of each fire alarm zone must be clearly indicated on each FAC panel and remote annunciator. This indication may be by engraved labels, framed directories, graphic display, or any combination of these. Use of label tape or handwritten labels for this purpose is not acceptable.

4.9 Detectors used for elevator capture must be on a separate (vertical) zone. The capture signal must come from the FAC panel and be triggered by elevator lobby smoke only. Detector-based relays must not provide this signal, but may be used to report the floor that has elevator lobby smoke by initiating an alarm in the adjacent corridor zone.

The operation of automatic fire-extinguishing systems should also be separately annunciated since this normally means that a substantial fire is in progress.

4.10 Systems are to be provided with a separate and independent source of emergency standby power. Switching to emergency standby power during alarm must not cause signal dropout. Batteries must meet the appropriate capacity requirements of the local code plus a 25 percent safety factor.

If NFPA standards are followed, limited standby battery capacity for the alarm system is still required when a single generator is used. Further information can be found in *NFPA 72®*, *National Fire Alarm Code®*.

4.11 A proprietary multiplex system must have Style "—" initiating device circuits, and Style "—" signaling line circuits in compliance with *NFPA 72®*, *National Fire Alarm Code®* (appropriate style letters should be added in this section).

4.12 All wiring must be appropriately color coded, and permanent wire markers must be used to identify the terminations for each zone at the control panel. The number of splices must be held to an absolute minimum. All junction boxes that are visible or accessible must be painted red, unless in finished areas.

Note: The engineer can specify wire markers at all junction boxes instead of color-coded wiring if necessary on large or complex systems. However, both color coding and wire markers are desired.

4.13 Wiring must be in metal conduit or surface metal raceway, and copper conductors must be used. The connectors should be sized in accordance with the equipment manufacturer's recommendations, but in no case are alarm-initiating circuits to be less than 16 AWG stranded/14 AWG solid or alarm-indicating circuits less than 14 AWG.

Stranded wire must comply with Section 760-16(a) of NFPA 70, *National Electrical Code®*.

4.14 Initiating-device or indicating-appliance circuits must not be included in raceways containing ac power or ac control wiring. Within the FAC panel, any ac control wiring must be properly separated from other circuits. The enclosure must have an appropriate warning label to alert service personnel to the hazard.

4.15 To prevent insulation damage, the following requirements apply:

Any threaded rigid conduit terminating at sheet metal boxes and cabinets must be provided with insulating bushings.

Any electrical metallic tubing (EMT) connectors must be of the all-steel compressing type, with insulated throats. Exception: The indenture type can be used if run exposed in finished areas, to obtain a tighter fit to the surface.

Any surface metal raceway (*wiremold*) must be carefully reamed to remove all burrs and sharp edges. Plastic bushings must be used at all raceway terminations.

4.16 All wiring must be checked for grounds, opens, and shorts prior to termination at panels and installation of detectors. The minimum resistance to ground or between any two conductors must be 10 megohms, verified with a megger.

Note: The owner must be given advance notice of these tests so that he or she or a representative can be present.

4.17 The system must be electrically supervised for open or (+/−) ground fault conditions in the alarm-initiating circuits, the alarm-indicating circuits, and the system alarm and trouble relay coils. Removal of any detection device, alarm appliance, system module, or standby battery connection must result in a trouble signal. A fire alarm signal must override trouble signals, but any prealarm trouble signal must reappear when the panel is reset.

4.18 If the FAC panel is in an unoccupied area or an area not normally occupied, a remote annunciator with an audible–visible trouble signal must be provided. The FAC panel area must be protected by a minimum of one smoke detector within 15 linear feet (4.5 m) of panel location.

4.19 Manual fire alarm boxes must require the use of a key or Allen-type wrench for reset.

If the specification concerns a school, dormitory, or other place with many occupants, the following note should also be included:

Note: False alarms at manual fire alarm boxes can be a serious problem in schools, dormitories, and other occupancies. In such cases, each manual box should be covered with a metal box that has a glass panel that can be broken to gain access. The optional key lock feature is recommended for convenient test or glass replacement.

5.0 *Requirements for Detectors*

5.1 Detectors must be the plug-in type, with a separate base to facilitate replacement and maintenance. Each detector or detector base must incorporate a light to indicate whether it is in alarm.

5.2 All air duct/plenum detectors and underfloor detectors must have a remote alarm indicator light, located in the nearest corridor or public area, and identified with a permanent label affixed to the wall or ceiling. Air duct/plenum detectors and underfloor detectors must also be installed in a manner that provides suitable access for periodic cleaning and calibration.

Note: Interlocking acoustical ceiling tile does not provide suitable access. For more information, refer to NFPA 90A, *Standard for the Installation of Air-Conditioning and Ventilating Systems.*

5.3 Open-area-type detectors mounted within 12 ft (3.7 m) of a walking surface must have their built-in locking device activated to prevent unauthorized removal.

5.4 HVAC ducts, lighting fixtures, and other factors affecting air flow must be considered in establishing the exact location of the detectors.

5.5 Open-area-type detectors should not be within 3 ft (0.9 m) of an air supply.

5.6 Unless suitably protected against dust, paint, and other material, detectors must not be installed until the final construction cleanup has been accomplished.

Note: A detector that is surface mounted rather than flush mounted responds better to smoke. Therefore, surface mounting must be used unless the detector has been listed for flush mounting.

5.7 CAUTION: Smoke detectors vary widely in their electric characteristics and must be carefully matched with a suitable control panel to ensure proper performance. The contractor must verify and certify the compatibility of the detectors to the owner's satisfaction. (See Chapter 5, Signal Transmission, for further information on compatibility.)

5.8 CAUTION: The limited power available from an individual smoke zone module generally cannot be relied on to alarm more than one to three detectors in the zone, depending on whether they have remote alarm lamps and/or auxiliary control relays. HVAC shutdown and other control functions must be accomplished from the FAC panel rather than by individual smoke detector relays.

5.9 CAUTION: Duct smoke detector mounting position and air sampling tube orientation are critical for proper detector operation. The equipment manufacturer's detailed installation instructions must be followed.

5.10 A unique identification number must be assigned to each detector (identification should be by zone number and device number within the zone). This number must be noted on the plans and also be permanently mounted adjacent to the detector or affixed to its base.

5.11 Detector trouble contacts (if any) must be series-wired electrically between the last alarm-initiating device and the end-of-line device.

5.12 The location of detectors must be dimensioned on the plans.

5.13 The temperature rating of the heat detectors should be indicated on the plans and bill of materials.

6.0 *Motorized Fire Dampers (Smoke Dampers)*

6.1 The interface between fire/smoke dampers and the fire alarm signaling system should be indicated here.

7.0 *Automatic Smoke Doors/Automatic Door Locks*

7.1 Automatic smoke doors must either have wall-mounted magnetic door holders and separate heavy-duty closers or combination door control/smoke detector units.

8.0 *Sprinkler System Connections*

8.1 Installation of sprinkler systems must comply with NFPA 13, *Standard for the Installation of Sprinkler Systems.*

8.2 Waterflow switches must be listed for the purpose.

8.3 Each waterflow switch must be provided with an integral 15-second time delay device to prevent nuisance alarms from surges in water pressure.

8.4 Permanent provision must be made for testing each switch by water flow equivalent to that from a standard ½-in. (12.7-mm) sprinkler.

8.5 Sprinkler gate valve supervision must not utilize the waterflow alarm circuit to indicate "trouble" on the waterflow alarm circuit.

9.0 *Verification of System Performance (Acceptance Test)*

9.1 Upon completion of the installation, the contractor and the manufacturer's authorized representative together must test every alarm-initiating device for proper response and zone indication, every alarm-signaling appliance for effectiveness, and all auxiliary functions such as capture of elevators and control of smoke doors/dampers, HVAC systems, and pressurization fans. This may be referred to as the "100 percent system test." Personnel other than the contractor who installed or furnished any device or appliance should be notified of the test and asked to supply appropriate information.

The owner and a designated representative must be given the opportunity to witness these tests. An itemized test report must then be submitted to the owner, detailing and certifying all results, including the measured sensitivity of each smoke detector. The data for each smoke detector must include the manufacturer's serial number (if any) plus specific location information adequate to quickly pinpoint the device (see Section 5.10 of this specification).

After completion of the 100 percent system test and submission of the certified test report, the contractor must set up a "pre-final" inspection with the owner or a designated representative. The system must operate for at least 30 days prior to this inspection.

9.2 The system must be inspected and functionally tested by the owner or designated representative on a sample basis.

9.3 The warranty period must begin after successful completion of the owner's inspections and tests. In the event of any system malfunctions or nuisance alarms, the contractor must take appropriate corrective action. This action can necessitate a repeat of the 100 percent system test if the owner so desires. Continued improper performance during warranty is cause to require the contractor to remove the system.

10.0 *System Documentation, Operator Training, and Maintenance*

10.1 The contractor must provide the owner with three copies of the following:
1. As-built wiring and conduit layout diagrams, incorporating wire color code and/or label numbers and showing all interconnections in the system.
2. As-built schematic wiring diagrams of all control panels, modules, annunciators, and communications panels.
3. As-built fire detector location drawings showing location dimensions of each detector and alarm box.
4. Copies of the manufacturer's technical literature on all major parts of the system, including detectors, manual stations, signaling appliances, alarm panels, and power supplies.

10.2 The manufacturer's authorized representative must instruct the owner's designated employees in proper operation of the system and all required periodic maintenance. This instruction must include two copies of a written summary in booklet or binder form that the employees can retain for future reference. Basic operating instructions for the system must be suitably framed and mounted at the FAC panel.

10.3 The contractor must have the manufacturer's authorized representative provide a quote to the owner for regular preventive maintenance. This quote provides the owner with information on internal versus contract cost for the maintenance. This quote must include the following services, after expiration of the standard warranty:

Maintenance every 2 months:
1. Test all waterflow alarm switches.

Maintenance every 6 months:
1. Inspect all FAC panels.
2. Operate all manual stations.
3. Test all restorable heat detectors functionally.
4. Test all smoke detectors functionally.

Maintenance every 12 months:
1. Clean all smoke detectors, including duct types.
2. Measure, record, and set sensitivity of each smoke detector.

3. Verify all auxiliary system functions.
4. Test standby batteries.
5. Verify operation of all audible and visible alarm appliances.

QUESTIONS

1. Design documents should NOT include:

   a. system objectives and criteria
   b. detailed description of spaces to be protected
   c. technical specifications of system components
   d. areas to be protected by suppression systems

2. The type of drawings that should be provided to the AHJ for acceptance testing are:

   a design drawings   b. as-built drawings
   c. riser drawings   d. electrical schematics

3. Typically, design objectives outline basic functions, such as:

   a. life safety   b. actuating   c. number of
                       suppression systems   smoke detectors
   d. b and c   e. a and b

4. Ceiling construction and configuration is most often found on the:

   a. site plan   b. mechanical drawings
   c. reflected ceiling plan   d. electrical drawings

5. Riser diagrams illustrate:

   a. wiring locations
   b. ceiling height
   c. the arrangement of fire detection devices, notification appliances, and alarm boxes related to control units
   d. the wiring terminal arrangement for each device

BIBLIOGRAPHY

NFPA Codes, Standards, Recommended Practices, and Manuals (see the latest *NFPA Catalog* for availability of current editions of the following documents):

NFPA 13, *Standard for the Installation of Sprinkler Systems.*

NFPA 70, *National Electrical Code®.*

NFPA 72®, National Fire Alarm Code®.

NFPA 90A, *Standard for the Installation of Air-Conditioning and Ventilating Systems.*

NFPA 170, *Standard for Fire Safety Symbols.*

11

Approvals and Acceptance

INTRODUCTION

The fire alarm system designer is responsible for obtaining proper approvals at several phases of system design and subsequent installation, acceptance testing, and periodic test and maintenance. This chapter describes the function of the authority having jurisdiction concerning a fire alarm system, types of standards and the standards development process, and listing tests for fire alarm system components. For related information see Chapter 8, Fire Alarm System Installation, and Chapter 12, Fire Alarm System Testing.

APPROVALS, AUTHORITIES, LABELING, AND LISTING

Standards for fire alarm systems, components, and related equipment contain the terms *approved, authority having jurisdiction, labeled,* and *listed.* The definitions of these terms are important to the fire alarm system designer because many applications of fire alarm systems are required to comply with requirements of public and private agencies. In addition, system manufacturers design and build systems in the context of these requirements. The usual definitions of these terms are as follows:

Approved means acceptable to the authority having jurisdiction.

Authority having jurisdiction is the organization, office, or individual responsible for approving equipment, an installation, or a procedure.

Labeled refers to equipment or materials to which has been attached a label, symbol, or other identifying mark of an organization that is acceptable to the authority having jurisdiction and is concerned with product evaluation, periodic inspection of the production of the labeled equipment or materials, and whose labeling indicates manufacturer compliance with appropriate standards or performance in a specified manner.

Listed refers to equipment, materials, or services included in a list published by an organization that is acceptable to the authority having jurisdiction and is concerned with evaluation of products or services, maintains periodic inspection of the production of the listed equipment or materials or periodic evaluation of the services, and whose listing states that the equipment, material, or service meets appropriate designated standards or has been tested and found suitable for the specified purpose.

The labeling or listing organization provides independent verification that the products meet appropriate standards for reliable operation when properly installed and maintained. Installation, testing, and maintenance standards (such as *NFPA 72®*, *National Fire Alarm Code®*) work with the product standards of the labeling or listing organization and are usually cited within the building codes (regulations). In granting approval, the authority having jurisdiction may base acceptance of a fire alarm signaling system installation, procedures, equipment, or materials on compliance with appropriate standards. Instead of using standards, the authority may require evidence of proper installation, procedure, or use, or refer to listing or labeling practices of a product evaluation organization. This organization will determine compliance of the system with appropriate standards for the current production of listed items.

The definition of *authority having jurisdiction* (AHJ) can vary. In NFPA documents, the AHJ is defined broadly, since jurisdictions and approval agencies vary depending on their responsibilities. Where public safety is the paramount consideration, the AHJ may be a federal, state, local, or other regional department, or an individual such as a fire chief, fire marshal, chief of a fire prevention bureau, head of the labor or health department, building official, electrical inspector, or other individual with statutory authority. The AHJ, for insurance purposes, could be an insurance inspection department, rating bureau, or other insurance company representative. At a government or military installation, the AHJ could be the command officer or designated departmental official.

Labeling and *listing* are not necessarily interchangeable terms since some organizations do not recognize equipment as listed unless it is also labeled. A listed product should be identified in accordance with the system employed by the listing organization. Most authorities having jurisdiction rely on the labeling or listing of organizations that it finds acceptable.

In many projects, a system designer is responsible for specifying the installation of a system and, unless he or she has delegated this responsibility, the designer in a sense is an authority having jurisdiction for all tradespersons responsible for implementation of the system design. Depending on terms of the contract with the facility owner, the designer may be directly responsible for system compliance with the requirements of the AHJ or may report to an owner.

The owner of the protected facility is ultimately responsible for a system being acceptable to a local authority, although the individual trades involved usually handle requests for approval at certain stages of a project.

It is important for the designer to know the identity of the AHJ since a request for approval at all stages of a project must be submitted to the proper organization or individual. Proceeding without approvals may cause work to be redone at later stages in the project and at greater expense.

Requests for an AHJ's approval in a project may take various forms. In general, standards requiring approval also require sufficient data to be submitted at the time the approval is requested so the AHJ can base approval on compliance with job requirements. The submission of data can be a simple certification that a system has been installed in a given building to a given standard or code. The submittal can also be more complex, with detailed data on equipment specifications, wiring diagrams, floor layouts, description of operation, dimensions, and so on. Complex submittals may have to be coordinated depending on the levels of project management and the detail of reporting required by project managers. Some jurisdictions may require design documents to be sealed by a licensed engineer or architect. Increasingly, jurisdictions are requiring third-party certification of the competency of designers and installation personnel. Often, such competency certifications for fire alarm technicians are issued by the National Institute for the Certification of Engineering Technicians (NICET).

QUALITY SYSTEMS

There is a growing use of quality systems in construction-related fields both in the United States and globally. Traditionally, quality assurance in construction has been centered on independent inspection, such as product approvals by third parties and periodic inspection of work by regulatory officials, and through licensing as an indication of qualifications, such as for engineers or specific trades.

Today these traditional approaches are being supplemented by quality systems centered on documentation and standardized procedures. These include the highly publicized International Standards Organization* (ISO) 9000 type systems. These systems have become important because on today's job site, skilled tradespeople are hard to find and work is performed by journeymen or general workers under the supervision of qualified individuals. In such an environment, clear communication of detailed work procedures is crucial and the existence of written procedures is the best way to accomplish this. Further, written procedures can be subjected to independent review and compliance can be monitored.

The National Electrical Contractors Association (NECA) has developed a standard for fire alarm system installation and maintenance, NECA 305, that includes detailed procedures for the installation of fire alarm system components in a workmanlike manner. NECA 305 also provides detailed procedures for maintenance on system components in a manner that is consistent with the *NFPA 72*.

Another quality system in place is the qualifications certification program operated by NICET. This is a five-level program of study and certification by exam that addresses knowledge and skills ranging from general installation to systems design and specification. With the growth in the number of NICET-certified individuals has come state and local requirements for NICET certification of fire alarm technicians.

CODES AND STANDARDS

Codes and standards fall into three broad categories: (1) codes, (2) installation standards, and (3) product standards.

Codes specify circumstances under which a given type of protection is required. For example, NFPA *101®*, *Life Safety Code®*, requires specific types of systems and devices for specific occupancies. NFPA *101* requires a corridor smoke detection system and one or more manual fire alarm stations for new hotels, depending on whether the hotel is sprinklered. An annunciator panel and an emergency voice/alarm communications system are also required, depending on building size and height. Codes are generally adopted into law by jurisdictions that regulate the built environment.

Installation standards detail how the protection specified by the code is to be achieved. In addition to details on installation and use, these standards include requirements for maintenance and periodic testing of the installed equipment. Installation standards are generally cited in regulations as required to be followed. Thus, standards achieve their legal status by reference.

Product standards specify which functions and capabilities are required of the hardware and conditions under which the equipment must operate. Standards developed by testing laboratories, such as Underwriters Laboratories Inc. (UL) and FM Global (FM), are examples of product standards. For the hotel example previously cited, UL 268, *Standard on Smoke Detectors,* describes tests for listing the smoke detectors used in hotel corridors. The standard includes requirements for basic sensitivity and the maximum acceptable change in sensitivity allowed when the detector is operated under various environmental conditions. Product standards are frequently not cited in regulations, but are required conditions for approval by the AHJ.

Codes can contain installation and product performance requirements, and installation standards can include product performance details. NFPA *101* contains installation and performance-type requirements for fire detection and alarm systems; *NFPA 72* contains performance requirements for specific devices. Requirements repeated in two or more standards are usually correlated; if conflicts occur, the designer should propose a solution that is acceptable to the authority having jurisdiction.

The Standards Development Process

Most standards used in the United States are developed under a consensus process. Initially, a sponsoring organization develops a draft standard. In some cases, public notice of the intent to develop such a standard is given, and proposals concerning the standard are solicited from interested parties, for example, manufacturers, building officials, government officials, private citizens, and fire service representatives.

This draft standard is then circulated to a broad group of interested parties for their review and comments. Comments on the draft are reviewed and considered by the originating body and are accepted, negotiated, or rejected. When the draft comments are negotiated or rejected, this process is carefully documented so the reasons for the action are clear.

The next step in some standards-making organizations is a formal voting procedure. Members of the National Fire Protection Association who attend one of its two member meetings each year vote to accept or reject the draft. Approval laboratories without a voting membership often submit their standards to a review and voting cycle administered by the American National Standards Institute (ANSI). ANSI recognition of a standard represents the completion of the consensus process.

Following completion of the consensus process, these standards may be adopted by federal, state, or local government as part of their building or fire regulations and become law. The standards may be adopted completely, in part, or with modifications. Code requirements for any system can vary from one jurisdiction to another depending on how they were adopted.

Final determination of how a standard is applied rests with the person or organization referred to in the codes as the authority having jurisdiction. Specific aspects of system design and operation may need to be approved by the authority before it is considered that the requirements of the standard have been satisfied.

Equivalency

Most standards in the United States contain an *equivalency* clause, which allows for the use of systems, methods, or devices of equivalent or superior quality, effectiveness, or safety that are different from those prescribed by the standard, as long as sufficient documentation is provided to demonstrate the equivalency to the satisfaction of the authority having jurisdiction.

Equivalency is critical to the fire alarm signaling system designer. Codes and standards are written in general terms and represent only the minimum acceptable requirements for general applications of systems and equipment. The designer should design a system to meet the intent and goals of safety for the specific application; in some circumstances, particular conditions or alternate methods are not implied by the code, and equivalent measures can be justified. A designer should use good engineering judgment when applying for acceptance under the equivalency provision.

The performance of and reference to fire tests or other available experimental data has been used to show equivalency. Computerized building fire simulation modeling and calculation methods are increasingly being used and allow for easier demonstration of equivalency. The Society of Fire Protection Engineers (SFPE), in cooperation with the NFPA, has published the *SFPE Guide to Performance-Based Fire Protection Analysis and Design of Buildings* to document a formal procedure that can be used to demonstrate performance or equivalency.

International commerce and the standards that foreign countries require U.S. manufacturers to meet may be more complex than U.S. requirements. Foreign product approval systems tend to also use a consensus process, but it can differ from one country to another. In most Western European countries, the process is developed by manufacturing, insurance, and government organizations and is similar to the U.S. system; in Japan, the system is administered by the government and is mandatory. The formation of the European Union (EU) has resulted in increased harmonization among member countries and the cooperative development of standards by the European Committee for Standardization (CEN) that preempts national standards.

A complicating factor in international product approval is that a large number of different approvals are often required and may include one approval per country or group of countries in which the product will be sold. This situation led to the creation of the ISO, which is working to correlate different standards and develop consistent international requirements. The activities of ISO Technical Committee 21, which develops test standards for fire detectors and fire alarm systems, is expected to result in a common set of fire detection and alarm standards. Subsequently, approval tests in international markets could be conducted in a single laboratory and a test report then submitted to other testing laboratories from which approval is desired. Any individual approval laboratory would be free to establish its own requirements for regular production testing and quality assurance. A system such as this has been instituted within the EU. Designated laboratories can authorize the use of the CEN mark, which is then accepted in any EU member country.

TESTING LABORATORY PROCEDURES

In the United States, testing laboratories test samples of products to standards developed by them or by other standards development bodies. A fire alarm signaling system designer should be familiar with the detailed inspection and quality assurance procedures that comply with any testing laboratory listing.

Once samples of the product that are representative of final production have passed all the tests, inspection procedures for the product are developed by the engineer who conducted the original investigation for the laboratory. Periodic unannounced plant inspections are conducted on production units to ensure that no changes are made that might result in unacceptable performance. These inspections include comparisons of the product being assembled to detailed descriptions and drawings of the originally tested item and a detailed evaluation of the manufacturer's quality assurance program and verification process. This inspection system mandates that any and all changes to the product be submitted to and approved by the testing laboratory before such changes are incorporated into production units. By contractual agreement, UL and FMRC, two of the major testing laboratories in the United States, prevent shipment of a product bearing their mark that may not meet the requirements of their individual standards. Numerous third-party testing laboratories exist to test products to UL, FM, or proprietary standards, who may use other follow-up procedures. AHJs usually examine these procedures to be comfortable with the listing lab before accepting products labeled by them.

Heat Detector Testing

In testing to list heat detectors, UL determines heat detector ratings such as spacings. Installation spacing ranging from 15 to 60 ft (4.5 to 18 m) is given based on actual tests (see Figure 11-1).

Heat detectors are tested by UL with an alcohol pan fire. In each of two tests, the time of operation is measured for 160°F (71°C)-rated automatic sprinklers that are installed on a 10 × 10 ft (3 × 3 m) spacing. When a fire of sufficient size is reached to operate the sprinklers in 2 min (±10 seconds) after ignition, the heat detector installed on its maximum spacing must have operated *before* the automatic sprinkler.

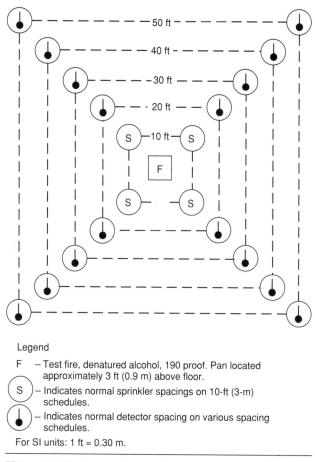

Legend

F   – Test fire, denatured alcohol, 190 proof. Pan located
    approximately 3 ft (0.9 m) above floor.

S   – Indicates normal sprinkler spacings on 10-ft (3-m)
    schedules.

   – Indicates normal detector spacing on various spacing
    schedules.

For SI units: 1 ft = 0.30 m.

Figure 11-1. Layout for UL heat detector test.

Operation of the sprinkler head is the point of comparison and the time–temperature curve is similar to the 15-ft (4.6-m) spacing curve shown in Figure 11-2. If earlier warning is desired, less-than-listed spacing will probably be necessary.

The distance that heat detectors are located from the fire is representative of the spacing for which the detector is being tested. This distance is also the maximum distance from the fire allowed for proper installation (which will also result in the slowest acceptable detector response).

The spacing curve shows an approximate air temperature at the detector of 206°F (97°C) when detection occurs (see Figure 11-2). This illustrates the effect of thermal lag, which is the difference between the set point of the detector

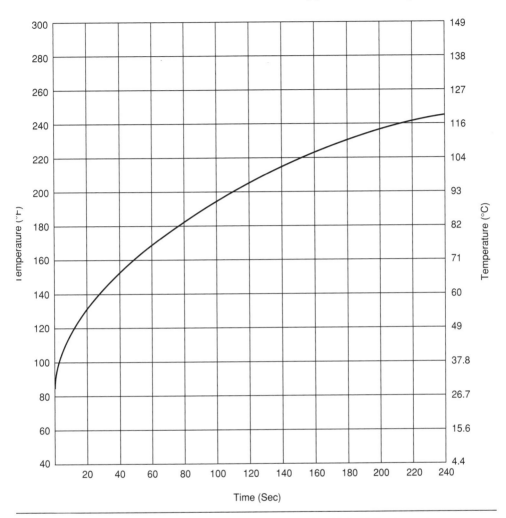

Figure 11-2. Heat detector time–temperature curve for 15-ft (4.6-m) spacings.

[perhaps as low as 135°F (57°C)] and the air temperature required for the detector to operate in a fire condition.

Smoke Detector Testing

A smoke detector is a device that detects visible or invisible particles of combustion. Numerous field tests have shown that heat detectors do not provide the speed of response to a fire needed for life safety. Two of the major field test programs were conducted during the 1970s in Indiana. The conclusions of the second test series state, "Supporting the first year results, the fixed temperature heat detectors rated for 50-ft (15-m) spacing [135°F (57°C)] used in this test series, in the room of fire origin, provided little life saving potential. These detectors failed to respond to a majority of the fires and when they did respond they were considerably slower than smoke detectors located remotely from the fire."[1]

Since smoke detectors respond much faster to a fire than a heat detector, the test fires they are required to detect are much smaller than the test fires used for the listing of heat detectors. Also, since smoke detectors operating on different principles of operation respond differently to smoke from different combustibles, UL has established five test fires, each with a different combustible that every detector must detect.

A further need for different small test fires for smoke detectors is summarized as follows: "The essential feature of smoke is its instability. Under the influence of a lively Brownian motion, the particles collide with one another and agglomerate. This process goes on continuously until the number of particles has been considerably diminished and the average size largely increased."[2] This effect can be seen in Figures 11-3 and 11-4, which are photos of fresh Douglas fir smoke particles and the same smoke after it has aged for a period of time, respectively.

Prior to conducting fire tests, smoke detectors are placed in a smoke box similar to the one shown in Figure 11-5. The detector is mounted in a horizontal position at the top of the upper chamber; smoke is then introduced and circulated across the lower chamber, up through the air straightener to the detector. The smoke density is gradually increased until the detector responds, establishing the sensitivity (set point) of the detector.

Additional tests are conducted with the oncoming smoke directed to each of four sides of the detector and also with the detector placed on end facing the oncoming smoke to determine a least favorable position of smoke entry (if any). If there is a least favorable entry position, the detectors are placed in this position with respect to the oncoming smoke when mounted in the fire test room. In all cases, the sensitivity of the detectors should be within the limits shown in Table 11-1. In addition to the above, the detectors selected for the fire tests are detectors set at the minimum production sensitivity (least sensitive). For the fire tests, all of the detectors except battery-operated single-station detectors are connected to a source of supply at their rated voltage. Battery-powered detectors are tested with the battery depleted to the trouble signal voltage level.

The five UL test fires used to list smoke detectors include two fires that produce a black smoke (*n*-heptane and polystyrene), two that produce a gray smoke (wood and newspaper), and one smoldering fire that produces gray smoke (wood is

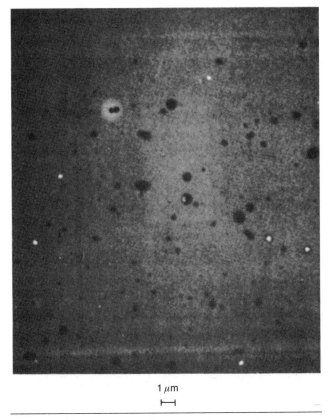

1 µm
⊢——⊣

Figure 11-3. Fresh Douglas fir smoke particles.

pyrolized on a hot plate). The first four are flaming fires and the fifth a smoldering fire, so a wide cross section of types of smoke are included in the listing process.

UL uses two methods of smoke measurement during the fire tests. The first is a light beam with the light source placed 5 ft (1.5 m) from the light-sensitive receiver (see Figure 11-6). An instrument measures the light being received (percentage of light transmitted) in clear air over this distance and the meter reading is adjusted to register 100 percent. Smoke entering the detector will obscure some of the light and decrease the meter reading. The meter readings for the five fire tests are then plotted on charts as shown in Figures 11-7 and 11-8.

Smoke obscuration of 2 percent/ft (0.3 m) is defined here as smoke that is of sufficient density to obscure 2 percent of the light over a distance of 1 ft (0.3 m). A 5-ft (1.5-m) beam is used instead of a 1-ft (0.3-m) beam to obtain a better resolution at the meter. Since a 5-ft (1.5-m) beam is used, as the light

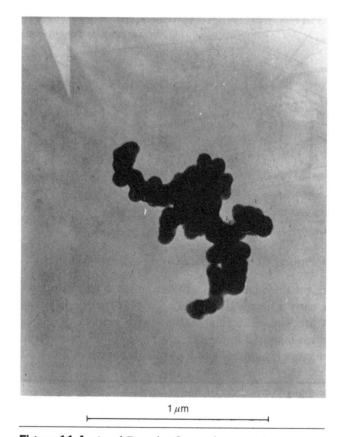

1 μm

Figure 11-4. Aged Douglas fir smoke particles.

travels through 1 ft (0.3 m) of smoke with a 2 percent/ft (0.3 m) obscuration, it cuts off 2 percent of the light at the 1-ft (0.3-m) position of the beam. During the next 1 ft (0.3 m) of travel, the 2 percent smoke cuts off 2 percent of the light that was left at the 1-ft (0.3-m) mark. As the light continues to travel through the 2 percent smoke, the light transmission is reduced by 2 percent of the light that was at each previous 1-ft (0.3-m) mark. So 2 percent smoke does not cut off 10 percent of the light at the 5-ft (1.5-m) mark, and that is why Figure 11-8 shows that it requires 2.1 percent obscuration per foot (0.3 m) to reduce the percentage light transmitted to 90; 4.4 percent for 80; 6.9 percent for 70; etc.

A second method of measuring smoke during test fires is through a measuring ionization chamber (MIC) device, developed by Cerberus Ltd., Switzerland. This instrument is preferred for testing the performance of ionization-chamber smoke detectors.

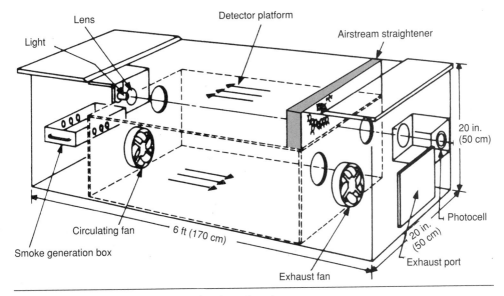

Figure 11-5. Smoke detector evaluation chamber.

The MIC characterizes the smoke present in the same reference terms as the ion chamber smoke detector under test. When used in conjunction with a light beam, the mean particle diameter of the generated smoke can be estimated. When a MIC is used with a light beam and a thermocouple, the combination of these three measuring instruments permits the measurement of the three most important parameters of the smoke detector test fires, and can be used to

Table 11-1. Visible Smoke Obscuration Limits*

Percent per Foot	Percent per Meter	OD per Foot	OD per Meter	
A. Visible Smoke Obscuration Limits (Gray Smoke)				
4.0	12.5	0.0177	0.0581	Maximum
0.2	0.65	0.00087	0.0029	Minimum
B. Visible Smoke Obscuration Limits (Black Smoke)				
10.0	29.2	0.046	0.150	Maximum
0.5	1.6	0.0022	0.0071	Minimum
C. Measuring Ionization Chamber (MIC) Measurement				
14.5 psi (100 Pa) (minimum) to 5.4 psi (37.5 Pa) (maximum)				

* OD, Optical density.

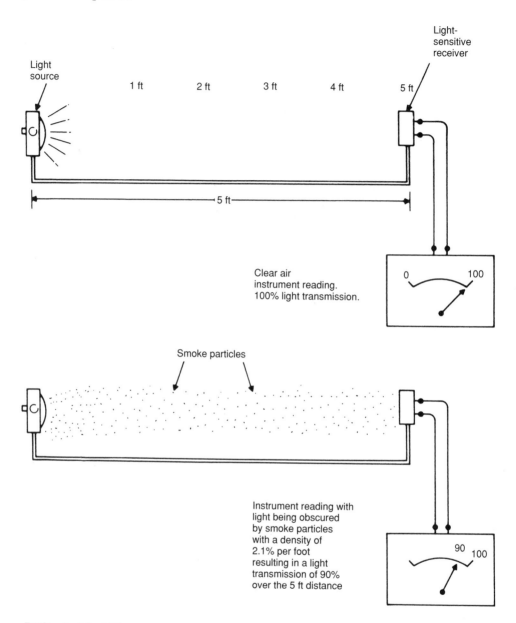

For SI units: 1 ft = 0.30 m.

Figure 11-6. Light beam measurement for UL smoke detector test.

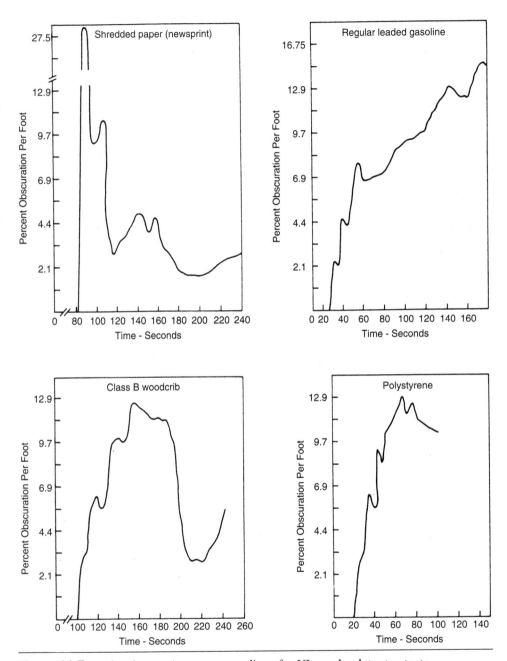

Figure 11-7. Light obscuration meter readings for UL smoke detector tests.

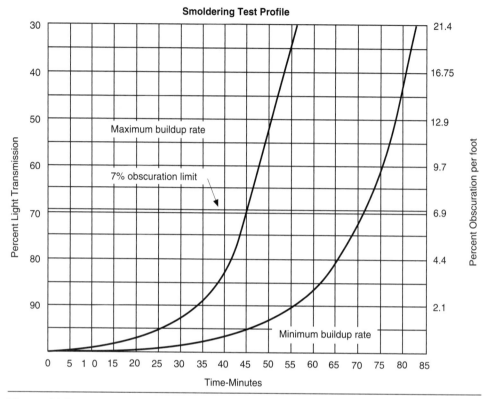

Figure 11-8. Percentage obscuration per foot (percentage light transmission).

determine whether repeatable fires are being produced (and to indicate the particle size of the smoke).

The MIC is designed to be a standard measuring chamber and has a parallel-plate electrode configuration in which the alpha-particle source (americium 241) is part of one of the electrodes. This configuration provides a measuring volume in which the ionization is uniform and approximately parallel to a constant electric field (see Figure 11-9).

The air is drawn mechanically through the chamber to reduce wind dependence. The air in the aspirated sample is drawn past the measuring volume, but not through it, since the flow would draw ions from the measuring volume, affecting the readings. Smoke is transferred from the air flow to the measuring volume by diffusion.

The current through the chamber is measured either directly (with an electrometer) or with a special amplifier. With an amplifier, an impedance transforming

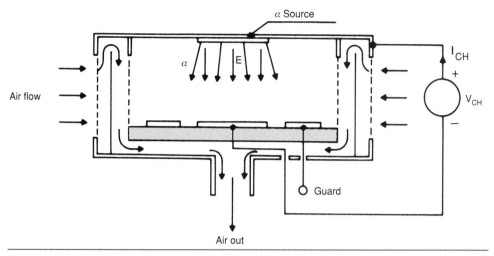

Figure 11-9. MIC configuration.

circuit is placed inside the MIC, transforming the high impedance level of the ionization chamber to a lower level. This is an advantage when the MIC device is used for a full-scale fire test because the length of the cable connecting the MIC to the amplifier is not critical. MIC test profiles for all five UL test fires are shown in Figures 11-10 through 11-14.

The five UL smoke detector test fires are conducted in a 36 × 22 × 10 ft (11 × 6.7 × 3 m) test room (see Figure 11-15). The test fires are shown in Table 11-2. The detectors are placed on the ceiling and side wall so they are located 17.7 ft (5.4 m) from the point on the ceiling that is directly over the test fire. Though this arrangement does not establish a listed spacing similar to the listed spacing for heat detectors, it does establish a criterion for pass/fail determination of smoke detector response to the smoke produced in the five test fires.

Classification (Grading) System for Automatic Fire Detectors

Listing of automatic fire detectors in the United States is based primarily on a pass–fail method of fire tests. In some applications, this approach does not provide sufficient data to select the most suitable detector needed. To assist users in detector selection, the CEN has proposed that the ISO consider adoption of a different approach to evaluate fire detectors. The CEN standard (EN54) includes a series of test fires using a variety of combustibles as shown in Table 11-3.

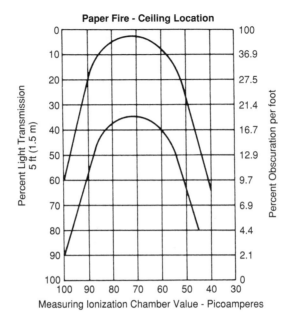

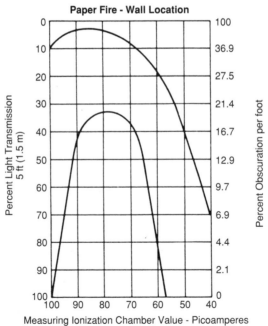

Figure 11-10. MIC profiles of the UL smoke detector test for paper fires, ceiling and wall location.

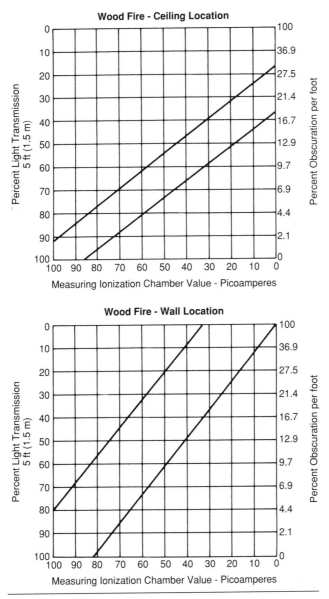

Figure 11-11. MIC profiles for the UL smoke detector test for wood fires, ceiling and wall location.

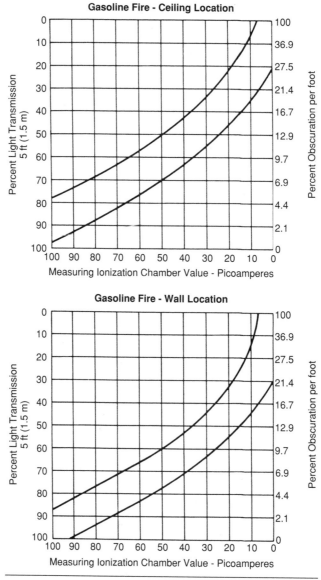

Figure 11-12. MIC profiles for UL smoke detector test for gasoline fires, ceiling and wall location.

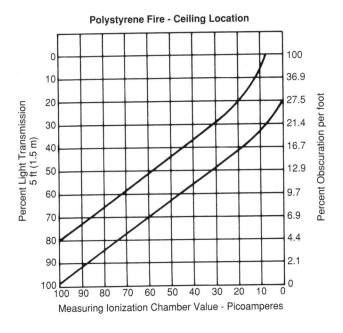

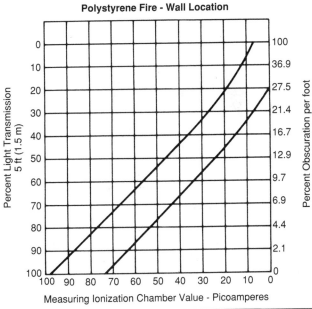

Figure 11-13. MIC profiles for UL smoke detector test for polystyrene fires, ceiling and wall location.

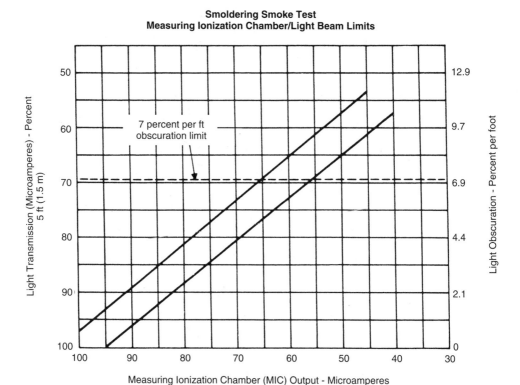

Figure 11-14. MIC profile for the UL smoke detector test for smoldering fires (measuring ionization chamber/light beam units).

The CEN test program classifies each type of fire detector, regardless of operation principle, as a Class A, B, or C in accordance with its response to each combustible. This classification indicates the sensitivity of a detector type in a particular fire situation. The classification is derived by using three limiting values in each fire test: (1) temperature T, (2) smoke density m (optical density measured with a light beam), and (3) smoke density y (ionization measured with a MIC) (see Table 11-4).

Four detectors are used in the fire tests and mounted in the fire test room as shown in Figure 11-16.

If the alarm points of all four detectors are reached before the measuring instruments record ΔT_1, m_1, and y_1, the detector is placed in Class A. If all alarm

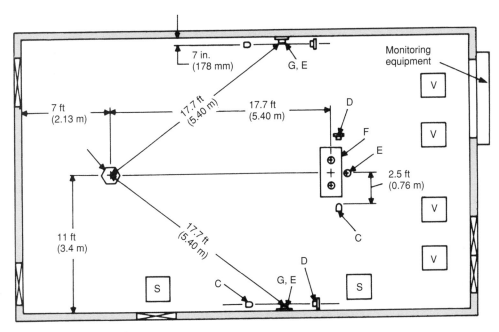

A. Fire test room dimensions
   1. Length - 36 ft (11 m)
   2. Width - 22 ft (6.7 m)
   3. Ceiling - height 10 ft (3.0 m) suspended type. Consists of 2 by 4 ft (0.6 by 1.2 m) by 5/8 in. (15.9 mm) thick noncombustible fissured mineral fiber layer in panels.

B. Test fire
   1. 3 ft (0.91 m) above floor for fire tests
   2. 8 in. (203 mm) above floor for smoldering smoke test

C. Lamp assembly - 4 in. (102 mm) below ceiling, 7 in. (178 mm) from each side wall.

D. Photocell assembly - spaced 5 ft (1.5 m) from lamp, photocell center 4 in. (102 mm) below ceiling 7 in. (178 mm) from each side wall.

E. Measuring ionization chamber (MIC)

F. Test panel, sidewall-mounted detectors

G. Test panel, sidewall-mounted detectors

S. Air supply

V. Exhaust vents

Figure 11-15. UL fire test room for smoke detector test.

Table 11-2. Smoke Detector Test Fires by Underwriters Laboratories

Test	Combustible	Time Detector Must Respond Within (min)	Typical Smoke Profile	Typical MIC Profile
A.	1.5 oz. shredded newsprint*	4	Figure 11-7	Figure 11-10
B.	6 × 6 × 2.5 in. layered fir wood strips*	4	Figure 11-7	Figure 11-11
C.	30 ml of *n*-haptane*	3	Figure 11-7	Figure 11-12
D.	1 oz. foam polystyrene-type packing material*	2	Figure 11-7	Figure 11-13
E.	Ponderosa pine sticks over hot plate†	70	Figure 11-8	Figure 11-14

* Flaming fire.
† Smoldering fire.
For SI Units: 1 oz. = 28 g; 1 ml = 0.034 oz; 1 in. = 25.4 mm.

Table 11-3. Test Fires by the European Committee for Standardization

Designation*	Type of Fire	Development of Heat
TF 1	Open cellulosic fire (wood)	Strong
TF 2	Smoldering pyrolysis fire (wood)	Can be neglected
TF 3	Glowing smoldering fire (cotton)	Can be neglected
TF 4	Open plastics fire (polyurethane)	Strong
TF 5	Liquid fire (*n*-heptane)	Strong
TF 6	Liquid fire (Methylated spiritis)	Strong

Upcurrent	Smoke	Aerosol Spectrum	Visible Portion
Strong	Yes	Predominantly invisible	Dark
Weak	Yes	Predominantly visible	Light, high scattering
Very weak	Yes	Predominantly invisible	Light, high scattering
Strong	Yes	Partially invisible	Very dark
Strong	Yes	Predominantly invisible	Very dark
Strong	No	None	None

* TF, Test fire.

Table 11-4. Limiting Values in European Committee for Standardization Classification Program

$\Delta T_1 = 15°C$	$y_1 = 1$
$\Delta T_2 = 30°C$	$y_2 = 2$
$\Delta T_3 = 60°C$	$y_3 = 4$
$m_1 = 0.3$ dB/m	
$m_2 = 0.6$ dB/m	
$m_3 = 1.2$ dB/m	

Plan View of Detectors, Fire Place, and Measuring Instruments. The Height of the Room is 12.5 ft - 14 ft (3.8 m - 4.2 m).

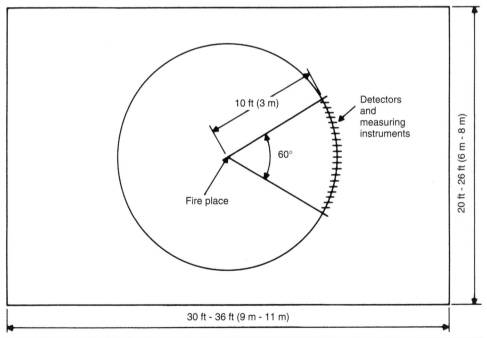

Figure 11-16. CEN/ISO fire test room.

Table 11-5. Smoke Detector Classification Chart

Test Fire	Class A	Class B	Class C	N
TF 1	X			
TF 2			X	
TF 3			X	
TF 4	X			
TF 5		X		
TF 6				X

before $\Delta T_2, m_2$, and y_2, the detector is placed in Class B; and if they alarm before $\Delta T_3, m_3$, and y_3, they are placed in Class C. If the detector does not respond to a specific test fire, it is classified as N for that combustible.

When the test results are placed on the classification chart, the user has a better picture of detector response to a specific combustible. However, the classification applies only to applications for representative test conditions.

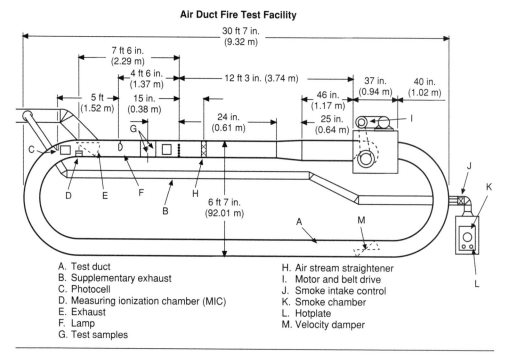

Air Duct Fire Test Facility

A. Test duct
B. Supplementary exhaust
C. Photocell
D. Measuring ionization chamber (MIC)
E. Exhaust
F. Lamp
G. Test samples
H. Air stream straightener
I. Motor and belt drive
J. Smoke intake control
K. Smoke chamber
L. Hotplate
M. Velocity damper

Figure 11-17. Air duct test facility. Two combustibles are used: Wood on a hot plate (L) in the smoke chamber produces gray smoke and *n*-heptane is burned for the black smoke test.

Table 11-6. Smoke Parameters

Gray Smoke:

| Duct Air Speed | | No. of Sticks | |
fpm	m/sec	Used in Test	Smoke Buildup Rate*
300	1.52	5	2.9 ± 0.5
1000	5.14	5	2.2 ± 0.8
2000	10.2	6	2.1 ± 0.5
3000	15.4	7	1.3 ± 0.4
4000	20.3	8	0.8 ± 0.2

Black Smoke:

| | | Receptacle Size | | | | | Amount of |
| Duct Air Speed | | Diameter | | Depth | | Combustible | Smoke Buildup |
fpm	m/sec	in.	mm	in.	mm	(ml)	Rate*
300	1.52	$3\frac{3}{8}$	85.7	$1\frac{7}{8}$	47.6	30	4 ± 1.5
1000	5.14	$3\frac{3}{8}$	85.7	$1\frac{7}{8}$	47.6	30	4 ± 1.5
2000	10.2	$3\frac{3}{8}$	85.7	$1\frac{7}{8}$	47.6	30	4 ± 1.5
3000	15.4	4	101.6	2	50.8	40	4 ± 1.5
4000	20.3	4	101.6	2	50.8	40	4 ± 1.5

* Given as percentage per foot obscuration per minute.

Table 11-7. Visible Smoke Obscuration Limits with Percentage Light Transmission*

Percent per Foot	Percent per Meter	OD per Foot	OD per Meter	Light Transmission (%)	
Gray Smoke					
4.0	12.56	0.0178	0.0583	81.5	Maximum
0.2	0.66	0.00087	0.0028	99.0	Minimum
Black Smoke					
10.0	29.26	0.0458	0.1504	59.0	Maximum
0.5	1.65	0.0022	0.0072	7.5	Minimum

* OD, Optical density.

A classification chart for a certain smoke detector might look similar to Table 11-5. Note that since methylated spirits do not produce smoke when burning, the detector did not respond to test fire 6.

Duct Detectors: One method of testing the performance of duct-type smoke detectors is with an air duct facility as shown in Figure 11-17 on page 338.

The smoke detectors are placed in the test facility at location *G*. At this point, the duct is 1 sq ft (0.09 m^2) and the detectors are mounted in or on the duct in the same way that they are mounted in the field. The smoke buildup rate at the various velocities to which they are tested is shown in Table 11-6. In each test, response to the smoke must be within the limits shown in Table 11-7.

QUESTIONS

1. Who can approve the acceptability of fire alarm system components and installation?

    a. building owner
    b. authority having jurisdiction
    c. insurance agent
    d. mortgage holder

2. Legal documents that specify the circumstances under which certain types of protection are required are:

    a. codes
    b. standards
    c. recommended practices
    d. manufacturers' instructions

3. Alternative methods or products can be accepted by the AHJ if shown to provide the same level of performance under what principle?

    a. similarity
    b. equivalency
    c. engineering judgment
    d. none of these

4. Space ratings are assigned in full-scale tests of:

    a. smoke detectors
    b. flame detectors
    c. heat detectors
    d. fire-gas detectors

5. *Approved* means:

    a. acceptable to the authority having jurisdiction
    b. acceptable to the owner
    c. acceptable to the general contractor
    d. acceptable to the architect

BIBLIOGRAPHY

NFPA Codes, Standards, Recommended Practices, and Manuals (see the latest *NFPA Catalog* for availability of current editions of the following documents):

NFPA *72*®, *National Fire Alarm Code*®.

NFPA *101*®, *Life Safety Code*®.

REFERENCES CITED

1. Harpe, S. W., Waterman, T. E., and Christian, W. J., "Detector Sensitivity and Siting Requirements for Dwellings—Phase 2," ITT Research Institute, Chicago, 1975.
2. Hassler, W. M., "Smoke Detection by Forward Light Scattering," *Fire Technology*, Vol. 1, No. 1, 1965, p. 43.

ADDITIONAL READING

UL 268, *Standard on Smoke Detectors*, Underwriters Laboratories Inc., Northbrook, IL.

CHAPTER

12

Fire Alarm System Testing

INTRODUCTION

After proper installation, a regular program of testing is a crucial factor in the reliable functioning of a fire alarm signaling system and its components. This chapter describes the importance of testing and the differences between acceptance testing and periodic testing. Test schedules for various types of systems and for automatic fire detectors are also provided. For additional information, see Chapter 8, Fire Alarm System Installation, and the sections on electrical supervision and on power supply requirements in Chapter 4, Fire Alarm System Components and Circuits.

SYSTEM RELIABILITY

All electronic components fail. Basic electronic components of a fire alarm system, then, will also fail at some time. Testing does not prevent failures, but identifies failures that have occurred so that repairs can be made promptly to minimize the time during which people and property are unprotected.

Failure rates for smoke detectors or other fire alarm system devices can be calculated based on the known failure rate of standard electronic components. The methods employed are used by the U.S. military and well documented in *Military Handbook 217*. Failure rates are generally expressed in failures per million hours of operation or its inverse, mean time between failure (MTBF). The failure rate is further defined as failure other than those caused by loss of power, physical damage, dirt or dust, electrical transients, or actual removal of the system or components. For smoke detectors, a failure rate of 4 failures per million hours of operation is the maximum design failure rate permitted in UL 268, *Standard for Smoke Detectors for Fire Protective Signaling Systems*. This standard cites a more conservative failure rate of 3.5 failures per 1 million hours when calculated by a simplified (parts count) method. Table 12-1 shows that at the rate of 3.5 failures per million hours, 3 percent of any lot of detectors will be expected to fail in 1 year. This failure rate progresses until at the end of 30 years, 60 percent of the detectors will have failed.

Failure rates are important especially when applied to the millions of smoke detectors in use in the United States. A failure rate of 3.5 per million hours of operation for every 10 million smoke detectors translates into 35 detectors failing every hour of every day; for every 100 million detectors, 350 are failing every hour of every day. Smoke detectors must be tested regularly to identify those detectors that have failed so that they can be promptly replaced.

With annual testing, a detector could be out of service for as long as 52 weeks without anyone being aware of the problem. According to Table 12-1, the probability for out-of-service time of annually inspected equipment is 28 weeks. This unprotected time grows to 273 weeks for testing once in 10 years, and approaches 900 weeks (about 17 years) for testing every 30 years (see Figure 12-1). By testing the detector twice a year, one can cut undiscovered downtime in half (see Figure 12-2). The out-of-service time with a 6-month test is, over a 30-year period, only 1/40th of the time with no test (see Table 12-1).

With a 1-year test interval, if the detector design failure rate is reduced from 4 to 2 failures per 1 million hours, the unprotected time drops from 45 to 36 weeks. However, testing a smoke detector with a failure rate of 4 failures per million hours once a month instead of once a year reduces the unprotected time from 45 to 6.7 weeks. Figure 12-3 clearly shows that regular testing of smoke detectors is the best way to identify a failed unit and, by promptly replacing it, reduce out-of-service time.

The failure rates discussed previously apply to both single-station residential smoke alarms and system smoke detectors.

SYSTEM MALFUNCTIONS

Testing and maintaining fire alarm systems ensures the proper operating characteristics of the system. Common system deficiencies can be corrected promptly through proper testing and maintenance. A study of fire alarm system effectiveness conducted by the Los Angeles County fire chiefs[1] found that, of the more than 46,000 systems surveyed, many connected to central stations did not function properly.

Table 12-1. Probability of Failure and Average Unprotected Time for Typical Ranges of Detector Failure Rates and Test Intervals for Various Service Life Periods

Unprotected Time (weeks)

Test Interval	1-Year Service				10-Year Service				20-Year Service				30-Year Service		
	2.0*	3.0	3.5	4.0	2.0	3.0	3.5	4.0	2.0	3.0	3.5	4.0	2.0	3.0	3.5
1 week	2.5	2.5	2.5	2.5	2.7	2.8	2.9	3.0	3.0	3.2	3.3	3.5	3.2	3.6	3.8
2 weeks	3.0	3.0	3.0	3.1	3.3	3.4	3.5	3.6	3.6	3.9	4.0	4.2	3.9	4.3	4.6
1 month	4.2	4.2	4.2	4.2	4.5	4.7	4.8	5.0	4.9	5.4	5.6	5.8	5.4	6.0	6.4
3 months	8.6	8.6	8.7	8.7	9.3	9.7	9.9	10.1	10.1	10.9	11.4	11.9	10.9	12.3	13.0
6 months	15.2	15.2	15.3	15.3	16.4	17.1	17.5	17.9	17.8	19.3	20.1	21.0	19.3	21.7	23.0
1 year	28.2	28.6	28.6	28.7	30.6	32.0	32.7	33.5	33.3	36.1	37.6	39.3	36.1	40.6	43.0
No test	28.2	28.6	28.6	28.7	268	271	273	276	550	565	573	581	848	881	898
Probability of failure during service period (%)	1.7	2.6	3.0	3.4	16.1	23.1	26.4	29.6	29.6	40.1	45.8	50.4	40.1	54.4	60.1

* Failures per million hours.

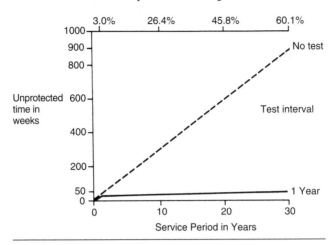

Figure 12-1. Probability of smoke detector failure with no test and 1-year test interval.

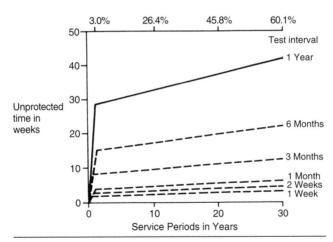

Figure 12-2. Probability of smoke detector failure with different test intervals during service period. Plot of Table 12-1 with 3.5 failures per million hours.

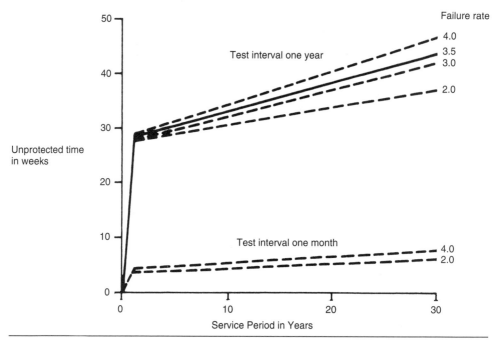

Figure 12-3. Failure rate for smoke detectors with different failure rates: monthly and annual tests.

Slightly more than 6000 fire alarms systems of those surveyed had failed or malfunctioned during the 12 months preceding the survey. The top 10 reasons for failure, ranked in order of frequency, were

1. Faulty flow switch (30 systems)
2. Lack of maintenance (29)
3. Water in conduit (23)
4. Power failure (18)
5. Telephone line trouble (17)
6. Detector failure (16)
7. Vandalism (not repaired) (11)
8. Battery failure (11)
9. Detectors too sensitive (10)
10. Poor installation (9)

Other failures were caused by central station problems, lack of detector maintenance, short circuit, equipment failure or incompatibility, technician-related issues, and so on.

Another study[2] concerned fire alarm signaling systems in 45 hotels; 20 hotels had 26 control panels less than 3 years of age and the remaining 25 had new control panels. The study included a visual inspection, functional test of control panels, and electric supervision test (wire disconnected).

Most control panels that did not operate properly had operational problems, such as burned-out lamps, dead batteries, or inoperative city connections. Many of the system smoke detectors in these hotels that would not alarm were not supervised or had other problems; some fire alarm-indicating appliances for the hotel systems were also not supervised, whereas others had burned-out lamps. Table 12-2 provides more details of the test failures in surveyed system components.

The hotel survey revealed further problems that could have been detected through a regular testing program. Additional system problems included

Table 12-2. Component Problems in Hotel Systems Survey (Percentage of Total Components)

Tested Component	System <3 Years Old	New System
Control Panels		
Alarm condition	19%	7%
Trouble signal	30%	14%
Operational problems	50%	39%
Total panels	26	27
Percentage with problems	99%	60%
System Smoke Detector		
No alarm	2%	0.9%
Not supervised	24%	0.5%
Mounted in wrong location	1%	0.07%
Other problems	3%	3.3%
Total detectors	942	1347
Percentage with problems	30%	3.51%
Alarm-Indicating Appliances		
Did not sound properly	6%	0.64%
Not supervised	11%	1.4%
Burned-out lamps	8%	3.5%
Total appliances	524	1720
Percentage with problems	25%	5.54%

unsealed riser piping between guest rooms, gaps in smoke doors, and improperly wired (unsupervised) sprinkler flow switches. Control panels had no trouble signal on transfer to battery power, or an inoperative resound circuit; system detector wires were looped around the terminal screw; single-station detectors were missing or had low-sounding horns; and some audible appliances were missing or improperly wired.

TESTING PROCEDURES

Testing procedures can be classified into two categories: acceptance and reacceptance testing procedures and periodic equipment and circuit testing procedures. *NFPA 72®, National Fire Alarm Code®*, contains test methods and minimum testing frequencies for every type of device that may be found in a fire alarm system. These are found in the mandatory portion of the Code and are required for any system that must comply with the Code.

Acceptance Testing or Commissioning

Every fire alarm system is required to be thoroughly tested after installation or significant alteration to the hardware, wiring, or site-specific software to ensure that the system meets its design objectives and is fully operational. Qualified personnel including a fire protection engineer, factory representative of the equipment manufacturer, and engineer of the installation company should participate in the acceptance test. A detailed testing procedure should be prepared prior to the test, with all parties becoming familiar with the procedure before the test is actually conducted.

After installation or alterations, satisfactory tests of the entire system should be made in the presence of a representative of the local authority. All functions of the system should be tested, including operation of the system in various alarm and trouble modes for which it is designed, for example, open circuit, grounded circuit, power outage, and so on.

As-built drawings and the manufacturer's manual should be available for verifying the agreement between the connected equipment and the as-built drawings. Prior to connection of the equipment, all conductors should be tested as follows. (*Note*: Equipment should not be wired during conductor testing, or equipment damage may result.)

1. Absence of unwanted voltages between circuit conductors and ground that would constitute a hazard or prevent proper system operation should be verified.

2. All conductors other than those intentionally and permanently grounded should be tested for isolation from ground using an insulation testing device such as a megger. All conductors other than those intentionally connected together should be tested for conductor-to-conductor isolation using an insulation testing device.
3. With each circuit pair short-circuited at the far end of the circuit, circuit resistance should be measured with an ohmmeter and recorded for each circuit shown on the as-built drawings.

The entire system should be tested as follows:

1. The control unit should be tested to verify it is in the normal supervisory condition as detailed in the manufacturer's manual.
2. Each initiating and indicating circuit should be tested to confirm that the integrity of installation conductors is being properly monitored by a suitable response at the control unit. One connection each should be opened at no less than 10 percent of all initiating devices and indicating appliances.
3. Each initiating device and indicating appliance of the system should be tested for alarm operation and proper response at the control unit. All intended functions should be tested in accordance with the manufacturer's manual, including all supplementary functions. Primary (main) power supplies and secondary (standby) power supplies should be tested.

A reacceptance test should be performed on all equipment and circuits affected after any addition, deletion, or mechanical or electric damage to the system has occurred or when conductors are added.

As part of the acceptance check, the local authority should be furnished a certificate of compliance (see Figure 12-4). All parts of the certificate should be completed after the system is installed and the installation wiring has been checked, except for Part 2, which should be completed after the operational acceptance tests have been completed. A preliminary copy of the certificate is given to the system owner and, when requested, to the authority or authorities having jurisdiction after completion of the installation wiring tests. A final copy of the certificate is distributed after completion of the operational acceptance tests.

NFPA 72 provides information for performing the installation wiring and operational acceptance tests required when completing the certificate of compliance (see Table 12-3).

Central Station Acceptance Testing: Many agencies can be involved in the approval of a central station supervising system installation because a central

---

FIRE ALARM SYSTEM
RECORD OF COMPLETION

Name of protected property: _____
Address: _____
Representative of protected property (name/phone): _____
Authority having jurisdiction: _____
Address/telephone number: _____

                    Organization name/phone   *Representative name/phone*

Installer _____
Supplier _____
Service organization _____
Location of record (as-built) drawings: _____
Location of operation and maintenance manuals: _____
Location of test reports: _____
A contract for test and inspection in accordance with NFPA standard(s)
Contract No(s): _____ Effective date: _____ Expiration date: _____

System Software
(a) Operating system (executive) software revision level(s): _____
(b) Site-specific software revision date: _____
(c) Revision completed by: _____
                     (name)   (firm)

1. Type(s) of System or Service

_____*NFPA 72*, Chapter 6 — Local
     If alarm is transmitted to location(s) off premises, list where received: _____
     _____

_____*NFPA 72*, Chapter 8 — Remote Station
     Telephone numbers of the organization receiving alarm:
     Alarm: _____
     Supervisory: _____
     Trouble: _____
     If alarms are retransmitted to public fire service communications centers or others, indicate location and telephone
     numbers of the organization receiving alarm: _____

     Indicate how alarm is retransmitted: _____

_____*NFPA 72*, Chapter 8 — Proprietary
     Telephone numbers of the organization receiving alarm:
     Alarm: _____
     Supervisory: _____
     Trouble: _____
     If alarms are retransmitted to public fire service communications centers or others, indicate location and telephone
     numbers of the organization receiving alarm: _____

     Indicate how alarm is retransmitted: _____

_____*NFPA 72*, Chapter 8 — Central Station
     Prime contractor: _____
     Central station location: _____

                                                                  (NFPA 72, 1 of 4)

Figure 12-4. Fire alarm system certificate and description. (*Continues*)

Means of transmission of signals from the protected premises to the central station:

_____ McCulloh   _____ Multiplex   _____ One-way radio

_____ Digital alarm communicator   _____ Two-way radio   _____ Others

Means of transmission of alarms to the public fire service communications center:

(a) _____

(b) _____

System location: _____

_____ NFPA 72, Chapter 9 — Auxillary

Indicate type of connection: _____ Local energy   _____ Shunt   _____ Parallel telephone

Location of telephone number for receipt of signals: _____

2. Record of System Installation

(Fill out after installation is complete and wiring is checked for opens, shorts, ground faults, and improper branching, but prior to conducting operational acceptance tests.)

This system has been installed in accordance with the NFPA standards as shown below, was inspected by
_____ on _____, includes the devices shown in 5 and 6, and has been in service since _____ .

____ NFPA 72, Chapters   1   2   3   4   5   6   7   8   9   10   11   (circle all that apply)

____ NFPA 70, *National Electrical Code*, Article 760

____ Manufacturer's instructions

____ Other (specify): _____

Signed: _____   Date: _____

Organization: _____

3. Record of System Operation

Documentation in accordance with Inspection Testing Form, Figure 10.6.2.3, is attached _____ .

All operational features and functions of this system were tested by_____ date _____
and found to be operating properly in accordance with the requirements of:

____ NFPA 72, Chapters   1   2   3   4   5   6   7   8   9   10   11   (circle all that apply)

____ NFPA 70, *National Electrical Code*, Article 760

____ Manufacturer's instructions

____ Other (specify): _____

Signed: _____   Date: _____

Organization: _____

4. Signaling Line Circuits

Quantity and class of signaling line circuits connected to system *(see NFPA 72, Table 6.6.1)*:

Quantity: _____   Style: _____   Class: _____

(NFPA 72, 2 of 4)

Figure 12-4. (*Continued*)

5. Alarm-Initiating Devices and Circuits

Quantity and class of initiating device circuits *(see NFPA 72, Table 6.5)*:

Quantity: _____ Style: _____ Class: _____

MANUAL

(a) Manual stations   Noncoded _____ Transmitters _____ Coded _____ Addressable _____
(b) Combination manual fire alarm and guard's tour coded stations _____

AUTOMATIC

Coverage: Complete _____   Partial _____

            Selective _____   Nonrequired _____

(a) Smoke detectors _____ Ion _____ Photo _____ Addressable _____
(b) Duct detectors _____ Ion _____ Photo _____ Addressable _____
(c) Heat detectors _____ FT _____ RR _____ FT/RR _____ RC _____ Addressable _____
(d) Sprinkler waterflow indicators: Transmitters _____ Noncoded _____ Coded _____ Addressable _____
(e) The alarm verification feature is disabled _____ or enabled _____ , changed from _____ seconds to _____ seconds.
(f) Other (list): _____

_____

6. Supervisory Signal-Initiating Devices and Circuits (use blanks to indicate quantity of devices)

GUARD'S TOUR

(a) _____ Coded stations
(b) _____ Noncoded stations
(c) _____ Compulsory guard's tour system comprised of _____ transmitter stations and intermediate stations
Note: Combination devices are recorded under 5(b), Manual, and 6(a), Guard's Tour.

SPRINKLER SYSTEM

Check if provided
(a) _____ Valve supervisory switches
(b) _____ Building temperature points
(c) _____ Site water temperature points
(d) _____ Site water supply level points

Electric fire pump:
(e) _____ Fire pump power
(f) _____ Fire pump running
(g) _____ Phase reversal

Engine-driven fire pump:
(h) _____ Selector in auto position
(i) _____ Engine or control panel trouble
(j) _____ Fire pump running

ENGINE-DRIVEN GENERATOR:

(a) _____ Selector in auto position
(b) _____ Control panel trouble
(c) _____ Transfer switches
(d) _____ Engine running
Other supervisory function(s) (specify): _____

_____
_____
_____

(NFPA 72, 3 of 4)

Figure 12-4. *(Continued)*

7. Annunciator(s)

Number: _____ Type: _____ Location: _____

8. Alarm Notification Appliances and Circuits

NFPA 72, Chapter 6 — Emergency Voice/Alarm Service

Quantity of voice/alarm channels: _____ Single: _____ Multiple: _____

Quantity of speakers installed: _____ Quantity of speaker zones: _____

Quantity of telephones or telephone jacks included in system: _____

Quantity and the class of notification appliance circuits connected to system *(see NFPA 72, Table 6.7)*:

Quantity: _____ Style: _____ Class: _____

Types and quantities of notification appliances installed:

(a) Bells _____ With Visible _____

(b) Speakers _____ With Visible _____

(c) Horns _____ With Visible _____

(d) Chimes _____ With Visible _____

(e) Other: _____ With Visible _____

(f) Visible appliances without audible: _____

9. System Power Supplies

(a) Fire Alarm Control Panel: Nominal voltage: _____   Current rating: _____

    Overcurrent protection: Type: _____   Current rating: _____

                    Location: _____

(b) Secondary (standby):

    Storage battery: _____ Amp-hour rating: _____

    Calculated capacity to drive system, in hours: _____

    Engine-driven generator dedicated to fire alarm system: _____

    Location of fuel storage: _____

(c) Emergency system used as backup to primary power supply: _____

    Emergency system described in NFPA 70, Article 700: _____

10. Comments

Frequency of routine tests and inspections, if other than in accordance with the referenced NFPA standard(s):

_____

System deviations from the referenced NFPA standard(s) are: _____

_____

_____

_____

(signed) for installation contractor/supplier   (title)   (date)

(signed) for alarm service company   (title)   (date)

(signed) for central station   (title)   (date)

Upon completion of the system(s) satisfactory test(s) witnessed (if required by the authority having jurisdiction):

(signed) representative of the authority having jurisdiction   (title)   (date)

(NFPA 72, 4 of 4)

Figure 12-4. *(Continued)*

Table 12-3. Frequencies of Testing*

	Initial and Reacceptance Tests	Monthly	Quarterly	Semiannually	Annually
1. Alarm Notification Appliances					
a. Audible Devices	X				X
b. Speakers	X				X
c. Visible Devices	X				X
2. Batteries: Central Station Facilities					
a. Lead–Acid Type					
1. Charger Test (replace battery as needed)	X				X
2. Discharge Test (30 min)	X	X			
3. Load Voltage Test	X	X		X	
4. Specific Gravity	X				
b. Nickel–Cadmium Type					
1. Charger Test (Replace battery as needed)	X		X		
2. Discharge Test (30 min)	X	X			X
3. Load Voltage Test	X	X	X		X
c. Sealed Lead–Acid Type					
1. Charger Test (Replace battery as needed)	X				
2. Discharge Test (30 min)	X	X			
3. Load Voltage Test	X	X			
3. Batteries: Fire Alarm Systems					
a. Lead–Acid Type					
1. Charger Test (replace battery as needed)	X			X	X
2. Discharge Test (30 min)	X			X	

3. Load Voltage Test	X		X	
4. Specific Gravity	X			
b. Nickel–Cadmium Type				
1. Charger Test (replace battery as needed)	X			X
2. Discharge Test (30 min)	X		X	X
3. Load Voltage Test	X			
c. Primary Type (Dry Cell)				
1. Load Voltage Test	X	X		
d. Sealed Lead–Acid Type				
1. Charger Test (replace battery every 4 yr)	X			X
2. Discharge Test (30 min)	X		X	X
3. Load Voltage Test	X			
4. Batteries: Public Fire Alarm Reporting Systems				
a. Lead–Acid Type	X (daily)			
1. Charger Test (replace battery as needed)	X			X
2. Discharge Test (2 hr)	X	X		
3. Load Voltage Test	X	X		
4. Specific Gravity	X		X	
b. Nickel–Cadmium Type				
1. Charger Test (replace battery as needed)	X			X
2. Discharge Test (2 hr)				X
3. Load Voltage Test	X	X	X	
c. Sealed Lead–Acid Type				
1. Charger Test (replace battery as needed)	X			X

(Continues)

Table 12-3. (Continued)

	Initial and Reacceptance Tests	Monthly	Quarterly	Semiannually	Annually
2. Discharge Test (2 hr)	X				
3. Load Voltage Test			X		X
5. Conductors/Metallic	X				
6. Conductors/Nonmetallic	X				
7. Control Equipment: Fire Alarm Systems Monitorerd for Alarm, Supervisory, Trouble Signals					X
a. Functions	X				X
b. Fuses	X				X
c. Interfaced Equipment	X				X
d. Lamps and LEDs	X				X
e. Primary (Main) Power Supply	X				X
f. Transponders	X				X
8. Control Equipment: Fire Alarm Systems Unmonitored for Alarm, Supervisory, Trouble Signals					
a. Functions	X		X		
b. Fuses	X		X		
c. Interfaced Equipment	X		X		
d. Lamps and LEDs	X		X		
e. Primary (Main) Supply	X		X		
f. Transponders	X		X		
9. Control Unit Trouble Signals	X				X
10. Emergency Voice/Alarm Communication Equipment	X				X

11. Engine-Driven Generator	X (weekly)		X
12. Fiber Optic Cable Power	X		X
13. Guard's Tour Equipment	X		
14. Initiating Devices			
a. Duct Detectors	X		X
b. Electromechanical Releasing Devices	X		X
c. Extinguishing System Switches	X		X
d. Fire-Gas and Other Detectors	X		X
e. Heat Detectors	X		X
f. Fire Alarm Boxes	X		X
g. Radiant Energy Fire Detectors	X		X
h. Smoke Detectors: Functional	X		X
i. Smoke Detectors: Sensitivity			One year after installation and alternate years thereafter
j. Supervisory Signal Devices	X	X	
k. Waterflow Devices	X	X	
15. Interface Equipment	X		X
16. Off-Premises Transmission Equipment	X	X	
17. Remote Annunciators	X		X
18. Retransmission Equipment	X		

(Continues)

Table 12-3. (Continued)

	Initial and Reacceptance Tests	Monthly	Quarterly	Semiannually	Annually
19. Special Hazard Equipment	X				X
20. Special Procedures	X				X
21. System and Receiving Equipment: Off-Premises					
a. Operational					
1. Functional: All	X				
2. Transmitters: Waterflow and Supervisory	X	X			
3. Transmitters: All Others	X				X
4. Receivers	X	X			
b. Standby Loading: All Receivers	X	X			
c. Standby Power					
1. Receivers: All	X	X			
2. Transmitters: All	X				X
d. Telephone Line: All Receivers	X	X			
e. Telephone Line: All Transmitters	X				X

*For testing addressable and analog devices, which are normally either affixed to a single molded assembly or affixed twist-lock type to a base, testing must be done utilizing the signaling style circuits (Styles 0.5–7). Analog-type detectors shall be tested with the same criteria.

station serves multiple properties under different ownerships. These agencies could include the local fire department or fire prevention bureau, insurers, state regulatory agencies, and so on.

All tests and maintenance for any fire alarm signaling system, including a central station system, should be supervised and performed by qualified personnel. Complete test and maintenance records should be kept for all systems and components.

Complete information regarding the central station system, including specifications, wiring diagrams, and floor plans, should be submitted for approval to the local authority prior to installation of equipment or wiring. All devices, combinations of devices, and equipment constructed and installed for a central station system should be approved for their intended purposes and a certificate prepared (see Figure 12-5). Upon completion of a system, a satisfactory test should be made in the presence of a representative of the local authority if so required.

Other Systems: Additional test and maintenance are required for emergency voice/alarm communication systems. After completion of acceptance tests, a set of reproducible as-built installation drawings, operation and maintenance manuals, test and maintenance schedules, and records for this system should be provided for the owner or a designated representative and kept current.

Audible appliances for the public operating mode of an emergency voice/alarm system should be tested and maintained in accordance with the manufacturer's instructions. Tests should be conducted using a portable sound level meter meeting the Type 2 requirements of ANSI S.1.4, *Specification for Sound Level Meters* (*A*-weighted scale).[3] Audible appliances in the private operating mode should be tested and maintained in accordance with the manufacturer's instructions.

Visible appliances for the public operating mode should be tested and maintained in accordance with the ratings, procedures, and component-parts replacement schedule recommended by the manufacturer for their intended use. Visible appliances for the private operating mode should be tested and maintained as rated and specified by the manufacturer or local authority having jurisdiction, including the recommended change schedule for replacement components.

Acceptance Testing for Automatic Fire Detectors: The owner should assign responsibility for inspections, tests, and maintenance programs to a designated person. People at all locations where the alarm system reports should be notified before testing to prevent unnecessary response and notified when testing has concluded. Any method or device used for testing in an atmosphere or process

Property name _____   Central station company_____

Address _____   Station location _____

To: _____   Address _____
    _(authority having jurisdiction)

1. A contract (dated _____) for test and inspection in accordance with NFPA standard(s) no.(s) _____ [edition date(s)] is in effect.

2. The system(s) has been installed in accordance with NFPA standard(s) no.(s) _____ [edition date(s) _____], includes the devices listed below, and has been in service since (date _____) .

3. Location   Building(s) — name, number, etc.

4. Manual fire   (a) Number of coded stations _____
   alarm service   (b) Number of non-coded stations _____ activating fire alarm transmitters
    (c) Number of combination manual fire alarm and guard's tour coded stations _____
    (d) Local annunciator — yes   no

5. Guard's tour   (a) Number of coded stations _____
   supervisory   (b) Number of non-coded stations _____ activating transmitters
   service   (c) Compulsory guard's tour system consisting of _____ transmitter stations and _____ interme- diate stations.
    Note: Record combination devices under 4(c) & 5(a)

6. Automatic fire   (a) Coverage   full   partial
   detection and   (If partial, indicate locations)
   alarm service   (b) Types of detectors and number of each (for line type, indicate number of circuits)
    (c) Local annunciator — yes   no
    (d) Local alarm   — yes   no
    (e) Number of coded fire signals _____
    Number of coded trouble signals _____

Figure 12-5. Certification of completion for a central station signaling system. (*Continues*)

classified as hazardous should be suitable for such use. Records of all inspections, tests, and maintenance should be kept on the premises for a minimum of 5 years.

After installation of the system, all detectors should be visually inspected to be sure that they are properly located and connected in accordance with the manufacturer's recommendations.

Heat Detectors: A restorable heat detector and the restorable element of a combination detector should be tested by exposing the detector to a heat source (such as a hairdryer or a shielded heat lamp) until it responds. The

7. Sprinkler sys-
tem water-
flow alarm &
supervisory
service

(a) Number of coded waterflow signaling
attachments _____
Number of waterflow switches _____ activating
transmitters
(b) Number of coded valve supervisory signaling
attachments _____
Number of valve switches _____
activating _____ transmitters
(c) Other supervisory service provided
Pressure: water   air
Temperature: water   room
Water level   fire pump: running   power
(d) Other fire service provided _____

8.

(a) Means of transmission of signals from the protected premises to the central station: McCulloh
multiplex   digital alarm communicator
two-way rf
(b) Means of retransmission of alarms to the public
fire service communications center: 1. _____
2. _____

9. Comments

(Signed) for central station company   Title   Date

Frequency of routine tests and inspections, if other than in accordance with
the referenced NFPA standard(s): _____

System(s) deviations from the referenced NFPA standard(s) are:

(Signed) for central station company   Title   Date

Upon completion of the system(s), satisfactory test(s) witnessed (if required by the
authority having jurisdiction)

By _____
     (representing the authority having jurisdiction)   Title   Date

Figure 12-5. (*Continued*)

detector should be reset after each heat test. Precautions should be taken to
avoid damage to the nonrestorable fixed-temperature element of a combination
rate-of-rise/fixed-temperature detector.

A pneumatic tube line-type detector should be tested with either a heat
source (if a test chamber is in the circuit) or pneumatically with a pressure
pump according to manufacturer's instructions. Line- or spot-type nonrestorable
fixed-temperature heat detectors should be tested mechanically or electrically

(not with heat) to verify alarm function. If required for proper performance, the loop resistance of line-type detectors should be measured to determine whether it is within acceptable limits for the equipment being used. The resistance should be recorded for future reference.

Detectors with a replaceable fusible alloy element should be tested by first removing the fusible element to determine that the detector contacts operate properly and then reinstalling the fusible element.

Smoke, Flame, and Other Fire Detectors: An alarm should be initiated with smoke or other aerosol at each smoke detector to demonstrate that smoke can enter the chamber and initiate an alarm.

Most residential smoke detectors have an integral test permitting an individual to check the system and its sensitivity. Underwriters Laboratories Inc. Standards 217, *Standard for Safety Single and Multiple Station Smoke Detectors,* and 268, *Standard for Safety Smoke Detectors for Fire Protective Signaling Systems,* require that each smoke detector be provided with either a means of measuring detector sensitivity or a built-in sensitivity test feature. These standards allow the test feature to be either electric or mechanical, but it is required to simulate a smoke level of no more than 6 percent/ft (20 percent/m). Though this exceeds the normal factory-set level of between 0.5 and 3 percent/ft (1.6–10 percent/m), it is felt that 6 percent/ft is a tolerable level. A more accurate measurement of smoke detectors in the field can be performed using a portable aerosol generator developed for this purpose.

Testing a detector, however, does not increase reliability. Tests will only identify detectors that have either failed or have shifted their sensitivity (set point).

Blowing smoke into a detector or using an aerosol spray to test detector sensitivity is not recommended since the density of either cannot be accurately determined. Using these methods could expose a detector to many times its alarm threshold and lead one to believe that the detector was still within calibrated limits when in fact the detector set point could have shifted to an unacceptable level.

One method used to simulate 6 percent smoke in a photoelectric detector uses a calibrated test wire built into the detector. This test wire passes through the light path when the test button is pressed. The test wire scatters the same amount of light onto the photosensitive device as when smoke of about 6 percent/ft density enters the smoke-sensing chamber.

In ionization detectors, various methods are used to simulate smoke. In some detectors, the electric conductivity of the smoke-sensing chamber is changed by placing a shunt across the chamber. In other detectors, the geometry of the chamber is physically changed or a voltage is impressed on the circuit in the smoke chamber that simulates the output voltage with a given density of smoke particles.

Because a specific level of smoke is needed for a proper test, only the factory calibration method or a calibrated particle generator should be used to test the sensitivity of smoke detectors.

To test a smoke detector sensitivity range, the detector should be tested using a calibrated test method, the manufacturer's calibrated sensitivity test instrument, listed control equipment arranged for the purpose, or another calibrated sensitivity test method acceptable to the local authority.

Detectors with a sensitivity outside the approved range should be replaced. Detectors listed as field adjustable may be either adjusted within an approved range or replaced.

Flame detectors, fire-gas detectors, and other fire detectors should be tested for operation in accordance with instructions supplied by the manufacturer or other test methods acceptable to the local authority. Inspection forms are also issued for automatic fire detectors in addition to system certification.

The inspection should include the following information on initial tests:

1. Date
2. Name of property
3. Address
4. Installer/maintenance company name, address, and representative
5. Name, address, and representative of approving agency or agencies
6. Number and type of detectors per zone for each zone
7. Functional test of detectors
8. Check of all smoke detectors
9. Loop resistance for all fixed-temperature line-type detectors, if any
10. Other tests as required by equipment manufacturers
11. Signature of tester and approval authority representative

Periodic Testing and Maintenance

Maintenance: Proper maintenance is as important as regular testing in a fire alarm signaling system. Automatic fire detector maintenance depends on the specific type of detector used, local environmental conditions, and manufacturer's recommendations. The manufacturer's recommendations should be implemented in the maintenance program to maintain system reliability.

Periodic Testing of Automatic Fire Detectors: Each detector should be visually inspected to ensure that it remains in good physical condition and that there are no changes such as building modifications, occupancy hazards, and environmental effects that would affect detector performance.

Detectors require periodic cleaning to remove dust or dirt that may have accumulated. The frequency of cleaning depends on the type of detector and the local ambient conditions. For each detector, cleaning, checking, operating, and sensitivity adjustment should be performed according to the manufacturer's instructions. These instructions should detail methods such as vacuuming to remove loose dust and insects and washing to remove heavy greasy and grimy deposits. The manufacturer may provide cleaning service at the factory or at field locations in lieu of these cleaning methods.

All automatic detectors suspected of exposure to a fire condition should be tested. Inspection forms should be issued each time a periodic test and inspection is performed. The form is identical to the initial inspection form except that it includes information on test frequency; name of person performing inspection, maintenance, and/or tests; that person's affiliation, business address, and telephone number; and designation of the detector(s) tested.

Smoke, Flame, and Other Fire Detectors: All smoke detectors should be visually inspected in place at least semiannually to identify missing detectors, detectors with impeded smoke entry, abnormally dirty detectors, and detectors no longer suitably located because of occupancy or structural changes. Smoke detectors should be tested annually.

Smoke detectors need to be cleaned for optimum detection results. If dust and dirt partially surround the smoke chamber of an ionization detector, smoke particulate may not reach the chamber and the detector may become less sensitive. If dust and debris insulate the radioactive foil, the rate of ion formation is reduced, making the ionization detector more sensitive and prone to false alarm. Accumulation of dust and film on photoelectric sensors such as on bulbs, lenses, and photocells diminishes the intensity of light within the detector. Light-scattering-type detectors become less sensitive as the light intensity is decreased. Light-attenuation-type detectors become more sensitive as the light intensity is decreased.

Detector sensitivity should be checked within 1 year after installation and every alternate year thereafter. Detectors that are abnormally sensitive should be replaced or cleaned and recalibrated.

Air duct detectors should undergo additional tests, consisting of visual inspection of the detector installation (including seals) to detect any abuse or modification of the device or installation and its intended operation. Also, the manufacturer's recommendations should be used to verify that the device will respond to smoke in the air stream; for example, measuring pressure drop or air flow through the detector for devices using sampling tubes is often recommended.

Flame detector sensitivity is affected by dust and dirt built up on the lens. The lens should be periodically cleaned for optimum detector performance; all flame detectors should be maintained in a clean condition for optimum detection results.

All flame detectors, fire-gas detectors, and other fire detectors should be tested at least semiannually as prescribed by the manufacturer, and more often if necessary.

Heat Detectors: For nonrestorable spot-type detectors, after the 15 years, at least 2 detectors of 100 (or fraction thereof) should be removed every 5 years, sent to a testing laboratory for testing, and replaced with new detectors. If a failure occurs on any of the detectors removed, additional detectors should be removed and tested until either a general problem involving faulty detectors or a localized problem involving 1 or 2 defective detectors is found.

For restorable heat detectors (except the pneumatic line type), one or more detectors on each signal initiating circuit should be tested at least semiannually and different detectors selected for each test. Each detector should have been tested within 5 years.

All pneumatic line-type detectors should be tested for leaks and proper operation at least semiannually. Nonrestorable line-type fixed-temperature detectors should also be tested for alarm function at least semiannually. The loop resistance should be measured, recorded, and compared with that previously recorded and any change in loop resistance investigated.

Fire Alarm Signaling Systems: In general, a fire alarm signaling system should be properly tested and inspected at prescribed intervals. For most systems, the local authority should be consulted on all system alterations and additions. The owner of the protected facility is responsible for providing proper system maintenance by either personnel at the facility or through a maintenance agreement with outside specialists.

All apparatus should be restored to normal as promptly as possible after each test or alarm and kept in normal condition for operation. Apparatus should be rewound, reset, or replaced as necessary. A complete record of the tests and operations of each system should be kept, in general, for at least 1 year. The record should be available for examination and reported to the local authority when required.

Central Station Systems: The central station is required to have a minimum of two persons on duty to ensure immediate attention to signals received. To provide prompt runner service to the protected premises, one of the two persons may leave the central station if provisions are made to contact the central station at least once every 30 minutes. This may be modified by the local authority when alarms are automatically transmitted to the public fire service communication center. Operation and supervision should be the primary functions of the operators and runners.

Necessary repairs to a central station should be commenced by competent maintenance personnel within 4 hours after notification of a need for service and should continue until completion. The central station should maintain a suitable stock of spare parts to allow updating and repair of equipment at the central station and the protected premises.

Tests of all circuits extending from the central station and tests of central station devices should be made at intervals not to exceed 24 hours. Complete and satisfactory tests should be performed monthly on all systems and devices except as follows:

1. Quarterly or more frequently for all transmitters, waterflow-actuated devices, automatic fire detection systems, and valve supervisory devices. After any vertically mounted sprinkler system control valve has been operated, the owner or occupant should be encouraged to perform a drain test to ensure that the valve has been fully reopened.
2. Annually for manual fire alarm boxes, combination night guard and fire alarm boxes, tank water level devices, building and tank water-temperature supervisory devices, and other sprinkler system supervisory devices.
3. Automatic fire detection devices on the system should be inspected and tested in accordance with *NFPA 72*.

In very large central station facilities, the test program should be arranged to maintain maximum protection and service at all times. Periodic testing is also necessary for protected-premises systems, which are connected to the central station.

Plant management and the central station should be notified just prior to testing and the name of the central station operator recorded. Each waterflow alarm device should be actuated by use of the inspector's test facilities. (Ringing of the local alarm bell is not an indication of transmitter operation.)

For sprinkler waterflow alarm tests, an actual waterflow, through the use of the inspector's test connection, should be used to test reliability of the sprinkler alarm unit as a whole. For a wet-pipe system, the test connection at the extremity of the system should be used.

Automatic fire detectors should be tested. Other alarm transmitters and supervisory switches and transmitters should be actuated in accordance with their frequency schedule. Circuits within the protected premises should be tested for proper operation in ground and open conditions. Primary power should be turned off and one fire alarm transmitter operated from batteries. The system and primary power should subsequently be restored to normal operating conditions. The central station and plant management should be notified of the test conclusion and the number and identity of signals checked. When several waterflow switches are connected to a single transmitter and test jacks are provided, the first and last tests should be made without the use of the test jack. For periodic testing of multiplex systems, the above procedure applies; *transponders* replace *transmitters* as described in the procedure.

Local System Testing: Periodic testing and inspection of local systems at prescribed intervals, in general, should be performed at the discretion of the qualified persons in charge of the system.

Auxiliary Systems: The auxiliary system should be inspected during monthly tests to observe the condition of all components and determine any changes in the property that may affect the protection. Reports of test results and changes should be furnished to the local authority. All initiating and transmitting devices on the system, including transmission of signals to the municipal communication center, should be tested monthly. (The local authority should be notified whenever an auxiliary system is not in service.)

Noncoded manual fire alarm boxes should be tested at least once every year and fire detectors tested according to *NFPA 72*.

If an engine-driven generator dedicated to the protective signaling system is used as a required power source, it should be tested by being operated weekly under load, disconnecting the normal supply to the system for a minimum of 1/2 hour in a continuous period.

Remote-Station and Proprietary Systems: All operator controls at the remote station designated by the authority having jurisdiction should be tested at each change of shift and drills conducted at regular intervals to satisfy the local authority.

Tests should be conducted annually for all automatic fire detection systems or other systems and devices except for waterflow-actuated devices and other equipment.

Waterflow-actuated devices should be tested at least every quarter. For sprinkler waterflow alarm tests, an actual waterflow, through the use of a test connection, should be used to test the reliability of the sprinkler alarm unit as a whole. Gate valve supervisory switches, manual fire alarm boxes, combination guard tour and fire alarm boxes, tank water level devices, building and tank water temperature supervisory devices, and other sprinkler system supervisory devices should be tested quarterly.

After any vertically mounted sprinkler system control valve has been operated, a drain test should be performed to ensure that the valve has been fully reopened. If possible, one of the supervisory switches tested at each inspection should be the most electrically remote device on the circuit being tested.

A flow through the alarm test bypass connection should be used to test the waterflow alarm of a dry-pipe, preaction, or deluge sprinkler system. For a wet-pipe sprinkler system, the test connection at the extremity of the system should be used.

Testing the waterflow alarm device for dry-pipe, deluge, or preaction systems using the bypass test valve does not require tripping of the dry-pipe, deluge, or preaction valve when all related equipment is maintained in proper operating condition. Trip tests of dry-pipe, deluge, or preaction valves are usually the responsibility of the property owner or lessee.

The inspector's test connection for a wet sprinkler system should have an orifice sized for the smallest sprinkler head in the system and be located in accordance with NFPA 13, *Standard for the Installation of Sprinkler Systems.*

If an engine-driven generator dedicated to the protective signaling system is used as a required power source, it should be tested by being operated weekly under load by disconnecting the normal supply to the system for a minimum of 1/2 hour in a continuous period.

Emergency Voice/Alarm Systems: Emergency voice/alarm communication systems should be maintained and tested by qualified personnel. Functional and operational testing of the voice/alarm signaling service should be conducted annually and should include the use of a representative number of reporting devices, such as two-way telephones for the fire service, telephones for fire warden use or general public emergency use, and automatic and manual voice paging systems in each zone. All elements of the system should be tested at least annually.

Notification Appliances: Notification appliances for protective signaling systems should be tested at the same interval. The notification appliances for the private operating mode should be tested once every 6 months. Notification appliances may require periodic cleaning to remove dust or dirt that has accumulated. The frequency of cleaning depends on the local ambient conditions. For each appliance, the cleaning, checking, operation, and volume adjustments should be attempted only after consulting the manufacturer's instructions.

A permanent record showing all details of the test, including the name of the inspector, the type number, the location, and the date, should be kept on the premises for at least 5 years.

System Identification Placard: Periodic testing verifies the ongoing operation of the system. A complete functional test of a protective signaling system includes testing of connections to any equipment monitored or controlled by the system. A permanently mounted identification placard should be located in or adjacent to the system control unit and should contain the following information:

1. Names, addresses, and telephone numbers of the installation and servicing contractors.
2. Reference to the standard (including date of issue) to which the system conforms.
3. Description of primary (main) and secondary (standby) power supplies. This should include physical location and identification of all overcurrent devices and control switches, type of secondary supply, standby battery specifications, standby generator, names of local inspection authorities (including references and dates), and location of the as-built drawings and system operating and maintenance instructions.

Sample placards are contained in *NFPA 72*. Placard information should be verified in the acceptance procedure as previously described in this chapter.

Periodic tests should be performed in accordance with recommended schedules or more frequently when required by the local authority. When less than a 100 percent test is being performed, a record should be maintained of the individual devices being tested so that different devices and appliances are used in subsequent tests.

Public Fire Service Communication Systems: Test procedures for an auxiliarized system cover only testing of the alarm equipment up to and including the trip coil in the master box. Additional test procedures are needed for the balance of the system, that is, the master box through to, and including, the public fire service communication center.

Testing facilities should be installed at the communication center and each satellite communication center (if used). Those facilities for systems leased from a nonmunicipal organization may be located elsewhere if approved by the local authority.

An emergency power source other than batteries should be operated to supply the system for a continuous period of 1 hr at least weekly. This test should require simulated failure of the normal power source.

Although public fire service communication systems are usually tested and maintained by municipal employees, this service can be performed by private contractors. In this case, complete written records of the installation, maintenance, test, and extension of the system should be forwarded to the responsible municipal employee as soon as possible. The municipal employee should also be notified of any failure and restoration of service.

Maintenance performed by other than the municipality or a municipal employee should be under written contract, guaranteeing performance acceptable to the local authority.

Telephone receiving equipment, that is, the power source, incoming circuits, and public safety answering point (PSAP) equipment, should also be tested. Circuits should be tested at a minimum of once a week, and PSAP equipment should be operated in the fail-safe mode at least once a week. All dispatch system apparatus should be restored to normal condition as promptly as possible after each test or alarm in which the apparatus functioned.

Where supervisory devices or tests indicate that trouble has occurred anywhere on the system, the operator should take appropriate steps to repair the fault or, if this is not possible, to isolate the fault and notify the official responsible for maintenance.

Manual test of dispatch circuit instruments must be made and recorded at least once in each 24-hour period. Circuits for graphic transmission of signals should be tested by a message transmission. Outside devices, radio, telephone, or other facilities for alerting volunteer and off-duty fire fighters should be tested daily, and all wired radio and voice amplification circuits should be subjected to a talking test at least twice daily.

In the communications center, manual tests of the power supply for wired dispatch circuits should be made and recorded at least once in each 24-hour period. Such tests should include:

1. Current strength of each circuit and voltage across terminals of each circuit inside protected devices. Changes in voltage or current of any circuit that amount to 10 percent of normal should be investigated immediately.
2. Voltage between ground and circuits. When this test shows a reading in excess of 50 percent of that shown in the voltage test, the trouble should be immediately located and cleared. Readings in excess of 25 percent should be checked. These readings should be taken with a voltmeter of not more than 100 ohms/V impedance to minimize false ground readings.

Note: Systems in which each circuit is supplied by an independent current source will require tests between ground and each side of each circuit. Common-current source systems will require voltage tests between ground and each terminal of each battery and other current source.

3. A ground current reading is acceptable in lieu of Item 2. When this method of testing is used, all grounds showing a current reading in excess of 5 percent of the normal line current should be checked immediately.
4. Voltage across the terminals of common battery on the switchboard side of fuses.
5. Voltage between common battery terminals and ground. Abnormal ground readings should be investigated immediately. If more than one common battery is used, each should be tested.

Dispatch circuit instruments should be manually tested at least once each 24-hour period.

For equipment in the fire station, the power supply for dispatch circuits should be manually tested and recorded at least once each 30 days. Such tests should include the voltage across terminals of any power source on the receiving device side of fuses and the voltage between any power source terminals and ground. Abnormal ground readings should be investigated immediately. Emergency generator equipment should be operated to supply the system for a continuous period of 1 hour at least weekly. This test should also require simulated failure of a normal power source. Batteries supplying dispatch equipment should be tested as described in the Power Supply Requirements section of Chapter 4.

Public reporting systems should also be tested and maintained. A complete record should be kept by the municipality of all test and alarm signals, all circuit interruptions and observations or reports of apparatus failure or derangements, and all seriously abnormal or defective circuit conditions indicated by test or inspection for public reporting systems. These records should include the date and time of all occurrences.

Tests should be conducted from the multiplex interface to verify trouble indications for common-mode failures, such as alternating current power failure. To test for the alternating current power, the alternating current power supply to the interface should be removed and at least one initiating device activated. The manufacturer's manual should be consulted for other common-mode failures and the described testing procedures conducted. Proper receipt and proper processing at the central supervising station should be verified.

Each fault condition the system is required to detect should be introduced on the signaling line circuit. Proper receipt and the proper processing of the signal at the central supervising station should be verified.

When multiplex protective signaling systems are equipped with various optional features unique to these systems, the manufacturer's manual should be consulted to determine the proper testing procedures for verifying the proper operation of these optional features.

Manual tests of box circuit instruments should be made and recorded at least once each 24 hours. Where applicable, all box circuit instruments should be tested by use of operators' keys. Where repeating facilities are necessary, the test of one box from every circuit from which no alarm was transmitted during the past month should be transmitted over the entire system. Boxes should be tested by operation under conditions simulating actual use. Test signals should be transmitted and recorded at the public fire service communication center. A periodic test should be performed on all fire alarm boxes at least once in each 60-day period and the boxes examined, cleaned, and all functions tested.

Records of boxes should include box identification, location address, circuit number (if applicable), physical mounting, description by manufacturer, model number, date of installation and power source (radio), and test dates and time. Field inspection forms for the boxes should include information on physical condition, paint, mounting, door function, drop wire or antenna; tests of all box functions; and maintenance.

Each coded radio box should automatically transmit a test message at least once in each 24-hour period. The test should include the operation of all message functions associated with each box tested. Such message functions should be transmitted to the respective communication center, received, and permanently recorded. Where solar charging of box battery or batteries is utilized, the solar cell associated with each box in the system should be examined and cleaned at least once in each 60-day period. Receiving equipment associated with coded radio-type systems should be tested at least once each hour. The receipt of test messages is considered sufficient, provided at least one such message is received each hour.

Manual tests of box and dispatch circuit instruments for telephone reporting systems should be performed and recorded at least once each 24-hour period. If a telephone (parallel) reporting system is used, the person testing a voice box should furnish identification and request the calling location.

RECORDS

Complete records sufficient to ensure reliable operation of all alarm system functions should be maintained in a satisfactory manner. *NFPA 72* contains an example of an inspection and testing form for fire alarm systems. This form or one similar to it should be used to ensure complete records are maintained.

A complete record should be kept by the municipality of all test and alarm signals, all circuit interruptions and observations or reports of apparatus failures or derangements, and all seriously abnormal or defective circuit conditions indicated by test or inspections. These records should include the date and time of all occurrences.

Additionally, an as-built (record) drawing cabinet locked and keyed in the same way as the FACU should be installed at each project that has a new fire alarm system and located where acceptable to the AHJ. The cabinet shall contain the following:

- A complete set of as-built drawings
- A copy of the AHJ inspection report
- A copy of the original test printout
- A hard-copy printout of the system configuration if it is a programmable system
- A copy of each maintenance inspection report
- Operating instructions
- A copy of the equipment submittals/specifications.
- Voltage drop calculations
- Battery calculations
- Any keys required for the system
- Sequence of operation

When a combination of leased/owned facilities exists, records that must be maintained by the lessor for the municipality should be specified. A report of operations summarizing important statistics should be prepared annually. Records of wired circuits (box and dispatch) should include outline plans showing terminals and box sequence, diagrams of office wiring, and materials including trade name, manufacturer, and year of purchase or installation. Emergency generating equipment periodic test records should include date and time; fuel, electric, coolant, and exhaust system conditions; and operating time.

MULTIPLEX SYSTEMS AND CIRCUIT STYLES

Additional acceptance and reacceptance test procedures are needed to verify multiplex-type protective signaling system performance. The manufacturer's manual and the as-built drawings provided by the supplier should be used to verify proper operation after the initial testing phase has been performed by the supplier or a designated representative.

General inspection	Yes	No
1. Roof:		
Is roof covering noncombustible?	____	____
Are scuppers and drains unobstructed?	____	____
Are lightning arrestors in good condition?	____	____
Are skylights protected by screens?	____	____
Is access to fire escapes unobstructed?	____	____
Do fire escape stairs appear to be in good condition?	____	____
Are fire escape stairs unobstructed?	____	____
Are standpipe and sprinkler roof tanks and supports in good condition?	____	____
Are standpipe and sprinkler control valves secured in proper position?	____	____
2. All floors (inspect from top floor to basement):		
Are self-closing fire doors unobstructed and properly equipped with closing devices?	____	____
Are fire exits and directional signs properly illuminated?	____	____
Is emergency lighting system operable?	____	____
Are corridors and stairways unobstructed?	____	____
Are fire exits unlocked and unobstructed?	____	____
Are sprinklers unobstructed?		
Are standpipe hose outlets properly marked and unobstructed?	____	____
Are sprinkler control valves properly labeled and unobstructed?	____	____
Are recorded weekly inspections made of all sprinkler control valves to make certain they are open?	____	____
Are dry-pipe valves (for sprinklers in areas exposed to freezing) in service, with air pressure normal?	____	____
Are all fire-detection and fire-suppression systems in service and tested regularly?	____	____
Are sufficient fire extinguishers present?	____	____
Are extinguishers of the proper type? (See NFPA 10, *Standard for Portable Fire Extinguishers*.)	____	____

Figure 12-6. Fire safety self-inspection form. (*Continues*)

Starting from the unpowered condition, the system should be initialized in accordance with the manufacturer's manual. One or more tests should be conducted to verify that communication exists between the central processing unit and the connected central supervisory station peripheral devices.

Each initiating device circuit should be tested for its alarm-reporting capability by operating at least one of the initiating devices connected to it. Upon completion of this test, an open-circuit trouble condition should be made to verify open-circuit fault detection. Additional tests should be conducted to verify

Are extinguishers properly hung
and labeled?   ____   ____
Are smoking regulations enforced with
employees and visitors?   ____   ____
Are aisles to exit routes unobstructed
and visible?   ____   ____
Is housekeeping properly maintained?   ____   ____
Are cleaning supplies safely stored?
Are all trash receptacles emptied at
least daily?   ____   ____

3. Ground floor:
Do entrance and exit doors provide
unobstructed egress?   ____   ____
Is safe egress uncompromised by
security measures?   ____   ____

Exterior Inspection

1. Evacuation:
Do all exits, emergency exits, and
fire escapes have unobstructed passage
to safe areas?   ____   ____

2. Environment:
Are grounds clear of accumulations of
flammable material?   ____   ____
Have neighboring occupancies minimized
exterior fire hazards?   ____   ____
Is fire service access clear?   ____   ____
Are standpipe and sprinkler system siamese
connections unobstructed and operable?   ____   ____
Are hydrants unobstructed?   ____   ____

Personnel Inspection

Training:
Do all staff members know how to transmit
a fire alarm?   ____   ____
Do all staff members know their assigned
duty in evacuating the building?   ____   ____
Do all staff members know how and when
to use portable fire extinguishers?   ____   ____
Do all staff members know their
responsibilities in fire prevention?   ____   ____

Figure 12-6. (*Continued*)

all possible status modes that the initiating-device circuit should provide. Proper receipt and processing of each test signal by the central supervisory station equipment should be verified.

When the style of an initiating-device circuit or signaling line circuit requires alarm signaling capability in the presence of a fault condition, that capability should be verified by the proper receipt and the proper processing of the alarm signal at the central supervising station. This test should be repeated for all fault conditions for the style of the circuit involved.

TESTING SCHEDULES

Table 12-3 lists acceptance and periodic tests unless otherwise noted. The recommended test means are not intended to exclude equivalent means, such as self-diagnostic testing. Tests conducted on equipment and wiring in hazardous locations can require special testing procedures.

CONCLUSION

Whereas testing and maintenance are important to a fire alarm signaling system and its components, it is equally important to periodically inspect an entire protected facility. Figure 12-6 (on page 374), shows some of the typical fire safety items that should be checked on a regular basis.

QUESTIONS

1. Acceptance testing is required after:
   a. initial installation
   b. significant alterations to the system
   c. changes to site-specific system software
   d. a, b, and c

2. Smoke detector sensitivity is tested:
   a. annually
   b. semiannually
   c. after 1 year and alternate years thereafter
   d. never

3. System components found defective must be replaced:
   a. immediately   b. within 24 hours
   c. within 30 days   d. at the earliest opportunity

4. Operator controls in supervising stations should be tested:
   a. at each shift change   b. daily
   c. weekly   d. monthly

5. Nonrestorable heat detectors should be tested:
   a. daily   b. monthly
   c. annually   d. sampled beginning in the 15th year

BIBLIOGRAPHY

NFPA Codes, Standards, Recommended Practices, and Manuals (see the latest *NFPA Catalog* for availability of current editions of the following documents):

NFPA 13, *Standard for the Installation of Sprinkler Systems.*

NFPA 72®, *National Fire Alarm Code®.*

REFERENCES CITED

1. Fire Chiefs of Los Angeles County, "Study on Central Stations Effectiveness," December 30, 1983.
2. Drouin, A., "Hotel Fire Alarm Life Safety Systems Audit," 1983.
3. ANSI S.1.4, *Specification for Sound Level Meters,* American National Standards Institute, New York.

ADDITIONAL READING

NBS, *"An Instrument to Evaluate Installed Smoke Detectors,"* NBSIR 78-1430, National Bureau of Standards, Washington, DC.

NFPA 10, *Standard for Portable Fire Extinguishers.*

NFPA 70, *National Electrical Code®.*

NFPA 101®, *Life Safety Code®.*

NFPA 1221, *Standard for the Installation, Maintenance, and Use of Emergency Services Communications Systems.*

UL 217, *Standard for Safety Single and Multiple Station Smoke Detectors,* Underwriters Laboratories, Inc., Northbrook, IL.

UL 268, *Standard for Safety Smoke Detectors for Fire Protective Signaling Systems,* Underwriters Laboratories, Inc., Northbrook, IL.

13

Public Fire Service Communication Systems

INTRODUCTION

The public fire service communication center receives notification of a fire condition initiated by any fire alarm signaling system, including local, central station, auxiliary, remote-station, proprietary, and household warning systems. Notification must be direct, via an auxiliary or remote-station system, or by phone or other means of retransmission. The communication center must be properly maintained and operated so the links are maintained among the fire department, the alarm system, and other points within the communication system.

This chapter describes installation, maintenance, and use of a public fire service communication system (and the facility in which it is located), and communication center operation, including retransmission of alarms and reporting and dispatch systems. For additional information, see Chapter 5, Signal Initiation; Chapter 6, Signal Transmission Methods and Procedures; Chapter 7, Fire Alarm Notification; and Chapter 12, Fire Alarm System Testing.

Public fire service communication systems and facilities include (but are not limited to) public reporting, dispatching, telephone, and both two-way and microwave radio systems. All of these fulfill two principle functions: receipt of fire alarms or other emergency calls from the public and retransmission of these alarms and emergency calls to fire companies and other interested agencies.

A public fire service communication center is defined as the building or portion of a building that houses the central operating part of the fire alarm system. The center is also usually the place where the necessary testing, switching, receiving, retransmitting, and power supply devices are located.

COMMUNICATION CENTER LOCATION AND CONSTRUCTION

If the fire service communication center building is located within 150 ft (46 m) of another structure, special attention should be given to guard against damage by protecting openings and constructing the roof to resist damage that might be caused by falling walls. The communication center should not be located below grade unless the structure is specifically designed for such a location. Floor elevation should be above the 100-year flood plan prediction if the location is below grade.

Seismic and wind loads prevalent in the geographic location of the building should be considered in design and construction of the communication center. Applicable building codes should always be followed. The building housing the center should be of fire-resistive construction or protected noncombustible/limited-combustible construction.

If the building is unprotected noncombustible/limited-combustible or ordinary construction, it should have a Class A fire-resistive roof covering and a sprinkler system in all areas of the building except the communication center and power room. The sprinkler system should be completely supervised by the communication center.

A fire service communication center can be located in other than a fire department building. In these buildings, the communication center should be separated from the other portions of the building by vertical and horizontal separations with minimum 2-hour fire resistance rating. Openings should be protected by self-closing or automatic fire doors or other assemblies with a minimum 1½-hour fire resistance rating. If spaces adjoining the communication center are ordinary-hazard occupancies (as defined in NFPA 13, *Standard for the Installation of Sprinkler Systems*), they should have an automatic fire alarm system. If such spaces are occupied by extra-hazard occupancies as defined therein, they should have an automatic sprinkler system. Interior finish material should have a maximum flame spread rating of 25.

The communication center and other buildings that house essential operating equipment should be secure. Entryways leading directly from the exterior should be protected by two doors and a vestibule. Entry to the communication center should be restricted to authorized persons only, and door openings should be protected by not less than a Class B self-closing fire door assembly.

Warm air heating, ventilating, and air conditioning services should be provided to the communication center by independent systems. Main water, sewer, storm sewer, or sprinkler lines should not pass through the communication center or its equipment rooms. An automatic fire alarm system connected to an audible and visible warning device at a continually manned location should protect the entire communication center.

Fire extinguishers should be provided for the communication center with at least two extinguishers with 2-A or greater rating and two extinguishers with a combined rating of 20-B:C or greater. If it is not possible to have these, two multipurpose extinguishers with 2-A:10-B:C rating may be provided.

The communication center should also be equipped with an emergency lighting system that can be immediately placed in service and powered by an independent source. Illumination should be sufficient to permit all necessary operations. At least one self-charging, battery-pack lantern should be available that lights automatically when power is interrupted.

Two sources of power should be provided to operate the communications network and its related supporting systems and equipment under all conditions. Three recommended power sources for the communication center that are considered acceptable are as follows:

1. One circuit from a utility distribution system and a second from an engine-driven generator plus a 4-hour-capacity standby storage battery.
2. Two circuits from separate utility distribution systems, serviced or connected so that normal supply to one will not be affected by trouble in the other. This requires supply from two building services on entirely separate distribution networks from independent generating stations.
3. Two engine-driven generators with one unit supplying normal system power and a standby unit that activates within 30 seconds. A standby storage battery having a 4-hour capacity should be provided.

All standby storage batteries incorporated into a power source network should be equipped with suitable float or trickle chargers. If two engine-driven generators are used as a second power source, the fire alarm system 4-hour battery can be omitted.

OPERATION OF THE COMMUNICATION CENTER

Reports of fires and other emergencies to the fire service originate from three principle sources: (1) the general public, (2) the business community (industrial, institutional, commercial, and mercantile), and (3) other public service/safety agencies. The reporting process can be made through any of the following means (individually or in combination): public commercial telephone systems, public emergency reporting/fire alarm systems, privately operated automatic alarm systems, and two-way radio communication systems.

Commercial Telephone Facilities

The conventional commercial telephone network is the most common method of reporting a fire emergency. The installation of outdoor telephone booths increases the availability of telephones for emergency reporting.

The *enhanced 911* system automatically provides both the calling number and the location of the related telephone instrument directly to the dispatcher. This indication is presented visually on a cathode ray tube (CRT) and can be automatically recorded if the communication center includes computer-aided dispatch or computerized logging support systems. The expanding implementation of 911 systems has eliminated the necessity of public reporting stations (street boxes), at least in those areas where publicly accessible telephones are well distributed and functional.

Although there are some disadvantages to total dependency upon telephones, they are often the only means of communication available in cities, suburbs, and rural areas. The public telephone system, in general, is a good means of reporting fires and other emergencies.

Municipal Fire Alarm Systems

Any municipal fire alarm system, whether a coded, voice, or code–voice alarm system, must provide a means by which an alarm can be transmitted from a street alarm box to the communications center.

Use of a public reporting station (fire alarm box) eliminates the difficulty of the dispatcher (who receives the alarm) in determining the location from where the alarm is being transmitted. When actuated, each device should transmit a distinct numerical code in addition to any other functions or capabilities provided.

A Type A public reporting system is one in which an alarm from a fire alarm box is received and is retransmitted to fire stations either manually or automatically (see Figure 13-1). A Type B system is one in which an alarm from a fire alarm box is automatically transmitted to fire stations and, if used, to outside alerting devices (see Figure 13-2).

The Type A system is usually permissible in any size of municipality or area and should be provided when there are more than 2500 emergency calls from boxes per year or where more than 2500 alarms are transmitted over the dispatch circuits.

Automatic retransmission of alarms from boxes by use of electronic equipment can be allowed if reliable facilities and override capability are provided. A public emergency reporting system can transmit other emergency signals or calls as long as they do not interfere with the proper handling of fire alarms.

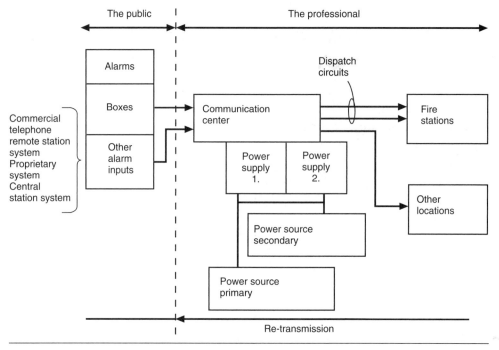

Figure 13-1. Type A public reporting system.

Dispatch Circuits and Equipment

A dispatch circuit is the means by which the fire alarm dispatcher notifies fire companies to respond to an alarm. The location from where the alarm was received is the minimum information that should be transmitted.

Two separate means of transmitting alarms to fire stations should be provided at the communication center. (Only one means of transmission is necessary when fewer than 600 alarms per year are received.) Each alarm transmitted, the date, and the time should be automatically recorded. Devices for transmitting coded or other types of signals should be arranged for manual setting and operation.

Computer-Aided Dispatch

Computer-aided dispatch (CAD) is a process by which a computer and its associated terminal(s) is used to provide relevant dispatch data (running assignments, address locations, equipment status, utility locations, special hazards, etc.) to the dispatcher or operator concerned. CAD is used to assess related response

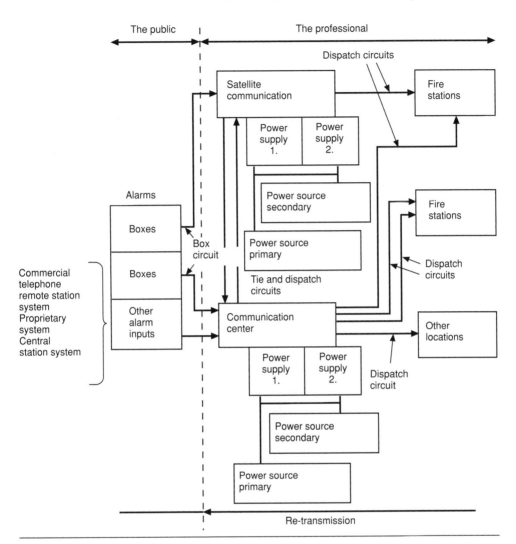

Figure 13-2. Type B public reporting system.

information and it identifies companies normally assigned and their current status (see Figure 13-3).

The degree to which a CAD system can be implemented to dispatch fire units depends upon the size of the fire department. A small department with one to three stations does not require a complex CAD system with its mobile digital terminals, station terminals, and printers. Small departments can use a personal computer with a dispatch software package to dispatch fire units.

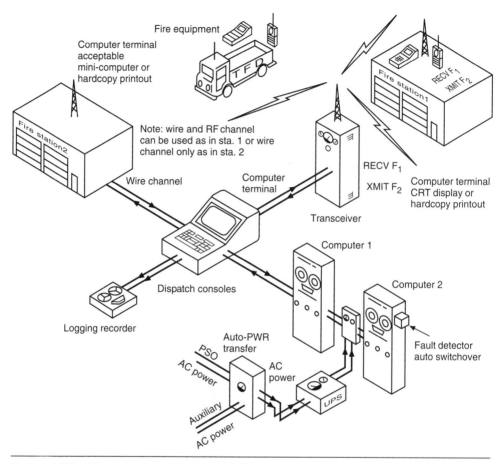

Figure 13-3. Diagram of a computer-aided dispatch system. RECV, receiver, XMIT, transmitter; UPS, uninterrupted power supply.

Personal computers should be carefully selected and used and matched to the proper application. The dispatch computer should not be used for other department applications. It should be kept free of time-consuming processes to allow for rapid retrieval of dispatch information. CAD systems for medium to large departments are more complex and must be custom designed for each department.

There are three major classes of CAD services for fire communication systems. A Class 1 CAD system is one in which computer technology and equipment selects and dispatches fire service personnel and equipment and emergency service assistance. A Class 2 CAD system is used as a support dispatch operation to voice- or graphic-type-operated dispatch systems. A Class 3 CAD system primarily supports fire service dispatching and is limited to status and logging information.

QUESTIONS

1. The reporting process can be made through any of the following means (individually or in combination):

   a. public commercial telephone systems
   b. public emergency reporting/fire alarm systems
   c. privately operated automatic alarm systems
   d. two-way radio communication systems
   e. all of the above

2. The communication center should also be equipped with an emergency lighting system:

   a. powered by an independent source
   b. powered by a battery pack
   c. powered by city power
   d. powered by 120 V ac

3. A Type A public reporting system is one in which an alarm from a fire alarm box is received and is retransmitted to fire stations:

   a. manually   b. automatically
   c. either manually or automatically   d. by telephone

4. A Type B reporting system is one in which an alarm from a fire alarm box is:

   a. automatically transmitted to fire stations
   b. manually transmitted to fire stations
   c. if used, transmits to outside alerting devices
   d. a and c
   e. b and c

5. The Type A system is usually permissible in any size of municipality or area and should be provided when there are more than:

   a. 1000 emergency calls from boxes per year
   b. 2500 emergency calls from boxes per year
   c. 1500 emergency calls from boxes per year
   d. 3500 emergency calls from boxes per year

BIBLIOGRAPHY

NFPA Codes, Standards, Recommended Practices, and Manuals (see the latest *NFPA Catalog* for availability of current editions of the following documents):

NFPA 10, *Standard for Portable Fire Extinguishers.*

NFPA 13, *Standard for the Installation of Sprinkler Systems.*

NFPA 72®, *National Fire Alarm Code®.*

NFPA 220, *Standard on Types of Building Construction.*

NFPA 1221, *Standard for the Installation, Maintenance, and Use of Emergency Services Communications Systems.*

14

Code Requirements

INTRODUCTION

Local code requirements for fire alarm systems can often be confusing since different codes may all apply to a particular system. The fire alarm signaling system designer must have a thorough understanding of a locality's current code or standard for fire alarm signaling systems and all pertinent requirements. Working knowledge of code requirements is a prime factor in cost-efficient fire alarm signaling system design.

This chapter describes general requirements contained in codes and standards; the role of testing laboratories is briefly described. A tabular listing of some NFPA codes and standards referenced throughout this text is also included. For related information on code requirements, see Chapter 11, Approvals and Acceptance.

TESTING LABORATORIES

It is important that all equipment in a fire alarm signaling system be listed by an independent testing agency for the appropriate signaling purpose. Use of listed equipment is often mandated by local codes, even though the code may not specify a particular testing agency's listing.

In the listing process, a sample of each fire alarm system component is tested to determine its compliance with one or more of the testing laboratory's standards. Subsequent follow-up inspections are conducted at the manufacturer's facilities to ensure continued compliance with the test standard. Two major well-known testing laboratories in the United States are Underwriters Laboratories Inc. (UL) and FM Global Technologies (FM). UL and FM both determine the suitability of fire alarm equipment for its intended service and test equipment for compliance with appropriate NFPA codes and standards.

Table 14-1. Testing Laboratory Standards for Fire Alarm Systems

Underwriters Laboratories Inc. (UL) Standards

UL 217	*Single and Multiple Station Smoke Detectors*
UL 268	*Smoke Detectors for Fire Protective Signaling Systems*
UL 268A	*Smoke Detectors for Duct Application*
UL 464	*Audible Signals*
UL 521	*Heat Detectors for Fire Protective Signaling Systems*
UL 864	*Control Units for Fire Protective Signaling Systems*
UL 985	*Household Fire Warning System Units*
UL 1480	*Speakers and Amplifiers*
UL 1481	*Power Supplies*
UL 1638	*Visual Signals*
UL 1730	*Annunciator Systems*

FM Global Technologies (FM) Standards

FM 3210	*Thermostats ("Heat Detectors") for Automatic Fire Detection*
FM 3230–FM 3250	*Smoke-Actuated Detectors for Automatic Fire Alarm Signaling*
FM 3260	*Flame Radiation Detectors for Automatic Fire Alarm Signaling*
FM 3820	*Electrical Utilization Equipment for Fire Alarm Control Equipment*
[No number]	*Facilities and Procedures Audit; Program Manual**

*This describes the FM's listing process.

Some UL and FM test standards that apply to fire alarm systems are shown in Table 14-1.

Fire alarm signaling system requirements in a particular code or standard depend on the occupancy for which the system is intended. A local occupancy code will specify the type of system and detection that is required. For example, minimum protection for a nursing home could be (1) a fire alarm system that provides evacuation signals, (2) smoke detectors in building corridors and certain other areas, and (3) system connection to the local fire department. NFPA *101®*, *Life Safety Code®*, describes types of fire alarm signaling systems for a variety of occupancies. A summary of the protection requirements in NFPA *101*, by occupancy is shown in Table 14-2.

Table 14-2. NFPA 101®, Life Safety Code®, Requirements by Type of Occupancy

	Hotels/Dormitories		Health Care		Apartments		Mercantile		Business	
	New	*Existing*	*New*	*Existing*	*New*	*Existing*	*New*	*Existing*	*New*	*Existing*
1. Detection, alarm, and communications systems in accordance with Section 7-6	16-3.4.1	17-3.4.1	12-3.4.1.1, 12-6.3.4.1	13-3.4.1, 13-6.3.4.1	18-3.4.1$^{\#}$	19-3.4.1$^{\#}$	24-3.4.1*, 25-4.4.3.1, 24-4.5.3.1	25-3.4.1*, 25-4.4.3.1, 25-4.5.3.1	26-3.4.1$^{\#}$	27-3.4.1$^{\#}$
2. System smoke detectors in corridors, connected to the fire alarm system, installed per NFPA 72®, National Fire Alarm Code®	16-3.4.4.1**			13-3.4.5.1¶		Option 2 only				
3. AC-powered single-station smoke detectors installed per NFPA 72										
a. In all dwelling units					18-3.4.4.1	19-3.4.4.1				
b. On each floor level (in all habitable rooms or in corridors)										

(*Continues*)

Table 14-2. (Continued)

	Hotels/Dormitories		Health Care		Apartments		Mercantile		Business	
	New	Existing	New	Existing	New	Existing	New	Existing	New	Existing
c. In each individual guest or sleeping room	16-3.4.4.2	17-3.4.4			18-3.4.4.2					
4. Manual pull stations, connected to the fire alarm system										
a. Within 200 ft of any point on any given floor	16-3.4.2	17-3.4.2‡	12-3.4.2§ 12-6.3.4.2	13-3.4.2§ 13-6.3.4.2	18-3.4.2.1	19-3.4.2.1	24-3.4.2$^{‡#}$	25-3.4.2$^{‡#}$	26-3.4.2$^{‡#}$	27-3.4.2$^{‡#}$
b. At a continuously supervised control point	16-3.4.2	17-3.4.2								
c. Near each required exit in the natural path of escape	16-4.2	17-3.4.2‡	12-3.4.2§ 12-6.3.4.2	13-3.4.2,§ 13-6.3.4.2	18-3.4.2.1	19-3.4.2.1	24-3.4.2$^{†#}$	25-3.4.2$^{‡#}$	26-3.4.2$^{‡#}$	27-3.4.2$^{‡#}$
5. Sprinkler supervisory equipment required to be connected to the fire alarm system	16-3.4.2	17-3.4.2	12-3.4.2, 12-6.3.4.2	13-3.4.2, 13-6.3.4.2	18-3.4.2.2	19-3.4.2.4				

Requirement	16	17	12	13	18	19	24/25	26	27
6. Occupant notification activated by the fire alarm system in accordance with Section 7-6.3	16-3.4.3.1	17-3.4.3.1	12-3.4.3.1, 12-6.3.4.3	13-3.4.3.1, 13-6.3.4.3	18-3.4.3.1	19-3.4.3.1	24-3.4.3.2, 24-4.4.3.4, 24-4.5.3.4; 25-3.4.3.2, 25-4.4.3.4, 25-4.5.3.4	26-3.4.3.2	27-3.4.3.2
7. Voice communication system if high rise, in accordance with Section 30-8.3	16-4.1				18-4.2			26-4.2	
8. Central annunciator panel connected to the fire alarm system	16-3.4.3.2		12-3.4.3.3		18-3.,4.3.2#	19-3.4.3.2#			
9. Fire alarm system connection to the fire department	16-3.4.3.3†	17-3.4.3.2†	12-3.4.3.2, 12-6.3.4.4	13-3.4.3.2, 13-6.3.4.4			24-3.4.3.3, 24-4.4.3.5, 24-4.5.3.4; 25-4.4.3.5, 25-4.5.3.5		

* Applies to Class A mercantile only.
† Fire department notification by telephone permitted.
‡ Exempted if automatic initiation means is provided.
§ Exempted if manual pull stations are provided at nurses' control stations.
¶ Required only in limited-care facilities.
# See exceptions.
** Exempted in sprinklered buildings.
For SI Units: 1 ft = 0.30 m.

385

Once the type of system has been determined, installation requirements can be found in *NFPA 72®, National Fire Alarm Code®*.

REQUIREMENTS IN LOCAL BUILDING AND FIRE CODES

After well-publicized tragic fires in residential and commercial occupancies in the 1970s and 1980s, many cities, counties, and states reviewed their building and fire codes, strengthening the protective aspects of the codes to prevent loss of life and property.

Fire safety codes and standards that are developed by the National Fire Protection Association (NFPA) can be adopted by a local community. The standards can be adopted by reference (title and publishing information on the NFPA standard are mentioned only in the local code) or by transcription (printing of the standard in the local code). Similarly, model building codes are developed by private associations for modification and adoption by local communities. Model building codes are developed by the following organizations:

1. Building Officials and Code Administrators International (BOCA) develops the *Basic/National Building Code*.
2. Southern Building Code Congress International (SBCCI) develops the *Standard Building Code*.
3. International Conference of Building Officials (ICBO) develops the *Uniform Building Code*.

Since 1994, these three model code organizations have collaborated on the development of a single set of codes under the banner of the International Codes Council (ICC). Known as the I-Codes (or International Codes), they will eventually replace the codes developed by the individual groups.

The NFPA has also committed to the development of a coordinated set of consensus codes, including a building code, *NFPA 5000™, Building Construction and Safety Code™*. Other codes in the NFPA set originate from NFPA and from other organizations through strategic partnerships with NFPA. These include the *Uniform Plumbing Code* and *Uniform Mechanical Code* from the International Association of Plumbing and Mechanical Officials (IAPMO), the *Uniform Fire Code* from the Western Fire Chiefs Association, and others.

Code requirements for fire alarm signaling systems generally define:

1. Occupancy descriptions and classifications
2. Locations for smoke detectors

3. Function of the fire alarm and emergency communication system (if an emergency communication system is determined necessary for the occupancy protected)
4. Operation of the voice/alarm function of a system
5. Provision for a fire department communication system
6. Components of the fire command station
7. Emergency power requirements and service
8. Manual fire alarm station location and use
9. Exit-door-unlocking components

EXAMPLE OF CODE REQUIREMENTS

Table 14-2 is a composite of life safety provisions in model building codes and NFPA *101* (*note:* Occupancies in a community that are administered by the federal government are usually exempt from local code requirements.)

General

These recommendations cover basic functions of complete automatic fire alarm and emergency communication systems. These systems are primarily intended to provide early indication of emergency conditions.

Applicability

These recommendations apply to all occupancies of the following use groups, as typically defined by the building code in effect in the locality.

Use Group B, Businesses: *Businesses* are defined to include all buildings and structures or parts thereof that are used for the transaction of business, the rendering of professional services, or other services that involve stocks of goods, wares, or merchandise in limited quantity for incidental office use or sample purposes. Businesses also include offices, banks, civic administration facilities, outpatient clinics, professional buildings, testing and research laboratories, radio stations, telephone exchanges, and similar establishments.

Use Group R1, Residential and Hotel Occupancies: *Residential and hotel occupancies* are defined to include all hotel, motel, and dormitory buildings that are arranged for shelter and sleeping accommodations of more than 20 persons.

Use Group R2, Residential Multifamily Occupancies: *Residential multifamily occupancies* are defined to include all multifamily dwellings having more than two dwelling units. This occupancy classification also includes dormitories, boarding, and lodging houses that are arranged for shelter and sleeping accommodations of more than 5 but not more than 20 persons.

When required by the authority having jurisdiction, the specified occupancies contain a listed fire alarm system for life safety. This system should be installed, tested, maintained, and used in accordance with applicable requirements. These requirements can be found in NFPA *101*.

Smoke Detection Systems

An approved smoke detection system is to be installed in the previously defined occupancies in the following locations:

1. Boiler and furnace rooms.
2. Return air ducts and plenums of HVAC systems serving floors other than the floor on which the HVAC equipment is located. Detectors should be located at each opening into the vertical return air ducts or shafts.
3. Corridor areas.
4. Elevator lobby areas.
5. Elevator penthouse areas.
6. Other areas that may be deemed necessary.

The detection system must be designed to activate the voice/alarm system on a selective basis. This selective activation must be dependent upon the compartmentation design of the system. The detection system, upon activation, must also place into operation all equipment necessary to prevent the spread of smoke.

In use group R1 and R2 occupancies as defined by NFPA *101*, an approved single-station smoke detector must be installed in each room and residential dwelling.

Fire Alarm and Emergency Communication Systems

An approved fire alarm and emergency communication system must be installed for all occupancies in accordance with applicable requirements of NFPA *101*.

Voice/Alarm Systems

Voice alarm systems are generally required in buildings where the fire emergency plan does not contemplate the immediate evacuation of all occupants. The operation of any smoke detector, sprinkler waterflow device, or manual fire alarm station in the occupancy must automatically activate a voice/alarm system. Activation of such a system must automatically sound an alert signal to desired areas in the occupancy protected.

The voice/alarm system must provide a predetermined message on a selective basis to the area where the alarm originated. The message must also provide information and give directions to the occupants. The alarm must also be so designed as to be clearly heard by all hearing-able occupants within all designated areas.

Voice/alarm system controls must be located in the fire command station and operated from that station. Voice/alarm system controls should be located so that a selective or general voice alarm can be initiated by the station operator. Communication must be established on either a selective or general basis to the following terminal areas: elevators, elevator lobbies, corridors, exit stairways, rooms and tenant spaces exceeding 1000 sq ft (92.936 m^2), dwelling units in apartment houses, and hotel guest rooms or suites.

Installation wiring for a voice/alarm system must be electrically supervised on a continuous basis. This supervision must detect opens, shorts, and ground conditions that could impair the function of the system.

Fire Department Communication System

A two-way fire department communication system must be provided for fire department use. This system must operate between the fire command station and every elevator, elevator lobby, and entry-enclosed exit stairway.

Fire Command Station

A fire command station for fire department operations must be provided in a location approved by the fire department. The fire command station must have the following components:

1. A voice alarm system panel.
2. A fire department communication panel.

3. Fire detection and alarm system annunciator panels.
4. Status indicators for elevators and annunciator. Such indicators must specify which elevators are operational.
5. Status indicators and controls for air-handling systems.
6. Controls for simultaneous unlocking of all stairway doors.
7. Sprinkler valve and waterflow detector display panels.
8. Status indicators for emergency power, light, and emergency system control.
9. A telephone for fire department use, which must have controlled access to the public telephone system.

Emergency Power

Emergency power provided must be capable of operating the emergency communication system during a fire or other emergency conditions. The emergency power must be derived by connection to an emergency power circuit of an emergency system.

Note: NFPA 70, *National Electrical Code®*, describes emergency power circuit requirements and is used as the local electrical code in many localities.

The emergency power requirements must provide service to the following emergency systems: voice/alarm system, fire department communication system, fire department elevator, mechanical air-handling system, fire detection and alarm system, fire protection equipment and devices, exitway and other emergency lighting, and exitway-door-unlocking system.

Manual Fire Alarm Stations

Manual fire alarm stations must be provided and used only for fire protective signaling purposes. These stations must be provided in the natural path of escape near the required exit from each area of the occupancy protected. Additional manual fire alarm stations must be located within 200 ft (61 m) horizontal from any point in the building.

Doors

All exit stairway doors that are to be locked from the stairway side must have an approved lockset. This lockset must be able to be unlocked from the fire command station by the operator and unlocked automatically upon actuation of the fire alarm system. The locks must be connected in a fail-safe manner so that, in event of a power failure, they will open automatically.

NFPA CODES AND STANDARDS

NFPA 72®, National Fire Alarm Code®, is the correlative standard to this text. However, other NFPA codes and standards with requirements applicable to a fire alarm system exist, as listed in Table 14-3.

Further, the following NFPA standards contain requirements for automatic detection, alarm, and release of extinguishing agents that may also be part of a fire alarm signaling system:

NFPA 11, Standard for Low-, Medium-, and High-Expansion Foam
NFPA 11A, *Standard for Medium- and High-Expansion Foam Systems*
NFPA 12, *Standard on Carbon Dioxide Extinguishing Systems*
NFPA 12A, *Standard on Halon 1301 Fire Extinguishing Systems*
NFPA 12B, *Standard on Halon 1211 Fire Extinguishing Systems*
NFPA 13, *Standard for the Installation of Sprinkler Systems*
NFPA 15, *Standard for Water Spray Fixed Systems for Fire Protection*
NFPA 17, *Standard for Dry Chemical Extinguishing Systems*
NFPA 17A, *Standard for Wet Chemical Extinguishing Systems*
NFPA 75, *Standard for the Protection Information Technology Equipment*
NFPA 80, *Standard for Fire Doors and Fire Windows*
NFPA 90A, *Standard for the Installation of Air-Conditioning and Ventilating Systems*
NFPA 170, *Standard for Fire Safety Symbols*
NFPA 231C, *Standard for Rack Storage of Materials*

Table 14-3. NFPA Requirements for Fire Alarm Systems*

NFPA Code or Standard	*NFPA Requirements*
NFPA 70	Wiring requirements for fire alarm signaling systems
NFPA *101*	System or protective functions required in various occupancies
NFPA 110	Installation and performance of emergency generator equipment
NFPA 1221	Municipal fire alarm and communication systems

* Codes and standards specify minimum acceptable, real-world protection in an occupancy. A fire alarm system designer is encouraged to exceed the minimum safety requirements in high-hazard occupancies, facilities with a heavy occupancy of transient or physically challenged occupants, or other facilities (and occupants therein) where extra protection is advised.

QUESTIONS

1. Requirements for the provision of a fire alarm system would be found in the:

   a. building code   b. fire code
   c. testing laboratory standard   d. a, b, and c

2. Requirements for the type(s) of systems to be installed generally depend on the:

   a. Opinion of the authority having jurisdiction
   b. occupancy type
   c. fuel load
   d. building height

3. Voice alarm systems are required:

   a. in all high-rise buildings
   b. in buildings that may be occupied by people with disabilities
   c. in buildings where the fire emergency plan does not contemplate the immediate evacuation of all occupants
   d. in nonsprinklered buildings

4. Smoke detectors are required in:

   a. electrical rooms   b. furnace rooms
   c. elevator lobbies   d. all of the above

5. A fire command center must have the following:

   a. a voice alarm panel   b. extra electric outlets   c. status indicators
                                                              for elevators

   d. a and c   e. a and b.

BIBLIOGRAPHY

NFPA Codes, Standards, Recommended Practices, and Manuals (see the latest *NFPA Catalog* for availability of current editions of the following documents):

NFPA 70, *National Electrical Code®*.

NFPA 72®, National Fire Alarm Code®.

NFPA *101®, Life Safety Code®*.

NFPA 1221, *Standard for the Installation, Maintenance, and Use of Emergency Services Communications Systems*.

ADDITIONAL READING

See Tables 14-1 and 14-3, and additional NFPA codes and standards cited in this chapter.

15

Basic Fire Alarm System Plans Review

INTRODUCTION

The process of plans review includes drawing submissions for the fire alarm system as well as equipment submittals from the installing contractor. The purpose of performing a plans review is to ensure that the fire alarm system as designed will meet the requirements of the applicable codes and standards as well as local ordinances. Generally speaking, an efficient plans reviewer will need experience or training in basic blueprint reading and familiarization with the symbols used on the architectural, mechanical, fire protection, and electrical plans. An understanding of basic math and algebra and skill in the use of architectural and engineering scale rulers will also be useful. A thorough plans review is the first step to a reliable and code-compliant fire alarm system installation.

PLANS REVIEW TOOLS

In order to be proficient in any endeavor, one must obtain and use the correct tools of the trade. The plans review process is no different in that respect. A plans reviewer is expected to have in his or her possession or access to the following:

- A calculator
- A large magnifying glass
- An architect's scale ruler
- An engineer's scale ruler
- Multiple colored pens or pencils
- Multiple colored highlighters

- The fire and building codes and standards in force in the jurisdiction
- Basic reference books and applicable code handbooks

Submission Documents Necessary

The documents needed for a thorough plans review include the following:

- Occupancy classification
- Architect's drawings (sealed by a registered architect)
- Floor plans showing the design of the fire alarm system (electric or dedicated fire alarm) sealed by a licensed/registered professional engineer
- Information showing the location of all obstructions
- Mechanical/HVAC plans indicating location of air supply and return registers and locations of duct smoke detectors and access doors for these detectors
- Cross-section details showing elevations
- Reflected ceiling plans
- Construction specification package containing the technical and installation details
- Fire alarm equipment supplier or manufacturer submittals
- Voltage drop calculations
- Standby battery calculations
- Fire alarm system riser diagrams
- Annunciator description information
- Name and address of monitoring station (if applicable)

The plans review department of the authority having jurisdiction should require specific details to ensure ease of review and completeness for plans and documents. Generally two sets of design drawings (DDs) signed and sealed by the designer of the fire alarm system plus at least one set of specifications are required to be submitted to the plans review department.

The following is a typical checklist of requirements that will provide enough detail for a thorough and efficient plans review:

Drawings: General

- All sheets should be of the same size
- All drawings must be scaled with compass points shown
- Each drawing must have a drawing number, revision, and date (if applicable)
- Name and address of project

- Name and address of installing contractor
- Name and address of building owner
- Name and telephone number of installation company contact person
- Contractor license classification and number
- Name and address of general contractor (if applicable)
- Name and address of electrical contractor (if applicable)
- Type of system: fire alarm, conventional hardwired, wireless, addressable, analog addressable, Class A, Class B
- Building occupancy type
- Sheet title
- Key plan for building sections
- Box for AHJ approval stamp
- Wiring legend
- Symbol list with manufacturer, part number, and back box
- Square footage of each building and total

Drawings: Floor Plan

- Building floor plan
- Location of as-built drawing cabinet
- Location of lock box
- Location of all obstructions exceeding 6 ft above finished floor (AFF)
- Intended use of each room (e.g., storage, classroom, restroom, vestibule)
- Location of all air supply and return registers
- Ceiling heights, ceiling details, and configuration
- Full height cross section of building; may be shown on an additional sheet
- Mounting heights of devices
- Location of main fire alarm control panel
- Location and source of emergency standby power
- Location of all annunciator panels, sub fire alarm panels, booster panels
- Location of all power sources, panel numbers, and breaker numbers for each piece of equipment
- Location of all fire alarm detection and notification devices, along with temperature ratings and candela ratings if applicable
- Location of all ancillary devices: door holders, door closers, gas shutoff, fan shutdown, smoke dampers, shunt trips, and so on
- Location of all fire sprinkler risers, waterflow switches, and tamper switches
- Locations of all fire pumps and fire pump controllers
- Location of all access doors and signage for duct detector access
- Location of all remote indicators for hidden devices

- Show zoning if a conventional system
- Wiring type, size, number of conductors and approximate wiring layout
- All rated walls

Drawings: Riser Diagram

- Riser diagram showing all devices as connected in the circuit
- Device addresses, room numbers, and/or names

Drawings: Details

- Circuit wiring diagram
- Typical device and ancillary device wiring

Specification Package

- Table of contents (each section shall have tabs and be separated accordingly)

1. Fire alarm control panels, power supplies, and annunciators
2. Detection devices
3. Audio/visual devices
4. System components, modules, and relays
5. Battery calculations and cut sheets
6. Voltage calculations
7. Compatibility listings (matrix, table, or info showing device compatibility)
8. Operating instructions for entire system
9. Manufacturers inspections instructions

FIRE ALARM SYSTEM NARRATIVE

The narrative report is an important piece of information. It states exactly what the system is, whether it is in existence, how it will function, and how it will interface with other equipment. The narrative should explain:

- The scope of work
- Type of system
- How the system will function
- Interface with other fire protection systems
- System sequence of operations

The fire protection construction documents should include a narrative report that describes the basis of the system design, the sequence of operations, and testing criteria. The basis (methodology) of design for the protection of the occupancy and hazard should be outlined showing compliance with the building code and applicable NFPA standards in the form of a narrative report. The portion of the narrative report that describes the sequence of operation entails the specific operation of system devices and equipment and their related integration. The testing criteria portion of the narrative report is broken down into the following three sections:

- Section 1. Testing Criteria
- Section 2. Equipment and Tools
- Section 3. Approval Requirements

The narrative report is typically divided into the following sections:

- **Section 1. Building Description**

  a. Building use group
  b. Total square footage of building
  c. Building height
  d. Number of floors above grade
  e. Number of floors below grade
  f. Square footage per floor
  g. Type(s) of occupancies (hazards) within the building
  h. Type(s) of construction
  I. Hazardous material usage and storage
  j. High storage of commodities within a building usually over 12 ft
  k. Site access arrangement for emergency response vehicles

- **Section 2. Applicable Laws, Regulations, and Standards**

  a. Building code sections, fire protection system requirements
  b. NFPA standards and edition used for design of each specific fire protection system
  c. Applicability of sections of the fire prevention code
  d. Applicability of local fire prevention regulations
  e. Applicability of approved local by-laws or ordinances
  f. Applicability of specialized codes (plumbing, elevator, electrical, architectural, access)
  g. Applicability of federal laws (OSHA, ADA, etc.)

- ## Section 3. Design Responsibility for Fire Protection Systems

This section identifies the accountability for a specific fire protection system design and the accountability for the integration of the fire protection systems constituting a building life safety system.

- ## Section 4. Fire Protection Systems to Be Installed

This section identifies key performance design criteria and features for each specific fire protection system such as:

a. Water supply, fire mains, and hydrants
b. Automatic sprinkler systems and components
c. Standpipe systems and components
d. Fire alarm systems and components
e. Automatic fire-extinguishing systems
f. Manual suppression systems
g. Smoke control/management systems
h. Kitchen cooking equipment and exhaust systems
i. Emergency power equipment
j. Hazardous-material monitoring equipment
k. Seismic considerations

- ## Section 5. Features Used in the Design Methodology

This section identifies the designer's intent in the overall design and criteria development of either a required or a nonrequired system.

- ## Section 6. Special Consideration and Description

This section identifies the designer's intent to deviate from prescriptive requirements of regulatory codes and standards with alternative methods.

EXAMPLE OF A FIRE ALARM SYSTEM NARRATIVE

Fire Protection Narrative Report

Building Name _____

Address _____

City, State _____

Date _____

1. Fire Alarm System Construction Documents

A. Basis of Design: The building contains [insert business use groups] use groups as defined by [insert building code]. The building is approximately [insert size, square footage, number of floors]. The fire alarm system is connected to [describe off-premises connection]. The building is protected by an automatic sprinkler system.

The scope of work includes [insert scope of work].

The following are the governing codes for this construction project:

- State building code.
- Fire safety code.
- Referenced standard: *NFPA 72®*-[insert applicable date], *National Fire Alarm Code®*.
- Local fire department requirements.

[Insert design engineer name] is providing the fire alarm system design services to the business owner.

B. Sequence of Operation: Fire Alarm System Control Unit (FACU). When the FACU receives a signal from any manual or automatic fire detection device, an alarm signal is sent to all notification devices in the building and transmitted to the off-site monitoring service via the digital alarm communicator transmitter (DACT). Additionally, all supervisory and trouble signals shall be transmitted to the off-site monitoring service via the DACT.

Smoke Detectors. When smoke detectors sense smoke, a signal is sent to the FACU, which starts the alarm sequence throughout the building. Smoke detectors are located in the areas identified on the fire alarm system layout.

Manual Pull Stations. Pull stations are located at the exits of the building. When these are used, a signal is sent to the FACU, which starts the alarm sequence throughout the building.

Horns and Strobes. Located throughout the building, the notification appliances contain a combination strobe light and horn, horn only, or strobe only. The devices are located on the walls of the spaces identified on the plans. During an alarm, the strobes will flash and the horns will broadcast the alarm signal. The alarm signal is a distinctive three-pulse temporal pattern, synchronized signal.

Tamper Switches. Tamper switches are located on the building's sprinkler system. If a valve is improperly closed, a signal is sent to the FACU. The FACU annunciates the audible supervisory signal. Upon the actuation of a sprinkler tamper switch, an alarm signal is transmitted to the off-site monitoring service and actuates the master box. This sequence shall not broadcast the alarm signal throughout the building.

Flow Switches. Flow switches are located on the building's sprinkler system. If a sprinkler head fuses due to a fire, the flow switch actuates. A signal is sent to the FACU. The FACU broadcasts the alarm signal throughout the building.

Fire Alarm System Wiring. Fire alarm system shall be connected using Class A circuits.

C. System Testing: The fire alarm equipment shall have a 100 percent acceptance test performed by the installing contractor and approved testing company for the building. Request for acceptance test by the fire department shall be made in writing. The letter shall be received by the fire department at least 5 working days prior to the desired date of acceptance testing.

The fire alarm system shall be tested in accordance with *NFPA 72*-[insert applicable date], Section 7-1.6.1, *Initial Acceptance Testing,* and any other special requirements of the fire department.

Equipment and Tools. During the preliminary and final testing of the systems the following equipment will be provided at the site by the contractors:

a. Manufacturer's instructions
b. Construction documents and as-built drawings
c. Approved narrative description
d. Sound-level meters
e. Voltage meters
f. Test magnets and or test spray for smoke detectors
g. Communication radios (if necessary)
h. Any necessary special tools, ladders, and so on
i. Fire alarm testing notification announcements

2. Building and Site Access: [Insert how the fire department will have access to the site and building.]

3. Fire Hydrants and Water Supply System: [Insert design engineer name] is not performing any work pertaining to the fire hydrant(s) or water supply system.

4. Automatic Sprinkler System Layout: [Insert design engineer name] is not performing any work pertaining to the automatic sprinkler system.

5. Automatic Sprinkler System Control Equipment: [Insert design engineer name] is not performing any work pertaining to the automatic sprinkler system.

6. Automatic Standpipe Systems: [Insert design engineer name] is not performing any work pertaining to automatic standpipe systems.

7. Standpipe System Hose Valves: [Insert design engineer name] is not performing any work pertaining to automatic standpipe sprinkler system or hose valves.

8. Fire Department Siamese Connections: [Insert design engineer name] is not performing any work pertaining to the automatic sprinkler system or fire department siamese connections.

9. Type/Description and Design Layout of the Signaling System: The fire alarm system is designed for [insert size of building and occupancy classification]. The system will comply with reference standards. The fire alarm system is designed to notify the fire department and building occupants of a fire, supervise the system, and provide trouble notification. The system is composed of new fire detection and notification appliances, new analog addressable FACU and new DACT. Combination horn and strobe light, horns-only, and strobe-only appliances are connected to the FACU via notification appliance circuits. The layout of the signaling system is shown on the accompanying fire alarm system design layout.

10. Location of Fire Protective Signaling System(s) Control Equipment: The fire protective signaling system control equipment is integral to the fire alarm control unit. It is located as shown on the fire alarm system design layout. There is no remote alarm annunciator connected to the fire alarm system. The electrical contractor shall provide an 8½ in. × 11 in. graphic map that details all of the detection devices and the accompanying addresses. Each address detection device shall be labeled on the graphic map. The graphic map shall be placed adjacent to the FACU.

11. Smoke Control System: There is no smoke control system for this building.

12. Location of Smoke Control System Equipment: There is no smoke control system for this building.

13. Integrated Building Life Safety and Signaling System: When the FACU receives a signal from any manual or automatic fire detection device, an alarm signal is sent to all notification devices in the building, actuates the master box, and is transmitted to the off-site monitoring service via the DACT. Additionally, all supervisory and trouble signals shall be transmitted to the off-site monitoring service via the DACT.

If the FACU receives a signal from a duct smoke detector, then a signal is sent to the corresponding control module, which is interfaced to the air-handling unit for power interruption.

Upon the actuation of a sprinkler tamper switch, a signal is sent to the FACU. The FACU will annunciate an audible supervisory signal. Additionally, an alarm signal is transmitted to the off-site monitoring service and actuates the master box. This sequence shall not broadcast the alarm signal throughout the building.

If a sprinkler head fuses due to a fire, the sprinkler waterflow switch will actuate. A signal is sent to the FACU. The FACU will broadcast an alarm signal throughout the building. Additionally, an alarm signal is transmitted to the off-site monitoring service and actuates the master box.

14. Fire-Extinguishing Systems: Other than the automatic fire sprinkler system, there is no other fire-extinguishing system for this building.

15. Fire-Extinguishing System Controls: Other than the automatic fire sprinkler system, there is no other fire-extinguishing system for this building.

16. Location of Fire Protection System(s) Equipment Room: The location of the FACU, DACT, master box, document box, and exterior beacon are shown on the fire alarm system design layout.

[Insert design engineer name] is not performing any work pertaining to the automatic sprinkler system.

17. Fire Protection System(s) Equipment Identification and Operation Signs: The manufacturer labels the FACU, DACT, and master box on the respective equipment. The operation of the FACU for acknowledgment, silence, and reset are prominently displayed on the cover panel of the FACU. Depressing the respective button on the FACU performs these functions.

The words PULL, LIFT, FIRE, or combination thereof are prominently displayed on the manual pull stations. The letters FIRE are prominently displayed on the notification appliances.

[Insert design engineer name] is not performing any work pertaining to the automatic sprinkler system.

18. Alarm, Supervisory, and Trouble Signal Transmission: When the FACU receives a signal from any manual or automatic fire detection device, an alarm signal will be actuated and transmitted to the off-site monitoring service via the DACT. Additionally, all supervisory and trouble signals shall be transmitted to the off-site monitoring service via the DACT.

Approval Requirements: The following approvals are necessary prior to the start of work:

  a. City of [insert city] Fire Department, permit for fire protection system. This permit is for the installation of the new fire alarm system.
  b. City of [Insert city] building permit as part of the planned electrical work involved with the construction project.

The following approval is necessary during the fire alarm system installation:

  City of [Insert city], electrical Inspection.

The following approval is necessary after the fire alarm system installation is complete:

A fire alarm acceptance test witnessed by the fire department, city of [insert city] electrical inspector, owner, owner's designee, and fire alarm servicing company.

End of Report

Respectfully Submitted,   Reviewed By _____

Fire Protection Engineer   Designer of Record_____

QUESTIONS

  1. A plans reviewer is expected to have in his or her possession or access to:
     a. an engineer's scale   b. a calculator
     c. the applicable codes   d. all of the above

2. The documents needed for a thorough plans review include the following:

   a. date system is to be installed
   b. expected plans review completion date
   c. reflected ceiling plans
   d. cost information of the installed system

3. All drawings must be:

   a. in color   b. scaled
   c. in order   d. submitted on time

4. A riser diagram will show:

   a. FACU room location
   b. terminal connections
   c. access doors for duct smoke detectors
   d. device addresses

5. A fire alarm system narrative description explains the:

   a. reason the system is going to be installed
   b. the design intent and operation of the fire alarm system
   c. battery calculations and cut sheet descriptions
   d. basic reference books used

BIBLIOGRAPHY

NFPA 72®, National Fire Alarm Code®.

"Building Trades Printreading—Part 3";
Heavy commercial construction.
Leonard P. Toenjes,
Contractor—BOOKs.com

16

Cost Analysis

INTRODUCTION

Economics, that is, cost, is a prime motivator in determining the amount and type of fire protection that is installed in a facility. This chapter describes various methods of cost analysis that can be used to determine the economic feasibility of various fire alarm signaling systems.

In economic analysis, the economic impact of design, construction, and equipment installation is studied to determine the relative worth of net economic gains to be expected from alternative solutions relative to their net economic costs. Resources can be easily wasted if an economic analysis is not performed.

Not all fire protection decisions, however, require a detailed comprehensive economic analysis. Determining whether to place a smoke detector in one small room of a building, for instance, would not justify a detailed analysis.

BASIC PRINCIPLES OF ECONOMIC ANALYSIS

Sound economic analysis is based on an objective, factual study of all possible alternatives, independent of financing. The analysis period should not extend beyond a reasonable forecast period. Since the analysis is a study of the future, past events and investments (or "sunk costs") are irrelevant. When preparing an economic analysis, factors used should have identical time periods, be separated into market versus nonmarket, and be discounted to the same time/date. Common factors of equal magnitude can be omitted from the analysis.

In cost analysis, the process of achieving a rational decision is made by a logical method of analysis. Eight key elements are

1. Recognition of a problem: the realization that a problem exists.
2. Definition of the goal or objective to be accomplished. What is the task?

3. Assembly of relevant data. What are the facts? Are additional data needed? Is the additional information worth the cost of obtaining it?
4. Identification of feasible alternatives. What are the practical alternative ways of accomplishing the objective or task?
5. Selection of the criteria (e.g., political, economic, ecological) for judging the best alternative.
6. Construction of the various interrelationships, frequently called mathematical modeling.
7. Prediction of outcome for each alternative.
8. Choice of the best alternative for achieving the objective.

The decision process does not simply proceed from the first to the last element; in fact, it is often necessary to reexamine earlier elements.

One of the most important concepts in engineering economics is the time value of money, which is expressed in the form of the willingness of people to pay interest for its use. The economics of fire protection can be estimated correctly only by including this factor.

DEFINITIONS

- *Interest:* A charge for borrowed money.
- *Rate of interest:* The percentage (per unit of time) that is paid on money borrowed.
- *Compound interest:* Received at the end of each time period, this interest is based on the original borrowed amount plus the accumulated interest to that time.

LIFE CYCLE COST

The life cycle cost of a fire protection system includes all costs associated with the system during its lifetime. Life cycle costs can include engineering/design costs; costs of initial equipment, installation, and system startup; annual inspection, maintenance, repair, and operation costs; repair costs (other than scheduled annual repair); and salvage cost or value (see Figure 16-1).

Use of the life cycle cost is a method of evaluating expenditures that recognizes the sum total of all costs associated with the expenditure during the time the system is in use. Cost comparisons for economic decision making should include all appropriate costs during the life of a fire protection system.

Alternative fire protection solutions are evaluated by determining the present cost of each alternative and selecting the one at the least cost but most by using the effectiveness.

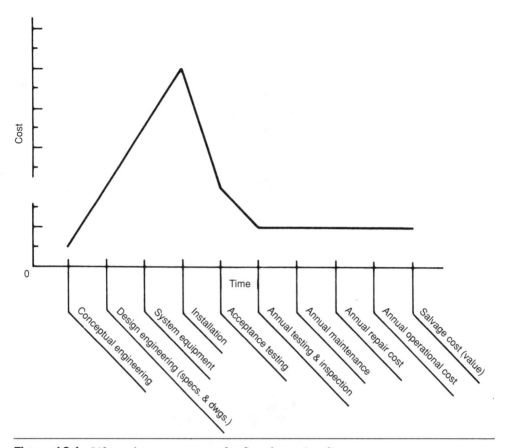

Figure 16-1. Life cycle system curve for fire alarm signaling systems.

Two of the easiest ways of comparing alternatives are by using the present cost and the series present cost. Simply stated, the analysis illustrates the maximum present worth of benefits and the minimum present cost. Three criteria for economic efficiency are presented in Table 16-1.

PRESENT COST ANALYSIS

To find the present cost of some future expenditure, the formula for the calculation is

$$P = \frac{F}{(1 + i)^n}$$

TABLE 16-1. Present-Worth Analysis

	Situation	*Criterion*
Fixed input	Amount of money or other input resource is fixed	Maximize present worth of benefits or other output
Fixed output	There is a fixed task, benefit, or other output to be accomplished	Minimize present costs or other input
Neither input nor output fixed	Neither amount of money or other input nor amount of benefits or other output is fixed	Maximize present worth of benefits minus present worth of costs, or, more simply, maximize net present worth

where F = future cost, P = present cost, n = number of periods, and i = interest rate per period.

The formula can be illustrated with a time line:

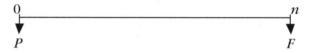

Example: A firm is trying to decide which of two alternative fire alarm signaling systems it should install to replace an outdated system. Alternative A will cost $110,000 to install in the second year, whereas Alternative B will cost $125,000 to install in the third year. Assuming all other costs are equal and the cost of borrowed money is 12 percent, which alternative should be selected?

Alternative	*Cost ($)*	*Periods*	*Interest (%)*
A	110,000	2	12
B	125,000	3	12

The time line for Alternative A is

The time line for Alternative B is

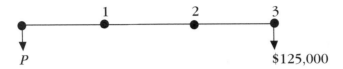

The present cost of Alternative A is

$$P = \frac{F}{(1 + i)^n}$$

$$= \frac{110,000}{(1 + 0.12)^2}$$

$$= \$87,691$$

The present cost of Alternative B is

$$P = \frac{F}{(1 + i)^n}$$

$$= \frac{125,000}{(1 + 0.12)^3}$$

$$= \$88,972$$

Since only the differences between the alternatives are relevant, the appropriate economic choice is Alternative A, the system with the minimum present cost.

SERIES PRESENT WORTH COST

To find the series present cost of some recurring future sum of money, the formula for the calculation is

$$P = A\frac{(1 + i)^n - 1}{i(1 + i)^n}$$

where A = uniform periodic payment, P = present cost, n = number of periods, and i = interest rate per period.

Sketching a time line of this formula shows

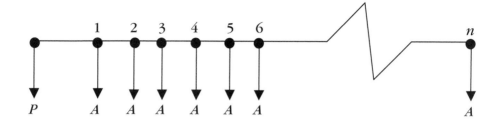

Example: A fire protection engineer working with an architectural engineering firm wishes to determine the present cost for annual expenditures to test and maintain fire detectors. The annual costs of functional testing and sensitivity testing for fire detectors are determined to be $1020 and $3380, respectively, or $4400 total. Assuming other annual costs are negligible and the interest on borrowed money is 12 percent, what is the present cost of the annual expenditure if the useful life of the fire detectors is 20 years?

The time line is

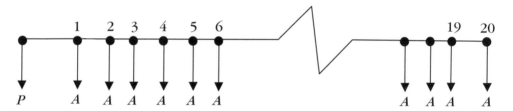

Using the series present cost formula ($A = \$4400$, $P =$ present cost, $n = 20$ years, and $i = 12$ percent), we obtain

$$P = A\left[\frac{(1 + i)^n - 1}{i(1 + i)^n}\right] = 4400\left[\frac{(1 + 0.12)^{20} - 1}{0.12(1 + 0.12)^{20}}\right]$$

$$= 4400\,(7.469)$$

$$= \$32,864$$

The series present cost factor CF (i, n),

$$\frac{(1 + i)^n - 1}{i(1 + i)^n}$$

TABLE 16-2. Data for Alternative Systems

Factor	Time (yr)	Alternative 1 ($)	Alternative 2 ($)
Engineering Design Cost	0	15,000	20,000
Equipment Cost	1	110,000	154,600
Installation Cost	1	62,500	41,000
Startup Cost	2	14,000	16,400
Inspection Cost	Annual	12,500	2,000
Testing Cost	Annual	17,400	3,200
Operation and Repair Cost	Annual	1,200	700
Salvage Value	20	6,500	2,000

does not need to be calculated each time since compound interest tables contain these data (interest tables assume that payments are made at the end of each period, so an appropriate adjustment should be made if payments are made at the beginning of each period).

The following example illustrates the cost analysis for a fire alarm system that is selected based on the lowest present cost.

Example: A fire protection engineer working with an architectural engineering firm and the facility's owner is trying to decide whether a multiplex fire alarm system or a hardwired fire alarm system is the more economical as determined by the life cycle cost of each system. Data for the alternative systems are provided in Table 16-2.

The total annual cost of Alternative 1 is $31,100 (total for inspection, testing, operation, and repair).

The time line for Alternative 1 is

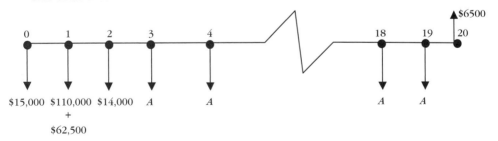

The present cost of Alternative 1 is
$$P = 15,000 + 172,500\text{CF}(12\%, 1) + 14,000\text{CF}(12\%, 2)$$
$$+ 31,100\text{CF}(12\%, 17) \times \text{CF}(12\%, 3) - 6500\text{CF}(12\%, 20)$$

Note that the annual cost payments are made during the end of the second year in which the fire alarm system operates. Use of an interest table gives:

$$P = 15{,}000 + 172{,}500(0.8929) + 14{,}000(0.7972)$$
$$+ 31{,}100(7.12) \times (0.7118) - 6500(0.1037)$$
$$= 15{,}000 + 154{,}025 + 11{,}161 + 157{,}615 - 674$$
$$= \$337{,}127$$

The present cost of Alternative 1 is $337,127.

For Alternative 2, the total annual cost A is $5900 (total for inspection, testing, operation, and repair). The time line for Alternative 2 is

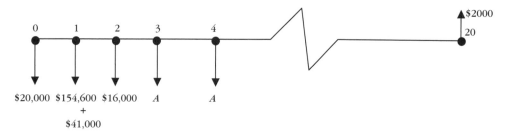

The present cost of Alternative 2 is found from

$$P = 20{,}000 + 195{,}600\text{CF}(12\%, 1) + 16{,}400\text{CF}(12\%, 2)$$
$$+ 5900\text{CF}(12\%, 17) \times \text{CF}(12\%, 3) - 2000\text{CF}(12\%, 20)$$

Using an interest table gives

$$P = 20{,}000 + 195{,}600(0.8929) + 16{,}400(0.7972)$$
$$+ 5900(7.12) \times (0.7188) - 2000(0.1037)$$
$$= 20{,}000 + 174{,}651 + 13{,}074 + 29{,}901 - 207$$
$$= \$237{,}419$$

The present cost of Alternative 2 is $237,833.

Alternative 2 should be selected since the criterion of selection is to minimize present cost.

OTHER COST ANALYSIS METHODS

Other techniques can be used to determine the most economical choice between comparable alternatives. These techniques are included here.

Annual Cash Flow Analysis

Annual cash flow analysis converts amounts of money to an equivalent uniform annual cost or benefit. Future sums and present cost are equated to equivalent uniform annual costs.

Example: Find A, given P:

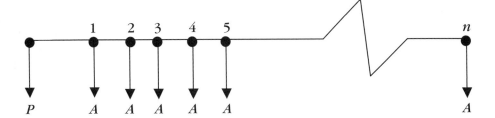

Capital recovery:

$$A = P \frac{i(1 + i)^n}{(1 + i)^n - 1}$$

Example: Find A, given F:

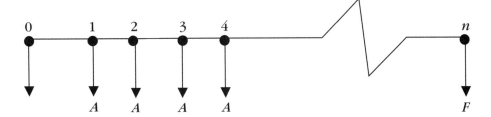

Sinking fund:

$$A = F \frac{i}{(1 + i)^n - 1}$$

Benefit/Cost Analysis

If the benefits of a project exceed the cost, then economically the project is worth undertaking:

$$\frac{\text{Benefits}}{\text{Cost}} = \frac{\text{Present worth of benefits}}{\text{Present cost}}$$

$$= \frac{\text{Equivalent uniform annual benefit}}{\text{Equivalent uniform annual cost}}$$

This is predicated upon a given minimum attractive rate of return.

Rate of Return: To determine the rate of return on an investment, the various consequences of an investment are converted to a cash flow. Then the cash flow is used to solve for the unknown value i, the rate of return. Two examples of cash flow equations are Present cost = Present worth of benefits and Equivalent uniform annual cost = Equivalent uniform benefits. Once the benefits and costs are known, the unknown rate of return can be calculated.

Payback Period

The payback period is the time required for the profit or other benefits from an investment to equal the cost of the investment; it is also the time required for savings on an investment to equal the cost of the investment.

The payback-period calculation is approximate, rather than exact. All costs and profits or savings of the investment prior to payback are included without considering differences in timing (when they were acquired). All the economic consequences beyond the payback period are ignored. Because it is an approximate calculation, payback period may or may not select the correct alternative. For example, both alternatives below have the same initial cost and payback period and benefits of $3000 each, so it would seem that either alternative is acceptable. But compare the time lines:

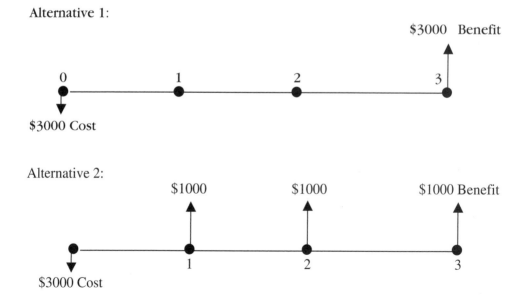

Alternative 1:

$3000 Benefit

$3000 Cost

Alternative 2:

$1000 $1000 $1000 Benefit

$3000 Cost

QUESTIONS

1. Life cycle cost is:

   a. the total cost over the life of the building
   b. the total cost over the life of the system
   c. the total cost over the life of the owner
   d. the sum of any periodic costs

2. The costs of alternative systems should be compared on the basis of:

   a. the present worth of their life cycle cost
   b. the present worth of their first cost
   c. their annual costs
   d. their interest rate

3. To determine the rate of return on an investment, the various consequences of an investment are converted to:

   a. costs   b. payback amounts
   c. cash flow   d. return on investments

4. The payback period is the time required for the profit or other benefits from an investment:

   a. to reach a satisfactory rate of return
   b. to amortize all costs
   c. to equal the cost of the investment
   d. to double its profits

5. A key element in cost analysis is:

   a. cost–benefit ratios
   b. choice of the best alternative to achieve the objective
   c. life cycle costs
   d. size of the investment

ADDITIONAL READING

Baresh, N. N., and Kaplan, S., *Economic Analysis for Engineering and Managerial Decision Making,* 2nd ed., McGraw-Hill, New York, 1978.

Grant, E. L., Ireson, W. G., and Leavenworth, R. S., *Principles of Engineering Economy,* 6th ed., Ronald Press, New York, 1976.

Newman, D. G., *Engineering Economic Analysis,* Engineering Press, San Jose, CA, 1977.

Ostwald, P. F., *Cost Estimating for Engineering and Management,* Prentice-Hall, Englewood Cliffs, NJ, 1974.

CHAPTER

17

Fire Warning Systems for Dwellings

INTRODUCTION

In 1999, 73.2 percent of structural fires occurred in residential occupancies (54 percent in one- and two-family dwellings). Also in 1999, residential fires were responsible for 80 percent of all fire deaths, 75.3 percent of civilian injuries, and 51 percent of total dollar losses from fire.

An effective household warning system integrates three elements: (1) the minimization of fire hazards, (2) the use of smoke detection equipment, and (3) the implementation of a family escape plan. This chapter covers types of warning systems and detectors and requirements for equipment, installation, maintenance, and testing. Additional detector information can be found in Chapter 6, Signal Transmission Methods and Processing; Chapter 4, Fire Alarm System Components and Circuits; and Chapter 12, Fire Alarm System Testing.

The primary function of fire warning equipment for dwelling units is to provide a reliable means of notifying the occupants of a dwelling unit of the presence of a threatening fire and the need to escape to a place of safety before such escape might be impeded by untenable conditions in the normal path of egress.

Life safety from fire in residential occupancies is based primarily on early notification to occupants of the need to escape, followed by the appropriate egress actions by those occupants. Fire warning systems for dwelling units are capable of protecting about half of the occupants in potentially fatal fires. Victims are often intimate with the fire, too old or too young, or too physically or mentally impaired to escape even when warned early enough that escape should be possible. For these people, other strategies such as protection-in-place or assisted escape or rescue may be necessary.

The performance of fire warning equipment for dwelling units discussed in *NFPA 72®*, *National Fire Alarm Code®*, depends on such equipment being properly selected, installed, operated, tested, and maintained in accordance with the provisions of this code and with manufacturers' instructions provided with the equipment.

GUIDELINES FOR FIRE WARNING SYSTEMS FOR DWELLINGS

A dwelling fire warning system consists of devices that produce an audible alarm signal within the dwelling unit to notify occupants of the presence of fire so they can escape. A dwelling fire warning system may also transmit signals and information to remote monitoring services, which that then notify first responders.

Advance warning of a fire condition is critical to escape; occupants may have only a few minutes to leave safely. Fire tests in residential structures have shown that detectable amounts of smoke precede detectable heat in almost all cases. Residential smoke alarms have lifesaving capabilities, but only if they are properly installed, regularly tested, and adequately maintained.

Smoke alarms are defined as devices responsive to smoke that incorporate a sensor, control components, and an alarm notification appliance in one unit operated from a power source either located in the unit or obtained at the point of installation. Smoke detectors lack the internal notification appliance and are usually powered from a remote source.

A typical dwelling fire warning system includes individual single-station smoke alarms or a system of interconnected alarms referred to as multiple-station alarms. Dwellings may also utilize systems consisting of a residential control panel to which are connected individual initiating devices and notification appliances. Such systems may also include burglar alarm, panic, medical emergency, carbon monoxide, flood, freezing, or other monitoring functions.

Residential smoke alarms were relatively expensive when they began to appear in the marketplace in the late 1960s. In 1970, however, the introduction of battery-operated smoke alarms, along with several new line-powered smoke alarms, began a period of public acceptance of single-station residential smoke alarms. Studies show that in the early 1970s, less than 1% of U.S. homes had any fire alarm equipment. By the mid-1990s, over 95% of U.S. homes had at least one smoke alarm. Although many of these were installed in newly constructed dwellings under building code provisions adopted in the late 1970s and 1980s, most were installed in existing homes voluntarily.

Current performance requirements and installation practices for residential smoke alarms derive from several test programs. The tests demonstrated that smoke alarms could provide a high level of safety due to their fast response to a fire, and that detectable quantities of smoke usually preceded detectable levels of heat showed that at least one smoke alarm on each story of a residence should be used as a minimum requirement (the so-called "every-level" smoke detection system). A *story* is defined as the portion of a building between the upper surface of one floor and the upper surface of the floor or roof next above. *NFPA 72* bases some of its fire detection criteria on these report.[1]

TYPES OF SYSTEMS

A dwelling fire warning system can range in size from a single-station smoke alarm in an apartment, manufactured home, or recreational vehicle to a group of multiple-station smoke alarms or a centrally wired fire alarm system containing numerous detectors and separate alarm signaling appliances. The fire warning system can include a burglar alarm system and an emergency medical alert system and can be connected to a central receiving headquarters by leased telephone line(s), dial-up connection, or digital alarm communicator.

When a smoke alarm is provided with one or more additional wires for interconnection with similar smoke alarms, it is referred to as a multiple-station alarm (see Figure 17-1). When interconnected according to the manufacturer's instructions, an alarm from any one of the interconnected devices sounds the

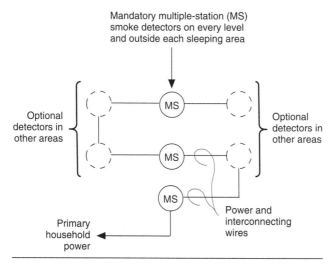

Figure 17-1. Typical household multiple-station smoke alarm system.

alarm in all the units. This is important because the smoke alarm near a bedroom should sound when a smoke alarm located in another area senses smoke, in order to provide sufficient noise levels to awaken sleeping occupants. The number of alarms that can be interconnected varies from one model to another. Some interconnection methods also allow for connection of heat alarms or manual fire alarm boxes to the interconnecting wire. These multiple-station smoke alarms can be used to form a fire warning system without installation of a separate control panel, but are limited to connections only within the individual dwelling unit.

Some smoke alarms are equipped with a transmitter that sends a signal to a receiver unit (usually located in a bedroom). If the device is an ac plug-in model, a line carrier transmitter impresses a high-frequency signal on the house wiring. The receiver unit detects the signal as long as both are on the same power company transformer. If the alarm is battery powered, it may contain a radio transmitter (similar to those used on garage door openers), with a range up to several hundred feet. A more detailed description of wireless transmission systems for residential and nonresidential use is found in Chapter 6, Signal Transmission Methods and Processing.

In addition to single- and multiple-station smoke alarms, there are dwelling fire warning systems similar in makeup and operation to a local fire alarm system. Such a system typically consists of a system control unit that derives main power from the ac house wiring and usually contains a rechargeable standby battery capable of operating the system for at least 24 hour (see Figure 17-2). In addition, this system uses either smoke alarms of the system-connected type or single-station smoke alarms, which may contain alarm contacts. The fire warning system could also have heat alarms and manual fire alarm boxes and separate alarm notification appliances such as bells, horns, or electronic sirens. The control unit for a fire warning system can also be connected to an automatic dialer, central station, or television cable to transmit the alarm to a point beyond the household. If a fire warning system is connected to some other alarm point, however, local code requirements for this type of alarm system must also be met.

Both wired and wireless dwelling fire warning systems often include a burglar alarm in addition to fire alarm functions. The fire alarm signal must take precedence over the burglar alarm signal in combination systems with distinctive audible alarms so the homeowner can immediately discern the difference between a fire and a burglary. A medical alert feature is sometimes included on a system and provides for transmission of alarms to a point beyond the household via leased lines, television cables, or other means (see Figure 17-3).

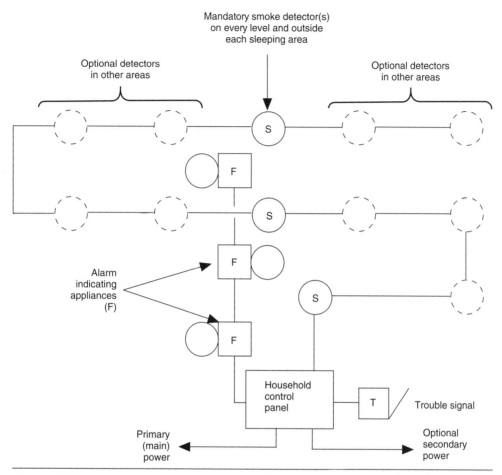

Figure 17-2. Typical dwelling fire alarm system, with separate control panel.

The single- and multiple-station smoke alarms used in dwelling fire warning systems operate on the same principle as the system-connected smoke detectors that are described in Chapter 5, Signal Initiation. The difference between them is that the single-station smoke alarm is self-contained with a built-in alarm notification appliance, whereas a multiple-station alarm is basically a single-station alarm with a method of interconnecting all of the smoke alarms together so that, when one senses smoke, all of the interconnected alarm-indicating appliances still sound. Typical single-station alarms are shown in Figure 17-4.

A system-connected smoke detector does not usually have a built-in alarm appliance, but (1) contains a relay that operates on alarm, (2) electronically impresses a signal on the alarm-initiating circuit, or (3) sends the alarm signal to

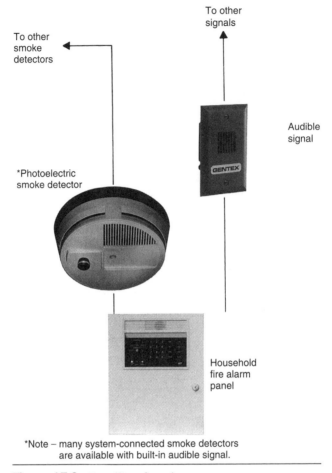

To other
signals

To other
smoke
detectors

Audible
signal

*Photoelectric
smoke detector

Household
fire alarm
panel

*Note – many system-connected smoke detectors
are available with built-in audible signal.

Figure 17-3. Dwelling fire alarm system.

the control panel by wireless transmission. See Chapter 6, Signal Transmission Methods and Processing.

FIRE WARNING SYSTEMS FOR ONE- AND TWO-FAMILY DWELLINGS

As defined by NFPA *101*®, *Life Safety Code*®, one- and two-family dwellings include buildings containing not more than two dwelling units in which each dwelling unit is occupied by members of a single family with no more than three outsiders, if any, accommodated in rented rooms. A dwelling unit is a single unit, providing complete, independent living facilities for the permanent use of

Single Station Smoke Detectors

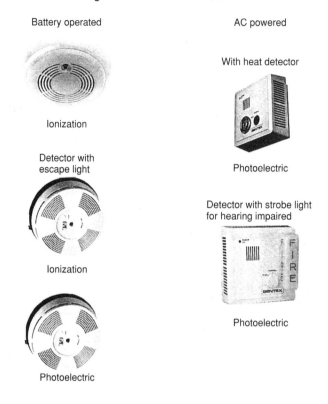

Battery operated

Ionization

Detector with
escape light

Ionization

Photoelectric

AC powered

With heat detector

Photoelectric

Detector with strobe light
for hearing impaired

Photoelectric

Note — many single station smoke detectors are available with
connections for multiple station operation.

Figure 17-4. Typical battery- and ac-powered single-
station smoke alarms.

one or more persons, including permanent provisions for living, sleeping, eating, cooking and sanitation. In *NFPA 72*, dwelling units do not include hotel or motel guest rooms or suites, dormitories, or sleeping rooms in nursing homes.

Detection

Approved single-station or multiple-station smoke alarms continuously powered by the house electrical service should be installed in accordance with *NFPA 72*. Many codes permit battery-powered alarms to be used in existing construction, but require ac-powered alarms with battery backup for new construction.

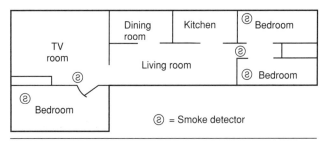

Figure 17-5. In dwellings with more than one sleeping area, a smoke alarm is provided to protect each separate sleeping area. In new construction, additional smoke alarms are required within each bedroom.

Single-station smoke alarms should be interconnected only within an individual living unit. Remote annunciation from single-station smoke alarms may be permitted.

A control panel and associated equipment, single- or multiple-station alarm device(s), or any combination thereof can be used to form a dwelling fire warning system in one- and two-family dwellings.

Requirements for fire warning systems including provision of and the number and location of devices is found in the occupancy chapters of NFPA *101*. Requirements for operation and performance, including sensitivity, audibility, power supplies, and so on, are found in *NFPA 72®*. For the convenience of the user, *NFPA 72* extracts the requirements from NFPA *101* and prints them along with its own requirements.

Smoke alarms are installed outside each separate sleeping area in the immediate vicinity of the bedrooms and on each additional story of the dwelling unit, including basement and excluding crawl spaces and unfinished attics (see Figures 17-5 through 17-8).

Smoke Alarm Location, Spacing, and Installation

The major threat from fire in a dwelling is at night when everyone is asleep. The principle threat to persons in sleeping areas comes from fires in the remainder of the unit; therefore, smoke alarm(s) are best located between the bedroom areas and the rest of the unit. In units with only one bedroom area on one floor, the smoke alarm is located as shown in Figure 17-7.

For dwellings with one or more split levels (i.e., adjacent levels with less than one full story separation between levels), a smoke alarm should suffice for an adjacent lower level, including basements (see Figure 17-8).

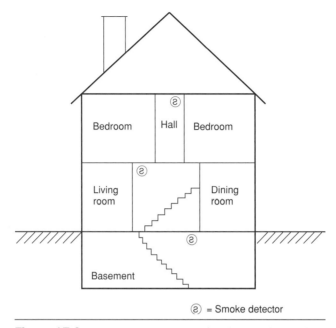

(2) = Smoke detector

Figure 17-6. As a minimum, a smoke alarm is located on each story.

In new construction, model building codes, NFPA *101,* and *NFPA 72* now require smoke alarms within sleeping rooms. These provide bedroom occupants protection from a fire starting within the bedroom where the door to that bedroom is closed. A smoke alarm outside the bedroom provides protection to others, but is ineffective for those in the room of origin when the door is closed.

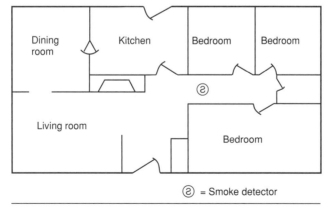

(2) = Smoke detector

Figure 17-7. A smoke alarm is located between the sleeping area and the rest of the dwelling unit.

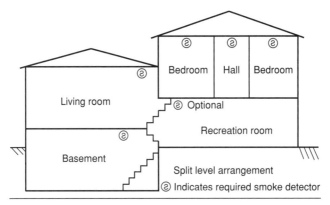

Figure 17-8. Required smoke alarms in a split-level arrangement. The indicated smoke alarmr is optional if no door is provided between split-level living and recreation rooms.

Furthermore, since smoke alarms in new construction are required to be interconnected, so that when one alarms, it sounds all of the smoke alarms, the audibility of the entire system is enhanced along with the ability of the system to awaken even sound sleepers.

Certain guidelines should be followed for locations of smoke alarms in certain areas of the dwelling. In rooms with ceiling slopes greater than 1 ft rise per 8 ft (0.3 m rise per 2.4 m) horizontally, smoke alarms should be located at the high side of the room. Smoke alarms should be located on or near the ceiling whenever practical. A smoke alarm installed in a stairwell should be located so smoke rising in the stairwell cannot be prevented from reaching the alarm by an intervening door or obstruction. A smoke alarm installed in the basement should be located in close proximity to the stairway leading to the floor above. A smoke detector installed to protect a sleeping area should be located outside the bedrooms but in the immediate vicinity of the sleeping area. In addition, a smoke alarm installed on a story without a separate sleeping area should be located in close proximity to the stairway leading to the floor above.

Smoke and heat alarms should be installed in those locations recommended by the manufacturer. When the space above the ceiling is open to the outside and little or no insulation is present over the ceiling, this results in ceiling temperatures that are cold in winter and hot in summer. Where the ceiling is significantly different in temperature from the air space below, smoke (and heat) has difficulty reaching the ceiling and to a smoke alarm that may be placed there. If a smoke alarm must be installed in these conditions, placement of the smoke alarm on a side wall, with the top 4 to 12 in. (0.1–0.3 m) from the ceiling, is preferred. It should be recognized that the condition of inadequately insulated ceilings and walls can also exist in multifamily housing (apartments), single-family housing, and manufactured homes. This situation can also exist (but to a lesser extent) with outside walls. Although a smoke alarm should be installed optimally on the ceiling and then on a side wall, if the side wall is an exterior wall with little or no insulation, the detector should be installed on an interior wall (one should ensure that the smoke alarm is listed for side-wall installation).

Smoke alarms should be located no closer than 3 ft (0.9 m) from heating or air-conditioning vents so that air issuing from the vent will not blow smoke away from the smoke alarm. Some smoke alarms are not suitable for location within kitchens because of false alarms from cooking vapors. Also, some smoke alarms are not recommended for garages (where automobile exhaust might cause alarms) or for attics or other unheated spaces where extremes of temperatures or humidity might affect smoke alarm operation. Before a smoke alarm is installed in any of these locations, its specifications should be checked to ensure it is appropriate for the intended area.

In dwellings employing radiant heating in the ceiling, the wall location is preferred (again, one should check to assure that the smoke alarm is listed for side-wall use). Radiant heating in the ceiling can create a hot-air boundary layer along the ceiling surface, which can seriously restrict the movement of heat and smoke to a ceiling-mounted smoke alarm.

Maintenance and Testing

Fire warning equipment in dwellings should be maintained in accordance with the recommendations of the equipment manufacturer. In general, this means little more than keeping the equipment clean and free of dust and replacing the batteries when needed. A maintenance agreement should be considered if the householder is unable to perform the required maintenance. *NFPA 72* requires that dwelling systems (with control panels) be serviced by a trained professional at least every 3 years. Smoke alarms should never be disconnected due to nuisance alarms (e.g., from cooking smoke). The smoke alarm should be relocated or replaced with a smoke alarm less sensitive to cooking smoke. A smoke alarm's sensitivity adjustment is normally set at the factory and should not be reset by an individual once the smoke alarm is in use. Some smoke alarms are available with a button to silence false alarms.

Tests or inspections as recommended by the manufacturer should be performed by the homeowner at least once a month (for other than battery-powered smoke alarms) and a minimum of once a week for battery-powered smoke alarms. It is also a good practice to establish a specific schedule for these tests. The importance of regular tests of smoke alarms and other electronic fire alarm equipment cannot be overstressed. Further information on testing fire alarm equipment is found in Chapter 12, Fire Alarm System Testing.

Proper care and maintenance of the smoke alarms is as important as the installation. Each residential smoke alarm is sold with an owner's booklet, which describes the necessary maintenance procedures. Fire reports involving death or serious injury where smoke alarms were installed have shown in almost all cases that the alarms were inoperative at the time of the fire because the homeowner had (1) failed to replace a worn-out battery, (2) failed to install the smoke alarm properly, (3) intentionally disconnected the power due to false alarms, or (4) failed to properly evacuate the dwelling. Because of this problem with battery-powered smoke alarms, *NFPA 72* mandates use of an ac primary power source with backup battery for smoke alarms in all new construction. This is a good recommendation even for existing smoke alarms in older construction.

SOUNDING APPLIANCES

Each detection device should cause the operation of an alarm. The alarm sound should be clearly audible in all bedrooms over background noise levels with all intervening doors closed. Audibility tests should be conducted under worst-case conditions (at night, with window air conditioners and room humidifiers operating). A single-station smoke alarm sounds an audible alarm at the device itself, and sound level output should be a minimum of 85 dBA at a distance of 10 ft (3 m).

For instance, there may be a noisy window air conditioner or room humidifier, which may generate an ambient noise level of 55 dBA or higher. The smoke alarms must be able to penetrate through the closed doors and be heard over the bedroom's noise levels with sufficient intensity to awaken sleeping occupants. Test data indicate that detection devices having sound pressure ratings of 85 dBA at 10 ft (3 m) and installed outside a bedroom can produce about 15 dBA over ambient noise levels of 55 dBA (approximately the minimum noise level of an air conditioner or other device) in the bedroom.

Test studies have shown that smoke alarms located remote from the bedroom area may not be loud enough to awaken the average person. In such cases, it is recommended that smoke alarms be interconnected so that operation of the remote smoke alarm will cause an alarm of sufficient intensity to penetrate the bedrooms. This interconnection can be accomplished by installation of a fire detection system, wiring together multiple-station alarm devices, use of line carrier or radio frequency transmitters/receivers, and so on.

An alarm may not be heard from a remote smoke detector. *NFPA 72* and many building codes now require that, where more than one smoke alarm is required to be installed within an individual dwelling unit, the smoke alarms should be wired in such a manner that the actuation of one alarm will actuate all the alarms. This ensures that smoke anywhere in the dwelling unit will activate the smoke alarm(s) near the bedroom(s).

A wired or wireless fire warning system normally has an alarm notification appliance circuit to connect various alarm signals, such as fire alarm bells, horns, or electronic sirens. These appliances can produce higher sound levels and can be located independent of the smoke alarms and can be clearly audible throughout the house (or even outside, where neighbors might hear the alarm).

VISIBLE APPLIANCES

NFPA 72 requires that homes occupied by one or more hearing-impaired persons be equipped with visible strobe lights or approved tactile (vibrating) devices. The light intensities are specified on the basis of research conducted by Underwriters Laboratories Inc. (UL) on a large number of hearing-impaired persons. They found that a light intensity of 15 candelas was sufficient to attract the attention of an awake person even if that person was not looking at the light. Further, *NFPA 72* requires that every-level smoke alarms be equipped with 15-candela strobes. Because of their low power consumption, these lights should be operated from a source of standby power when provided.

The UL study found that light of significantly greater intensity is required to awaken a sleeping person: 110 candelas minimum for about a 95 percent waking efficiency. Thus, rooms in which hearing-impaired persons sleep are required to be provided with a 110-candela light mounted on the wall, 24 in. (0.6 m) or more below the ceiling, where the light will not be attenuated by any smoke collected at the ceiling. If mounted on the ceiling or within 24 in., the light intensity must be 177 candelas to provide sufficient light at the pillow surface, based on the maximum expected smoke density at the ceiling necessary to activate the smoke alarm. These high-intensity lights cannot be powered practically from a battery supply, so standby power connection is not required.

One important note is that *NFPA 72* places the responsibility for the provision of these lights on the homeowner or tenant because it is often not possible for a landlord or builder to determine that a person is hearing-impaired.

A HOME FIRE SAFETY SYSTEM
Minimizing Fire Hazards

Since it is always easier to prevent a fire than to control one, minimizing fire hazards in and around the home is a key element in home fire safety. Minimizing hazards includes such things as removing all nonessential flammable liquids and other highly combustible items from the home, properly storing essential combustibles away from ignition sources, and not overloading electrical circuits. If a fire develops rapidly enough, however, residential smoke alarms may not provide sufficient time for the occupants of a dwelling to escape before exit routes become untenable.

Escape Planning

Another key element in residential fire protection is developing and practicing a family escape plan. Residential smoke alarms can only warn occupants of a fire in the home. There are three important elements of a home escape plan that should be followed by all occupants:

1. All occupants should leave the home immediately and call the fire department from a neighbor's house. Many deaths have been reported in fires where people have unwisely taken extra time to get dressed, gather valuables, or look for pets.
2. Everyone should have an alternate way out of each room in case the primary exit is blocked by fire.
3. A prearranged outside meeting place should be determined so everyone will know that the entire family has escaped the fire. Deaths have resulted because people have reentered a burning home to look for persons who had already escaped.

QUESTIONS

1. The primary function of fire warning equipment for dwelling units is to:
   a. notify the fire department of the fire
   b. notify the occupants of the need to escape
   c. supervise residential sprinklers
   d. meet code requirements

2. The primary protection for dwelling units is provided by detectors or alarms that respond to:
   a. smoke   b. heat   c. carbon monoxide   d. flame

3. Requirements for the number and location of devices in a dwelling fire warning system are found in:

 a. *NFPA 72®, National Fire Alarm Code®*
 b. NFPA *101®, Life Safety Code®*
 c. NFPA 74, *Household Fire Warning Equipment*
 d. NFPA 1, *Uniform Fire Code*™

4. Smoke alarms should not be installed in:

 a. attics   b. garages   c. kitchens   d. a, b, and c

5. Visible notification appliances are required:

 a. in all dwellings
 b. in dwellings occupied by hearing-impaired individuals
 c. in hotel guest rooms and suites
 d. outside the dwelling to notify the fire department

BIBLIOGRAPHY

NFPA Codes, Standards, Recommended Practices, and Manuals (see the latest *NFPA Catalog* for availability of current editions of the following documents):

NFPA 72®, National Fire Alarm Code®.

NFPA *101®, Life Safety Code®.*

REFERENCE CITED

1. Bukowski, R. W., Waterman, T. E., and Christian, W. J., "Detector Sensitivity and Siting Requirements for Dwellings: A Report of the NBS 'Indiana Dunes Tests,'" NFPA, Quincy, MA, 1975.

ADDITIONAL READING

U.S. Fire Administration, "Protecting Your Family from Fire," FA 130 (Spanish language version: FA 129), U.S. Fire Administration, Gaithersburg, MD.

Appendix

Fire Alarm and Emergency Communication Symbols

Referent (Synonym)	Symbol	Comments
Signal-Initiating Devices		
Manual Stations (call point)	□	General
Manual Alarm Box (pull station and pull box)	▣	
Telephone Station (telephone call point)		
Automatic detection and supervisory devices	○	General
Heat detector* (thermal detector)		Includes fixed-temperature, rate-compensation, and rate-of-rise detectors
Smoke Detector		Includes photoelectric and ionization-type detectors
Smoke Detector in Duct		
Gas detector	▲	
Flame Detector* (flicker detector)		Includes ultraviolet, infrared, and visible radiation-type detectors
Flow detector/Switch		
Pressure Detector/Switch*		Alternate term: pressure switch: air, water, etc.
Level Detector/Switch*		
Tamper Detector/Switch		Alternate term: tamper switch
Valve with Tamper Detector/Switch		

432

Referent (Synonym)	Symbol	Comments
Audible-type alerting devices (sounder)		
Speaker/Horn (electric horn)		
Bell (gong)		
Water Motor Alarm (water motor gong)		Shield optional
Horn with Light (horn with strobe)		
Visual Type		
Light (lamp, signal light, indicator lamp, strobe)		
Illuminated Exit Sign		
Illuminated Exit Sign with Direction Arrow		
Emergency Illumination Symbols		
Emergency Light, Battery Powered, One Lamp		
Emergency Light, Battery Powered, Two Lamps		
Emergency Light, Battery Powered, Three Lamps		
Control and Supervisory Devices		
Control Panel		
Door Holder		
Mains, Pipe		
Public Water Main		Indicate pipe size
Private Water Main		Indicate pipe size
Water Main Under Building		Indicate pipe size
Suction Main		Indicate pipe size

(Continues)

Referent (Synonym)	*Symbol*	*Comments*
Valves		
Post Indicator and Valve		Indicate valve size
Key-Operated Valve		Indicate valve size
OS & Y Valve (outside screw and yoke, rising stem)		
Indicating Butterfly Valve		Indicate valve size
Nonindicating Valve (nonrising-stem valve)		
Valve in Pit		Indicate valve size
Check Valve		Indicate valve size

*Symbol orientation must not be changed.

Glossary

Addressable Device. A fire alarm system component with discrete identification that can have its status individually identified or that is used to individually control other functions.

Alarm. A warning of fire danger.

Alarm Service. The service required following the receipt of an alarm signal.

Alarm Signal. A signal indicating an emergency requiring immediate action, as an alarm for fire from a manual box, a waterflow alarm, an alarm from an automatic fire alarm system, or other emergency signal.

Alarm Verification Feature. A feature of automatic fire detection and alarm systems to reduce unwanted alarms wherein smoke detectors report alarm conditions for a minimum period of time, or confirm alarm conditions within a given time period after being reset, in order to be accepted as a valid alarm initiation signal.

Annunciator. A unit containing two or more identified targets or indicator lamps in which each target or lamp indicates the circuit, condition, or location to be annunicated.

Authority Having Jurisdiction. The organization, office, or individual responsible for approving equipment, an installation, or a procedure.

Note: The phrase *authority having jurisdiction* is used in NFPA documents in a broad manner since jurisdictions and approval agencies vary as do their responsibilities. Where public safety is primary, the authority having jurisdiction may be a federal, state, local, or other regional department or individual such as a fire chief, fire marshal, chief of a fire prevention bureau, labor department, health department, building official, electrical inspector, or others having statutory authority. For insurance purposes, an insurance inspection department, rating bureau, or other insurance company representative may be the authority having jurisdiction. In many circumstances the property owner or a designated agent assumes the role of the authority having jurisdiction; at government installations, the commanding officer or departmental official may be the authority having jurisdiction.

Auxiliarized Local System. A local system that is connected to the municipal alarm facilities.

Auxiliarized Proprietary System. A proprietary system that is connected to the municipal alarm facilities.

Auxiliary Fire Alarm System. A connection to the municipal fire alarm system to transmit an alarm of fire to the municipal communication center. Fire alarms from an auxiliary alarm system are received at the municipal communication center on the same equipment and by the same alerting methods as alarms transmitted from municipal fire alarm boxes located on streets.

Auxiliary Trip Relay. A relay used to operate a municipal master box from an auxiliarized control panel.

Bell, Single-Stroke. A bell whose gong is struck only once each time operating energy is applied.

Bell, Vibrating. A bell that rings continuously as long as operating power is applied.

Box (or Station), Fire Alarm:

1. Noncoded. A manually operated device that, when operated, closes or opens one or more sets of contacts and generally locks the contacts in the operated position until the box is reset.

2. Coded. A manually operated device in which the act of pulling a lever causes the transmission of not fewer than three rounds of coded alarm signals. Similar to the noncoded type, except that, instead of a manually operated switch, a mechanism to rotate a code wheel is utilized. Rotation of the code wheel in turn causes an electric circuit to be alternately opened and closed, or closed and opened, thus sounding a coded alarm that identifies the location of the box. The code wheel is cut for the individual code to be transmitted by the device and can operate by clockwork or an electric motor. Clockwork transmitters can be prewound or can be wound by the pulling of the alarm lever. Usually the box is designed to repeat its code four times and automatically come to rest. Prewound transmitters must sound a trouble signal when they require rewinding. Solid-state electronic coding devices are also used in conjunction with the fire alarm control unit to produce coded sounding of the audible signaling appliances.

Breakglass Box (or Station). A box is one in which it is necessary to break a special element in order to operate the box.

Ceiling. The upper surface of a space, regardless of height. Areas with a suspended ceiling have two ceilings, one visible from the floor and one above the suspended ceiling.

Ceiling Height. The height from the continuous floor of a room to the continuous ceiling of a room or space.

Central Station System. A system or group of systems in which the operations of circuits and devices are signaled automatically to, recorded in, maintained, and supervised from an approved central station having competent and experienced observers and operators, who, upon receipt of a signal, take the required action. Such systems are controlled and operated by a person, firm, or corporation whose principal business is the furnishing and maintaining of supervised signaling service.

Channel. A path for signal transmission between two or more stations or channel terminations. A channel can consist of wire, radio waves, or equivalent means of signal transmission.

Chimes. A single-stroke or vibrating-type audible notification appliance that has a xylophone-type striking bar.

Circuit. The conductors or radio channel and associated equipment used to perform a definite function in connection with an alarm system.

Circuit Interface. A functional assembly that interfaces one or more of its initiating-device circuits with a signaling line circuit in a manner that permits the central supervising station to indicate the status of each of its individual initiating-device circuits.

Coded Signal. A signal pulsed in a prescribed code for each round of transmission. A minimum of three rounds and a minimum of three impulses is required for an alarm signal.

Combination Detector. A device that either responds to more than one of the fire phenomena, such as smoke, heat, flame, and fire gas, or employs more than one operating principle to sense one of these phenomena. Typical examples are a combination of a heat detector with a smoke detector and a combination rate-of-rise and fixed-temperature heat detector.

Combination System. A local fire alarm system for fire alarm, supervisory or guard's tour supervisory service, whose components may be used in whole or in part in common with a nonfire signaling system, such as a paging system, a burglar alarm system, a musical program system, or a process monitoring service system, without degradation of or hazard to the fire alarm system.

Communication Channel. A signaling channel (usually leased from a communication utility company) having two or more terminal locations and a suitable information handling capacity, depending on the characteristics of the system used. One terminal location is at the central supervising station and the other terminal location or locations are sources from which are transmitted alarm signals, supervisory signals, trouble signals, and such other signals as the central supervising station is prepared to receive and interpret.

Control Unit. A device with the control circuits necessary to (1) furnish power to a fire alarm system, (2) receive signals from alarm-initiating devices and transmit them to audible alarm notification appliances and accessory equipment, and (3) electrically supervise the system installation wiring and primary (main) power. The control unit can be contained in one or more cabinets in adjacent or remote locations.

Delinquency Signal. A signal indicating the need for action in connection with the supervision of guards or system attendants.

Emergency Voice/Alarm Communication System. A system that provides dedicated manual, automatic, or both types of facilities for originating and distributing voice instructions as well as alert and evacuation signals pertaining to a fire emergency to the occupants of a building.

End-of-Line Device. A device used to terminate a supervised circuit.

Fault. An open, ground, or short condition on any line(s) extending from a control unit that could prevent normal operation.

Fire Alarm System. A combination of compatible initiating devices, control units, and notification appliances designed and installed to produce an alarm signal in the event of fire.

Flame Detector. A device that detects the infrared, ultraviolet, visible radiation produced by a fire.

Frequency Division Multiplexing. A signaling method characterized by the simultaneous transmission of more than one signal in a communication channel. Signals from one or multiple terminal locations are distinguished from one another by virtue of each signal being assigned to a separate frequency or combination of frequencies.

Ground Fault. A condition in which the resistance between a conductor and ground reaches an unacceptably low level.

Ground Fault Detector. A device that detects the presence of a ground condition on system wiring.

Heat Detector. A device that detects abnormally high temperature or rate of temperature rise.

Horn. An audible signal appliance in which energy produces a sound by imparting motion to a flexible component, which vibrates at some nominal frequency.

Initiating Device. A manually or automatically operated device, the normal intended operation of which results in a fire alarm or supervisory signal indication from the control unit. Examples of alarm signal initiating devices are heat detectors, manual fire alarm boxes, smoke detectors, and waterflow switches. Examples of supervisory signal-initiating devices are water-level indicators, sprinkler-system valve-position switches, pressure supervisory transmitters, and water temperature switches.

Initiating Device Circuit (IDC). A circuit to which automatic or manual signal-initiating devices such as fire alarm boxes, fire detectors, and waterflow alarm devices are connected, where the signal received does not identify the individual device operated.

Labeled. Equipment or materials to which has been attached a label, symbol, or other identifying mark of an organization acceptable to the authority having jurisdiction and concerned with product evaluation, which maintains periodic inspection of the production of the labeled equipment or materials and by whose labeling the manufacturer indicates compliance with appropriate standards or performance in a specified manner.

Leg Facility. That part of a signaling line circuit connecting each protected building to the trunk facility or directly to the central supervising station.

Listed. Equipment or materials included in a list published by an organization acceptable to the authority having jurisdiction and concerned with product evaluation, which maintains periodic inspection of the production of listed equipment or materials and whose listing states either that the equipment or material meets appropriate standards or has been tested and found suitable for use in a specified manner.

Note: The means for identifying listed equipment may vary for each organization concerned with product evaluation, some of which do not recognize equipment as listed unless it is also labeled. The authority having jurisdiction should utilize the system employed by the listing organization to identify a listed product.

Local Alarm System. A local system sounding an alarm as the result of the manual operation of a fire alarm box or the operation of protection equipment or systems, such as water flowing in a sprinkler system, the discharge of carbon dioxide, the detection of smoke, or the detection of heat.

Local Energy Auxiliary Alarm System. An auxiliary alarm system that employs a locally complete arrangement of parts, initiating devices, relays, power supply, and associated components to automatically trip a municipal transmitter or master box over electric circuits that are electrically isolated from the municipal system circuits.

Local Supervisory System. A local system arranged to supervise the performance of guards' tours or the operative condition of automatic sprinkler systems or other systems for the protection of life and property against a fire hazard.

Local System. A system that produces a signal at the premises being protected.

Maintenance. Repair service, including periodic inspections and tests, required to keep the fire alarm system and its component parts in an operative condition at all times, together with replacement of the system and its components when for any reason they become undependable or inoperative.

Master Box. A municipal fire alarm box that may also be operated by remote means.

Multiplexing. A signaling method characterized by the simultaneous or sequential transmission, or both, and reception of multiple signals in a communication channel, including means for positively identifying each signal.

Municipal Communication Center. The building or portion of a building used to house the central operating part of the fire alarm system; usually the place where the necessary testing, switching, receiving, retransmitting, and power supply devices are located.

Municipal Fire Alarm Box. A specially manufactured enclosure housing a manually operated transmitter used to send an alarm to the municipal communication center.

Municipal Transmitter. A specially manufactured enclosure housing a transmitter that can only be tripped remotely, used to send an alarm to the municipal communication center.

Noncoded Signal. Signal from any notification appliance that is energized continuously.

Notification Appliance. Any audible or visible signal employed to indicate a fire, supervisory, or trouble condition. Examples of audible signal appliances are bells, horns, sirens, electronic horns, buzzers, and chimes. A visible indicator consists of a lamp, target, meter deflection, or equivalent.

Notification Appliance Circuit. A circuit or path directly connected to a notification appliance(s), such as bells, horns, chimes, etc.

Paging System. A system intended to page one or more persons by means such as a voice over loudspeaker stations located throughout the premises, or by means of coded audible signals or visual signals similarly distributed, or by means of lamp annunciators located throughout the premises.

Parallel Telephone Auxiliary Alarm System. An auxiliary alarm system connected by a municipally controlled individual circuit to the protected property to interconnect the actuating devices and the municipal fire alarm switchboard.

Parallel Telephone System. A telephone system in which an individual wired circuit is used for each box.

Permanent Visual Record (Recording). Immediately readable, not easily alterable print, slash, punch, etc., listing all occurrences of status change.

Proprietary Fire Alarm System. An installation of fire alarm systems that serves contiguous and noncontiguous properties under one ownership from a central supervising station located at the protected property, where trained, competent personnel are in constant attendance. This includes the central supervising station; power supplies; signal-initiating devices; initiating-device circuits; signal notification appliances; equipment for the automatic, permanent visual recording of signals; and equipment for the operation of emergency building control services.

Rectifier. An electrical device without moving parts that changes alternating current (ac) to direct current (dc).

Remote Station Fire Alarm System. An installation using supervised dedicated circuits to transmit alarm, supervisory, and trouble signals from one or more protected premises to a remote location at which appropriate action is taken.

Repeater Facility. Equipment needed to relay signals between the protected premises and the central supervising station.

Runner Service. Employees, other than the required number of operators, on duty at all times at the central supervising station at a runner station, or in a vehicle in constant radio contact with the central supervising station available for prompt dispatching, when necessary, to the protected premises.

Shunt Auxiliary Alarm System. An auxiliary alarm system electrically connected to an integral part of the municipal alarm system extending the municipal circuit into the protected property to interconnect the actuating devices. When operated, these devices open the municipal circuit shunted around the trip coil of the municipal transmitter or master box, which is thereupon energized to start transmission without any assistance whatsoever from a local source of energy.

Note: The shunt system runs municipal power wires into protected premises. Thus, the municipality may lose control of its circuit. In addition, an open circuit in this shunt loop will cause an alarm condition. The use of a shunt-type system is a matter of individual municipal policy.

Signaling Line Circuit (SLC). A circuit or path between any combination of circuit interfaces, control units, or transmitters (channel or trunk and leg) over which multiple system input signals or output signals, or both, are transmitted and received.

Signaling Line Circuit Interface. A system component that connects a signaling line circuit to any combination of initiating devices, initiating-device circuits, notification appliances, notification appliance circuits, system control outputs, and other signaling line circuits.

Smoke Detector. A device that detects visible or invisible particles of combustion.

Spacing. A horizontally measured dimension relating to the allowable coverage of fire detectors.

Supervision. The process of monitoring a circuit, switch, or device in such a manner that a trouble signal is received when a fault that would prevent normal operation of the system occurs.

Supervisory Service. The service required to monitor performance of guard patrols and the operative condition of automatic sprinkler systems and of other systems for the protection of life and property.

Supervisory Signal. A signal indicating the need for action in connection with the supervision of guards tours, sprinkler and other extinguishing systems or equipment, or the maintenance features of other protective systems.

Supplementary. Equipment or operation not required by signaling system standards and designated as such by the authority having jurisdiction.

Transmitter. A system component to which initiating devices or groups of initiating devices are connected. The component transmits signals to the central supervising station indicating the status of the initiating devices and the initiating-device circuits.

Trouble Signal. An audible signal indicating trouble of any nature, such as a circuit break or ground, occurring in the devices or wiring associated with a fire alarm system.

Trunk Facility. That part of a signaling line circuit connecting two or more leg facilities to the central supervising station or satellite station.

Two-Way Fire Department Communication System. An electrically supervised telephone system providing private voice communication capability between the command center or central control panel and designated remote locations.

Visible Signal. Response to the operation of an initiating device by one or more direct or indirect visible notification appliances. For a direct visible signal, the sole means of notification is by direct viewing of the light source. For an indirect visible signal, the sole means of notification is by illumination of the area surrounding the visible signaling appliance.

Waterflow Switch. An assembly approved for the service and so constructed and installed that any flow of water from a sprinkler system equal to or greater than that from a single automatic sprinkler of the smallest orifice size installed on the system will result in activation of this switch and subsequently indicate an alarm condition.

Zone. A designated area of a building. Commonly, zones within a building are annunciated to rapidly locate a fire.

ANSWERS TO CHAPTER QUESTIONS

Chapter 2:   1 (a), 2 (d), 3 (c), 4 (d), 5 (c).

Chapter 3:   1 (a), 2 (c), 3 (c), 4 (c), 5 (b).

Chapter 4:   1 (e), 2 (e), 3 (b), 4 (c), 5 (b).

Chapter 5:   1 (b), 2 (c), 3 (a), 4 (a), 5 (c).

Chapter 6:   1 (d), 2 (c), 3 (a), 4 (c), 5 (c).

Chapter 7:   1 (c), 2 (d), 3 (b), 4 (a), 5 (c).

Chapter 8:   1 (c), 2 (d), 3 (c), 4 (b), 5 (c).

Chapter 9:   1 (c), 2 (b), 3 (a), 4 (e), 5 (d).

Chapter 10: 1 (c), 2 (b), 3 (e), 4 (d), 5 (c).

Chapter 11: 1 (b), 2 (a), 3 (b), 4 (e), 5 (a).

Chapter 12: 1 (d), 2 (c), 3 (a), 4 (a), 5 (d).

Chapter 13: 1 (e), 2 (a), 3 (c), 4 (d), 5 (b).

Chapter 14: 1 (a), 2 (b), 3 (c), 4 (d), 5 (d).

Chapter 15   1 (d), 2 (c), 3 (b), 4 (d), 5 (b).

Chapter 16: 1 (b), 2 (a), 3 (c), 4 (c), 5 (b).

Chapter 17: 1 (b), 2 (a), 3 (b), 4 (d), 5 (b).

Index